Vibration of Mechanical Systems

This is a textbook for a first course in mechanical vibrations. There are many books in this area that try to include everything, thus they have become exhaustive compendiums that are overwhelming for an undergraduate. In this book, all the basic concepts in mechanical vibrations are clearly identified and presented in a concise and simple manner with illustrative and practical examples. Vibration concepts include a review of selected topics in mechanics; a description of single-degree-of-freedom (SDOF) systems in terms of equivalent mass, equivalent stiffness, and equivalent damping; a unified treatment of various forced response problems (base excitation and rotating balance); an introduction to systems thinking, highlighting the fact that SDOF analysis is a building block for multi-degree-of-freedom (MDOF) and continuous system analyses via modal analysis; and a simple introduction to finite element analysis to connect continuous system and MDOF analyses. There are more than 60 exercise problems and a complete solutions manual. The use of MATLAB® software is emphasized.

Alok Sinha is a Professor of Mechanical Engineering at The Pennsylvania State University (PSU), University Park. He received his PhD degree in mechanical engineering from Carnegie Mellon University. He has been a PSU faculty member since August 1983. His areas of teaching and research are vibration, control systems, jet engines, robotics, neural networks, and nanotechnology. He is the author of *Linear Systems: Optimal and Robust Control.*

He has served as a Visiting Associate Professor of Aeronautics and Astronautics at MIT, Cambridge, MA, and as a researcher at Pratt & Whitney, East Hartford, CT. He has also been an associate editor of *ASME Journal of Dynamic Systems, Measurement, and Control.* At present, he serves as an associate editor of *ASME Journal of Turbomachinery* and *AIAA Journal.*

Alok Sinha is a Fellow of ASME. He has received the NASA certificate of recognition for significant contributions to the Space Shuttle Microgravity Mission.

VIBRATION OF MECHANICAL SYSTEMS

Alok Sinha

The Pennsylvania State University

CAMBRIDGE UNIVERSITY PRESS
Cambridge, New York, Melbourne, Madrid, Cape Town,
Singapore, São Paulo, Delhi, Mexico City

Cambridge University Press
32 Avenue of the Americas, New York, NY 10013-2473, USA

www.cambridge.org
Information on this title: www.cambridge.org/9780521518734

First published 2010
Reprinted 2013

A catalog record for this publication is available from the British Library.

Library of Congress Cataloging in Publication Data

Sinha, Alok
Vibration of mechanical systems / Alok Sinha.
p. cm.
Includes bibliographical references and index.
ISBN 978-0-521-51873-4 (hardback)
1. Machinery – Vibration. I. Title.
TJ177.S56 2010
621.8′11–dc22 2010021143

ISBN 978-0-521-51873-4 Hardback

To
My Wife Hansa
and
My Daughters Divya and Swarna

CONTENTS

PREFACE

This book is intended for a vibration course in an undergraduate Mechanical Engineering curriculum. It is based on my lecture notes of a course (ME370) that I have been teaching for many years at The Pennsylvania State University (PSU), University Park. This vibration course is a required core course in the PSU mechanical engineering curriculum and is taken by junior-level or third-year students. Textbooks that have been used at PSU are as follows: Hutton (1981) and Rao (1995, First Edition 1986). In addition, I have used the book by Thomson and Dahleh (1993, First Edition 1972) as an important reference book while teaching this course. It will be a valid question if one asks why I am writing another book when there are already a large number of excellent textbooks on vibration since Den Hartog wrote the classic book in 1956. One reason is that most of the books are intended for senior-level undergraduate and graduate students. As a result, our faculties have not found any book that can be called ideal for our junior-level course. Another motivation for writing this book is that I have developed certain unique ways of presenting vibration concepts in response to my understanding of the background of a typical undergraduate student in our department and the available time during a semester. Some of the examples are as follows: review of selected topics in mechanics; the description of the chapter on single-degree-of-freedom (SDOF) systems in terms of equivalent mass, equivalent stiffness, and equivalent damping; unified treatment of various forced response problems such as base excitation and rotating balance; introduction of system thinking, highlighting the fact that SDOF analysis is a building block for multi-degree-of-freedom (MDOF) and continuous system analyses via modal analysis; and a simple introduction of finite element analysis to connect continuous system and MDOF analyses.

As mentioned before, there are a large number of excellent books on vibration. But, because of a desire to include everything, many of these books often become difficult for undergraduate students. In this book, all the basic concepts in mechanical vibration are clearly identified and presented in a simple manner with illustrative and practical examples. I have also attempted to make this book self-contained as much as possible; for example, materials needed from previous courses, such as differential equation and engineering mechanics, are presented. At the end of each chapter, exercise problems are included. The use of MATLAB software is also included.

ORGANIZATION OF THE BOOK

In Chapter 1, the degrees of freedom and the basic elements of a vibratory mechanical system are presented. Then the concepts of equivalent mass, equivalent stiffness, and equivalent damping are introduced to construct an equivalent single-degree-of-freedom model. Next, the differential equation of motion of an undamped SDOF spring–mass system is derived along with its solution. Then the solution of the differential equation of motion of an SDOF spring–mass–damper system is obtained. Three cases of damping levels – underdamped, critically damped, and overdamped – are treated in detail. Last, the concept of stability of an SDOF spring–mass–damper system is presented.

In Chapter 2, the responses of undamped and damped SDOF spring–mass systems are presented. An important example of input shaping is shown. Next, the complete solutions of both undamped and damped spring–mass systems under sinusoidal excitation are derived. Amplitudes and phases of steady-state responses are examined along with force transmissibility, quality factor, and bandwidth. Then the solutions to rotating unbalance and base excitation problems are provided. Next, the basic principles behind the designs of a vibrometer and an accelerometer are presented. Last, the concept of equivalent viscous damping is presented for nonviscous energy dissipation.

In Chapter 3, the techniques to compute the response of an SDOF system to a periodic excitation are presented via the Fourier series expansion. Then it is shown how the response to an arbitrary excitation is obtained via the convolution integral and the unit impulse response. Last, the Laplace transform technique is presented. The concepts of transfer function, poles, zeros, and frequency response function are also introduced.

In Chapter 4, mass matrix, stiffness matrix, damping matrix, and forcing vector are defined. Then the method to compute the natural frequencies and the mode shapes is provided. Next, free and forced vibration of both undamped and damped two-degree-of-freedom systems are analyzed. Last, the techniques to design undamped and damped vibration absorbers are presented.

In Chapter 5, the computation of the natural frequencies and the mode shapes of discrete multi-degree-of-freedom and continuous systems is illustrated. Then the orthogonality of the mode shapes is shown. The method of modal decomposition is presented for the computation of both free and forced responses. The following cases of continuous systems are considered: transverse vibration of a string, longitudinal vibration of a bar, torsional vibration of a circular shaft, and transverse vibration of a beam. Last, the finite element method is introduced via examples of the longitudinal vibration of a bar and the transverse vibration of a beam.

1

EQUIVALENT SINGLE-DEGREE-OF-FREEDOM SYSTEM AND FREE VIBRATION

The course on Mechanical Vibration is an important part of the Mechanical Engineering undergraduate curriculum. It is necessary for the development and the performance of many modern engineering products: automobiles, jet engines, rockets, bridges, electric motors, electric generators, and so on. Whenever a mechanical system contains storage elements for kinetic and potential energies, there will be vibration. The vibration of a mechanical system is a continual exchange between kinetic and potential energies. The vibration level is reduced by the presence of energy dissipation elements in the system. The problem of vibration is further accentuated because of the presence of time-varying external excitations, for example, the problem of resonance in a rotating machine, which is caused by the inevitable presence of rotor unbalance. There are many situations where the vibration is caused by internal excitation, which is dependent on the level of vibration. This type of vibration is known as self-excited oscillations, for example, the failure of the Tacoma suspension bridge (Billah and Scanlan, 1991) and the fluttering of an aircraft wing. This course deals with the characterization and the computation of the response of a mechanical system caused by time-varying excitations, which can be independent of or dependent on vibratory response. In general, the vibration level of a component of a machine has to be decreased to increase its useful life. As a result, the course also examines the methods used to reduce vibratory response. Further, this course also develops an input/output description of a dynamic system, which is useful for the design of a feedback control system in a future course in the curriculum.

The book starts with the definition of basic vibration elements and the vibration analysis of a single-degree-of-freedom (SDOF) system, which is the simplest lumped parameter mechanical system and contains one independent kinetic energy storage element (mass), one independent potential energy storage element (spring), and one independent energy dissipation element (damper). The analysis deals with natural vibration (without any external excitation) and forced response as well. The following types of external excitations are considered: constant, sinusoidal, periodic, and impulsive. In addition, an arbitrary nature of excitation is considered. Then, these analyses are presented for a complex lumped parameter mechanical system with multiple degrees of freedom (MDOF). The design of vibration absorbers is presented. Next, the vibration of a system with continuous distributions of mass, such as strings, longitudinal bars, torsional shafts, and beams, is presented. It is emphasized that the previous analyses of lumped parameter systems serve as building blocks for computation of the response of a continuous system that is governed by a partial differential equation. Last, the fundamentals of finite element analysis (FEA), which is widely used for vibration analysis of a real structure with a complex shape, are presented. This presentation again shows the application of concepts developed in the context of SDOF and MDOF systems to FEA.

In this chapter, we begin with a discussion of degrees of freedom and the basic elements of a vibratory mechanical system that are a kinetic energy storage element (mass), a potential energy storage element (spring), and an energy dissipation element (damper). Then, an SDOF system with many energy storage and dissipation elements, which are not independent, is considered. It is shown how an equivalent SDOF model with one equivalent mass, one equivalent spring, and one equivalent damper is constructed to facilitate the derivation of the differential equation of motion. Next, the differential equation of motion of an undamped SDOF spring–mass system is derived along with its solution to characterize its vibratory behavior. Then, the solution of the differential equation of motion of an SDOF spring–mass–damper system is obtained and the nature of the response is examined as a function of damping values. Three cases of damping levels, underdamped, critically damped, and overdamped, are treated in detail. Last, the concept of stability of an SDOF spring–mass–damper system is presented along with examples of self-excited oscillations found in practice.

1.1 DEGREES OF FREEDOM

*Degrees of freedom (**DOF**) are the number of independent coordinates that describe the position of a mechanical system at any instant of time.* For example, the system shown in Figure 1.1.1 has one degree of freedom x, which is the displacement of the mass m_1. In spite of the two masses m_1 and m_2 in Figure 1.1.2, this system has only one degree of freedom x because both masses are connected by a rigid link, and the displacements of both masses are not independent. The system shown in Figure 1.1.3 has two degrees of freedom x_1 and x_2 because both masses m_1 and m_2 are connected by a flexible link or a spring, and the displacements of both masses are independent.

Next, consider rigid and flexible continuous cantilever beams as shown in Figures 1.1.4 and 1.1.5. The numbers of degrees of freedom for rigid and flexible beams are 0 and ∞, respectively. Each continuous beam can be visualized to contain an infinite number of point masses. These point masses are connected by rigid links for a rigid beam as shown in Figure 1.1.2, whereas they are connected by flexible links for a flexible beam as shown in Figure 1.1.3. Consequently, there is one degree of freedom associated with each of the point masses in a flexible beam.

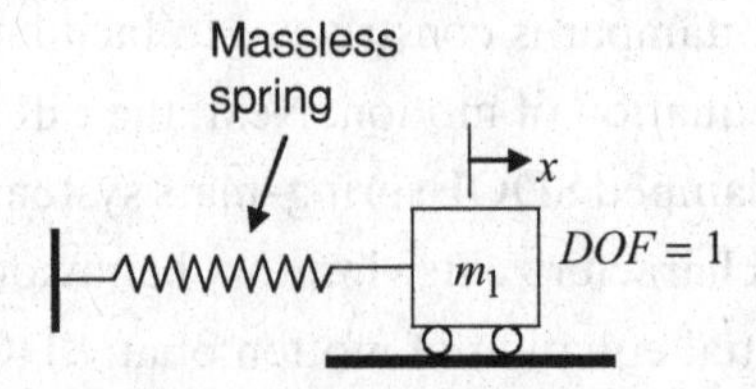

Figure 1.1.1 An SDOF system with a single mass

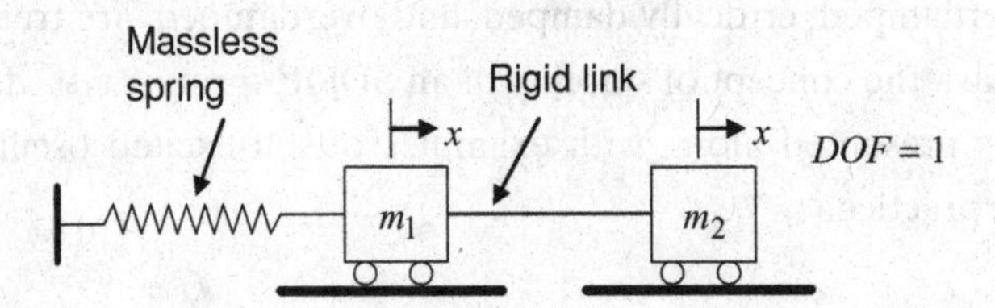

Figure 1.1.2 An SDOF system with two masses

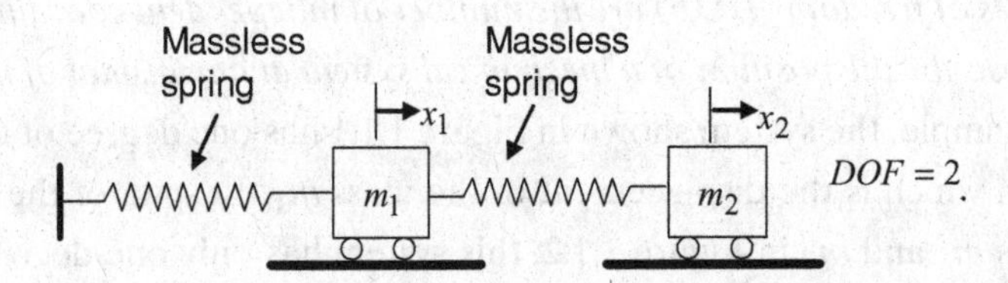

Figure 1.1.3 Two DOF systems with two masses

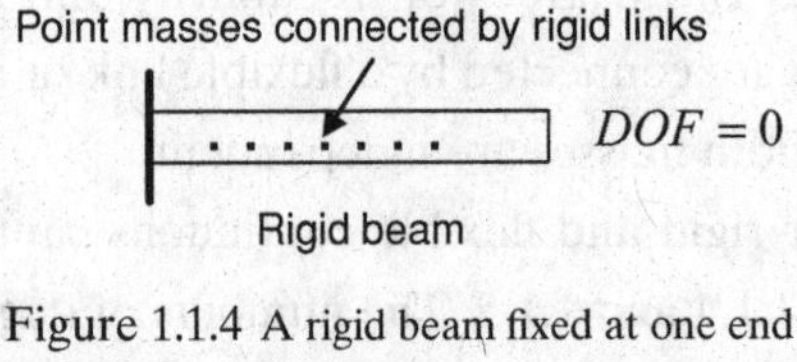

Point masses connected by rigid links

$DOF = 0$

Rigid beam

Figure 1.1.4 A rigid beam fixed at one end

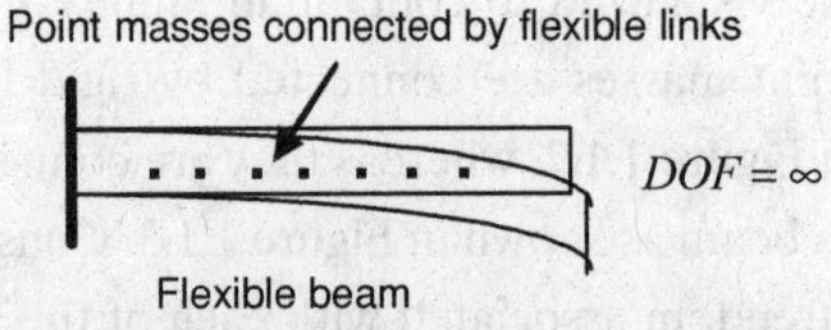

Point masses connected by flexible links

$DOF = \infty$

Flexible beam

Figure 1.1.5 A flexible beam fixed at one end

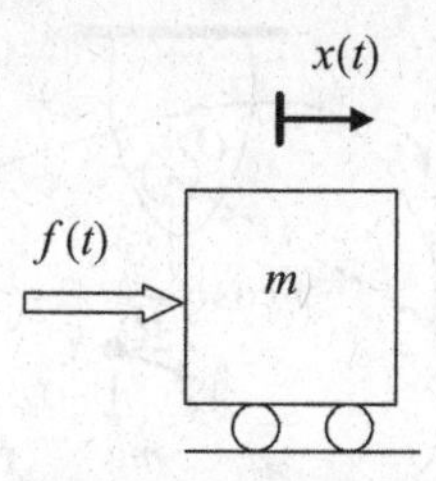

Figure 1.2.1 A mass in pure translation

1.2 ELEMENTS OF A VIBRATORY SYSTEM

There are three basic elements of a vibratory system: a kinetic energy storage element (mass), a potential energy storage element (spring), and an energy dissipation element (damper). The description of each of these three basic elements is as follows.

1.2.1 Mass and/or Mass-Moment of Inertia

Newton's second law of motion and the expression of kinetic energy are presented for three types of motion: pure translational motion, pure rotational motion, and planar (combined translational and rotational) motion.

Pure Translational Motion

Consider the simple mass m (Figure 1.2.1) which is acted upon by a force $f(t)$.

Applying Newton's second law of motion,

$$m\ddot{x} = f(t) \tag{1.2.1}$$

where

$$\dot{x} = \frac{dx}{dt} \quad \text{and} \quad \ddot{x} = \frac{d^2x}{dt^2} \tag{1.2.2a, b}$$

The energy of the mass is stored in the form of kinetic energy (KE):

$$\text{KE} = \frac{1}{2}m\dot{x}^2 \tag{1.2.3}$$

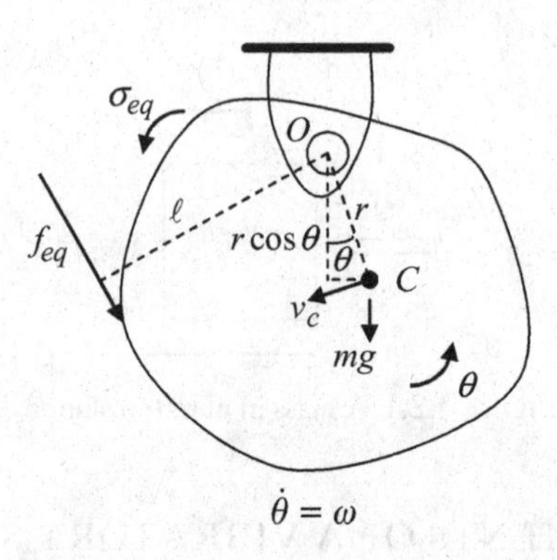

Figure 1.2.2 A mass in pure rotation

Pure Rotational Motion

Consider the mass m (Figure 1.2.2) which is pinned at the point O, and acted upon by an equivalent external force f_{eq} and an equivalent external moment σ_{eq}. This mass is undergoing a pure rotation about the point O, and Newton's second law of motion leads to

$$I_o\ddot{\theta} = -mgr\sin\theta + f_{eq}\ell + \sigma_{eq} \tag{1.2.4}$$

where I_o is the mass-moment of inertia about the center of rotation O, θ is the angular displacement, and ℓ is the length of the perpendicular from the point O to the line of force.

The KE of the rigid body is

$$\text{KE} = \frac{1}{2}I_o\dot{\theta}^2 \tag{1.2.5}$$

The potential energy (PE) of the rigid body is

$$\text{PE} = mg(r - r\cos\theta) \tag{1.2.6}$$

Planar Motion (Combined Rotation and Translation) of a Rigid Body

Consider the planar motion of a rigid body with mass m and the mass-moment of inertia I_c about the axis perpendicular to the plane of motion and passing through the center of mass C (Figure 1.2.3). Forces $f_i, i = 1, 2, \ldots, n_f$, and moments $\sigma_i, i = 1, 2, \ldots, n_t$, are acting on this

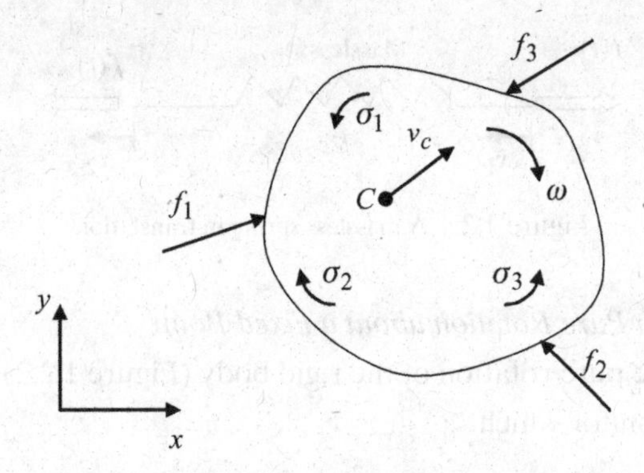

Figure 1.2.3 Planar motion of a rigid body

rigid body. Let x_c and y_c be x- and y- coordinates of the center of mass C with respect to the fixed $x-y$ frame. Then, Newton's second law of motion for the translational part of motion is given by

$$m\ddot{x}_c = \sum_i f_{xi}(t) \tag{1.2.7}$$

$$m\ddot{y}_c = \sum_i f_{yi}(t) \tag{1.2.8}$$

where f_{xi} and f_{yi} are x- and y- components of the force f_i. Newton's second law of motion for the rotational part of motion is given by

$$I_c\ddot{\theta} = I_c\dot{\omega} = \sum_i \sigma_i(t) + \sum_i \sigma_{fi}^c \tag{1.2.9}$$

where σ_{fi}^c is the moment of the force f_i about the center of mass C. And, θ and ω are the angular position and the angular velocity of the rigid body, respectively. The KE of a rigid body in planar motion is given by

$$\text{KE} = \frac{1}{2}mv_c^2 + \frac{1}{2}I_c\omega^2 \tag{1.2.10}$$

where v_c is the magnitude of the linear velocity of the center of mass, that is,

$$v_c^2 = \dot{x}_c^2 + \dot{y}_c^2 \tag{1.2.11}$$

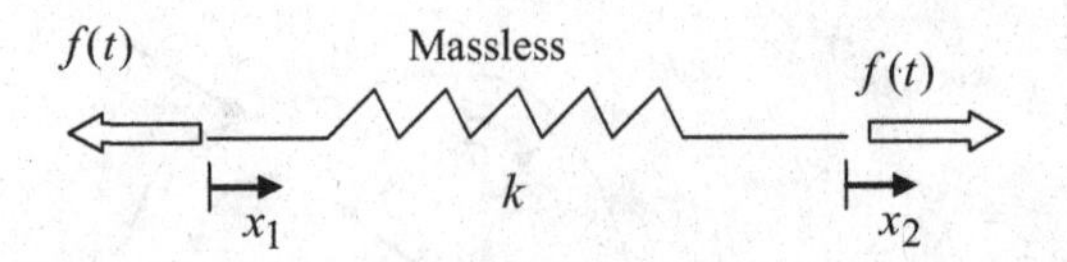

Figure 1.2.4 A massless spring in translation

Special Case: Pure Rotation about a Fixed Point

Note that the pure rotation of the rigid body (Figure 1.2.2) is a special planar motion for which

$$v_c = r\omega \tag{1.2.12}$$

and Equation 1.2.10 leads to

$$\mathrm{KE} = \frac{1}{2}(mr^2 + I_c)\omega^2 \tag{1.2.13}$$

Using the parallel-axis theorem,

$$I_o = I_c + mr^2 \tag{1.2.14}$$

Therefore, Equation 1.2.5 is obtained for the case of a pure rotation about a fixed point.

1.2.2 Spring

The spring constant or stiffness and the expression of PE are presented for two types of motion: pure translational motion and pure rotational motion.

Pure Translational Motion

Consider a massless spring, subjected to a force $f(t)$ on one end (Figure 1.2.4). Because the mass of the spring is assumed to be zero, the net force on the spring must be zero. As a result, there will be an equal and opposite force on the other end. The spring deflection is the difference between the displacements of both ends, that is,

$$\text{spring deflection} = x_2 - x_1 \tag{1.2.15}$$

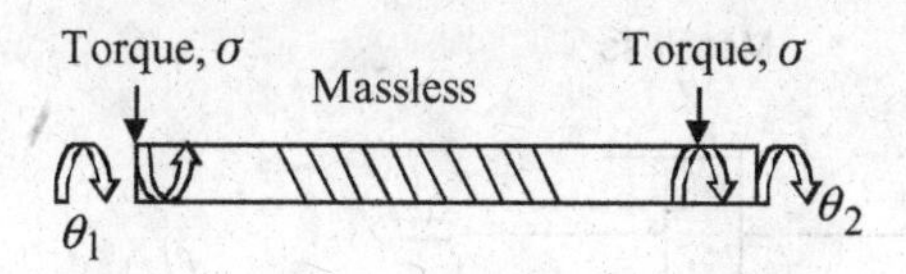

Figure 1.2.5 A massless spring in rotation

and the force is directly proportional to the spring deflection:

$$f(t) = k(x_2 - x_1) \tag{1.2.16}$$

where the proportionality constant k is known as the spring constant or stiffness.

The PE of the spring is given by

$$\text{PE} = \frac{1}{2}k(x_2 - x_1)^2 \tag{1.2.17}$$

It should be noted that the PE is independent of the sign (extension or compression) of the spring deflection, $x_2 - x_1$.

Pure Rotational Motion

Consider a massless torsional spring, subjected to a torque $\sigma(t)$ on one end (Figure 1.2.5). Because the mass of the spring is assumed to be zero, the net torque on the spring must be zero. As a result, there will be an equal and opposite torque on the other end. The spring deflection is the difference between angular displacements of both ends, that is,

$$\text{spring deflection} = \theta_2 - \theta_1 \tag{1.2.18}$$

and the torque is directly proportional to the spring deflection:

$$\sigma(t) = k_t(\theta_2 - \theta_1) \tag{1.2.19}$$

where the proportionality constant k_t is known as the torsional spring constant or torsional stiffness.

The PE of the torsional spring is given by

$$\text{PE} = \frac{1}{2}k_t(\theta_2 - \theta_1)^2 \tag{1.2.20}$$

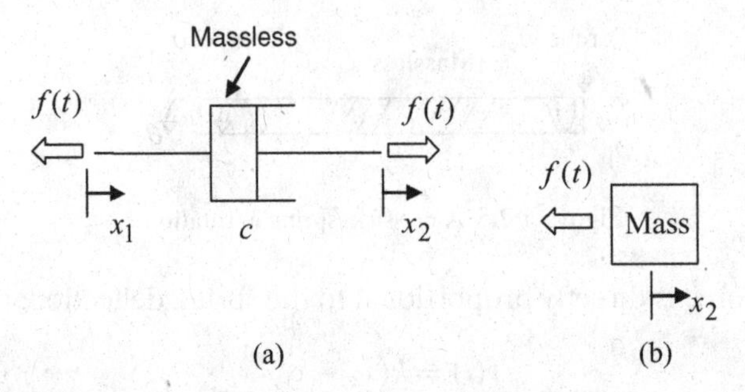

Figure 1.2.6 (a) A massless damper in translation; (b) A mass attached to the right end of the damper

It should be noted that the PE is independent of the sign of the spring deflection, $\theta_2 - \theta_1$.

1.2.3 Damper

The damping constant and the expression of energy dissipation are presented for two types of motion: pure translational motion and pure rotational motion.

Pure Translational Motion

Consider a massless damper, subjected to force $f(t)$ on one end (Figure 1.2.6a). Because the mass of the damper is assumed to be zero, the net force on the damper must be zero. As a result, there will be an equal and opposite force on the other end, and the damper force is directly proportional to the difference of the velocities of both ends:

$$f(t) = c(\dot{x}_2(t) - \dot{x}_1(t)) \tag{1.2.21}$$

where the proportionality constant c is known as the damping constant. The damper defined by Equation 1.2.21 is also known as the linear viscous damper.

If there is a mass attached to the damper at the right end (Figure 1.2.6b) with the displacement x_2, the work done on the mass against the damping force for an infinitesimal displacement dx_2 is

$$f(t)dx_2 = f(t)\frac{dx_2}{dt}dt = f(t)\dot{x}_2 dt \tag{1.2.22}$$

The energy dissipated by the damper equals the work done on mass against the damping force, that is, from Equation 1.2.22,

$$\text{energy dissipated by the damper} = f(t)\dot{x}_2 dt = c(\dot{x}_2 - \dot{x}_1)\dot{x}_2 dt \tag{1.2.23}$$

As an example, consider $x_1(t) = 0$ and $x_2(t) = A\sin(\omega t - \phi)$, where A and ϕ are constants, and ω is the frequency of oscillation. In this case, from Equation 1.2.23,

$$\text{energy dissipated by the damper} = c\dot{x}_2^2 dt = cA^2\omega^2\cos^2(\omega t - \phi)dt \tag{1.2.24}$$

Substituting $\upsilon = \omega t - \phi$ into Equation 1.2.24,

$$\text{energy dissipated by the damper} = cA^2\omega\cos^2\upsilon d\upsilon \tag{1.2.25}$$

As a result, the energy dissipated by the damper per cycle of oscillation is

$$\int_0^{2\pi} cA^2\omega\cos^2\upsilon d\upsilon = \pi c\omega A^2 \tag{1.2.26}$$

It should be noted that the energy dissipated by the viscous damper per cycle of oscillation is proportional to the square of the vibration amplitude.

Pure Rotational Motion

Consider a massless torsional damper, subjected to a torque $\sigma(t)$ on one end (Figure 1.2.7). Because the mass of the damper is assumed to be zero, the net torque on the damper must be zero. As a result, there will be an equal and opposite torque on the other end, and the damper torque is directly proportional to the difference of the angular velocities of both ends:

$$\sigma(t) = c_t(\dot{\theta}_2(t) - \dot{\theta}_1(t)) \tag{1.2.27}$$

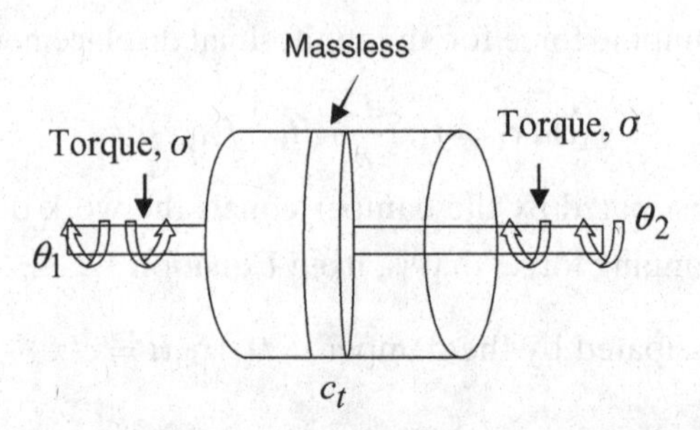

Figure 1.2.7 A massless torsional damper

where the proportionality constant c_t is known as the damping constant. The damper defined by Equation 1.2.27 is also known as the linear viscous damper.

If there is a mass attached to the damper at the right end with the angular displacement θ_2, the work done against the damping torque for an infinitesimal displacement $d\theta_2$ is

$$\sigma(t)d\theta_2 = \sigma(t)\frac{d\theta_2}{dt}dt = \sigma(t)\dot{\theta}_2 dt \tag{1.2.28}$$

The energy dissipated by the damper equals the work done on mass against the damping torque, that is, from Equation 1.2.28,

$$\text{energy dissipated by the damper} = \sigma(t)\dot{\theta}_2 dt = c_t(\dot{\theta}_2 - \dot{\theta}_1)\dot{\theta}_2 dt \tag{1.2.29}$$

The expression similar to Equation 1.2.26 can be derived for the energy dissipated by the torsional viscous damper per cycle of oscillation.

1.3 EQUIVALENT MASS, EQUIVALENT STIFFNESS, AND EQUIVALENT DAMPING CONSTANT FOR AN SDOF SYSTEM

In this section, equivalent mass, equivalent stiffness, and equivalent damping constant are derived for a rotor–shaft system, spring with

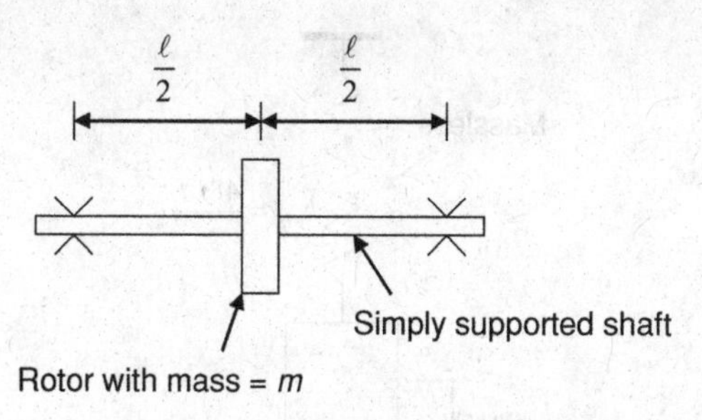

Figure 1.3.1 A rotor–shaft system

nonnegligible mass, parallel and series combinations of springs and dampers, and a combined rotational and translational system.

1.3.1 A Rotor–Shaft System

Consider a rotor with mass m which is supported at the mid-span of a simply supported shaft of length ℓ (Figure 1.3.1). The mass of the shaft is negligible in comparison with the mass of the rotor. For the purpose of transverse vibration modeling, the shaft is considered as a simply supported beam (Crandall et al., 1999). For a simply supported beam, when a load P is applied at the mid-span (Figure 1.3.2), the deflection δ of the shaft at the mid-span is obtained from the results provided in Appendix A (Equation A.2) by

$$\delta = \frac{P\ell^3}{48EI} \tag{1.3.1}$$

where E and I are the Young's modulus of elasticity and the area moment of inertia, respectively.

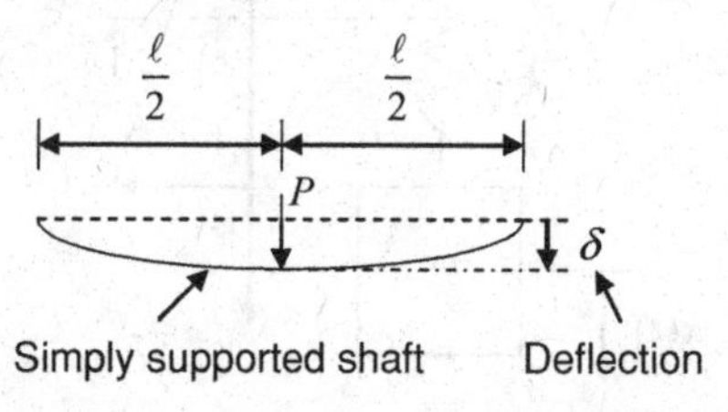

Figure 1.3.2 Deflection of a simply supported shaft

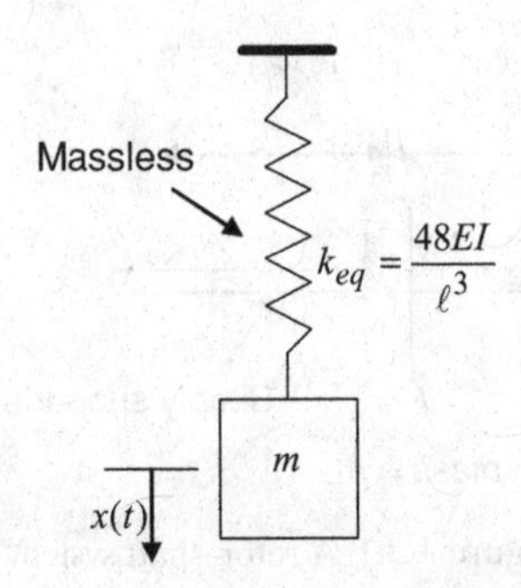

Figure 1.3.3 Equivalent SDOF system for a rotor–shaft system

Therefore, the equivalent stiffness of the shaft is defined as

$$k_{eq} = \frac{P}{\delta} = \frac{48EI}{\ell^3} \tag{1.3.2}$$

And, an equivalent SDOF system can be constructed as shown in Figure 1.3.3, where $x(t)$ is the displacement of the rotor mass.

1.3.2 Equivalent Mass of a Spring

Consider an SDOF system (Figure 1.3.4) in which the mass of the spring m_s is not negligible with respect to the main mass m.

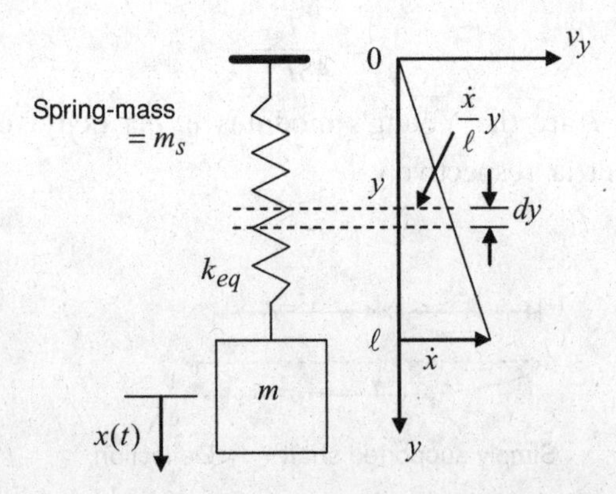

Figure 1.3.4 A spring–mass system with nonnegligible mass of spring

Then, an equivalent system with a massless spring can be obtained on the basis of the total KE. Let the length of the spring be ℓ, and assume that the mass of the spring is uniformly distributed over its length. Then, the mass of a spring strip of the length dy will be

$$dm_y = \frac{m_s}{\ell} dy \tag{1.3.3}$$

To determine the KE, the velocity v_y of the spring strip at a distance y from the base of the spring must be known. But, we only have the following information:

$$@y = 0,\ v_y = 0 \tag{1.3.4}$$

and

$$@y = \ell,\ v_y = \dot{x}\, v_y = \dot{x} \tag{1.3.5}$$

The velocity at an intermediate point ($0 < y < \ell$) is not known. Therefore, it is **assumed** that the velocity profile over the length of the spring is linear as shown in Figure 1.3.4, that is,

$$v_y = \frac{\dot{x}}{\ell} y \tag{1.3.6}$$

The KE of the spring strip of length dy at a distance y from the base of the spring is

$$dKE_s = \frac{1}{2} dm_y (v_y)^2 \tag{1.3.7}$$

Using Equations (1.3.3) and (1.3.7),

$$dKE_s = \frac{1}{2}\frac{m_s}{\ell}\left(\frac{\dot{x}}{\ell} y\right)^2 dy \tag{1.3.8}$$

Therefore, the total KE of the spring is given by

$$KE_s = \int_0^\ell \frac{1}{2}\frac{m_s}{\ell}\left(\frac{\dot{x}}{\ell} y\right)^2 dy = \frac{1}{2}\frac{m_s \dot{x}^2}{\ell^3}\int_0^\ell y^2 dy = \frac{1}{2}\frac{m_s}{3}\dot{x}^2 \tag{1.3.9}$$

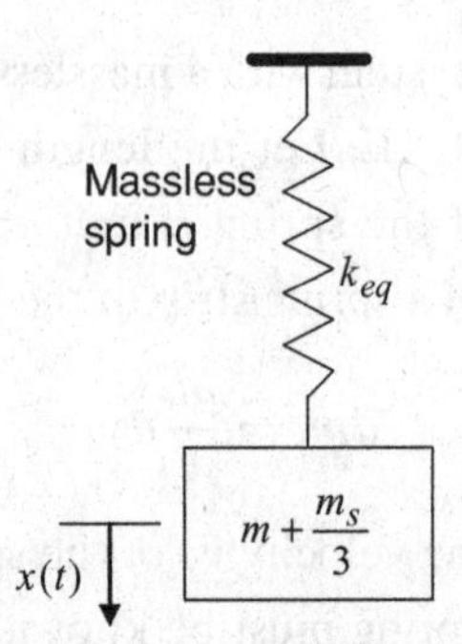

Figure 1.3.5 Equivalent SDOF system with a massless spring

Hence, the total KE of the system with a nonnegligible spring mass is given by

$$\text{KE} = \frac{1}{2}m\dot{x}^2 + \frac{1}{2}\frac{m_s}{3}\dot{x}^2 = \frac{1}{2}m_{eq}\dot{x}^2 \tag{1.3.10}$$

where the equivalent mass m_{eq} is given by

$$m_{eq} = m + \frac{m_s}{3} \tag{1.3.11}$$

And, the equivalent SDOF system can be created with a massless spring and a mass m_{eq} as shown in Figure 1.3.5. The systems shown in Figures 1.3.4 and 1.3.5 will have the same amount of KE under the assumption of a linear velocity profile for the spring.

1.3.3 Springs in Series and Parallel

Springs in Series

Consider a series combination of massless springs with stiffnesses k_1 and k_2 (Figure 1.3.6).

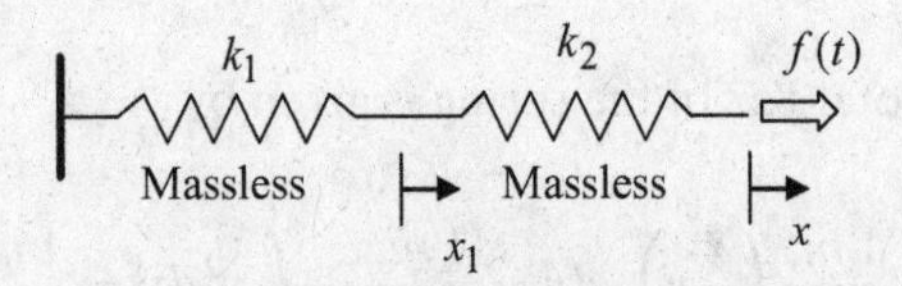

Figure 1.3.6 Springs in series

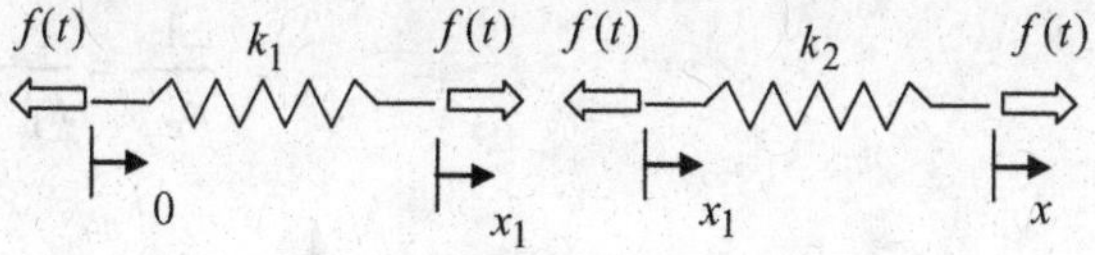

Figure 1.3.7 Free body diagrams for springs in series

The free body diagram of each spring is shown in Figure 1.3.7 for which the following relationships can be written:

$$f(t) = k_1 x_1 \tag{1.3.12}$$

and

$$f(t) = k_2(x - x_1) \tag{1.3.13}$$

It is important to note that both springs, which are in series, carry the same amount of force. From Equations 1.3.12 and 1.3.13,

$$x_1 = \frac{k_2}{k_1 + k_2} x \tag{1.3.14}$$

Substituting Equation 1.3.14 into Equation 1.3.13,

$$f(t) = k_{eq} x \tag{1.3.15}$$

where

$$k_{eq} = \frac{k_1 k_2}{k_1 + k_2} \tag{1.3.16}$$

Here, k_{eq} is the equivalent stiffness and the system in Figure 1.3.6 can be replaced by a system with only one spring with the stiffness k_{eq} as shown in Figure 1.3.8.

Equation 1.3.16 can also be written as

$$\frac{1}{k_{eq}} = \frac{1}{k_1} + \frac{1}{k_2} \tag{1.3.17}$$

Springs in Parallel

Consider a parallel combination of massless springs with stiffnesses k_1 and k_2 (Figure 1.3.9).

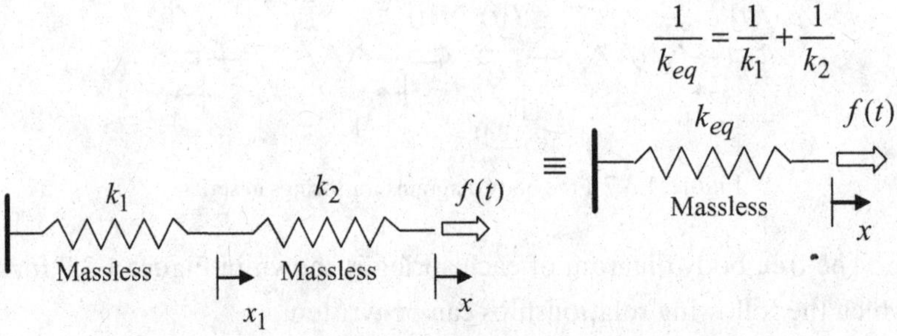

$$\frac{1}{k_{eq}} = \frac{1}{k_1} + \frac{1}{k_2}$$

Figure 1.3.8 Equivalent system with only one spring for a series combination

The free body diagram of each spring is shown in Figure 1.3.10 for which the following relationships can be written:

$$f_1(t) = k_1 x \tag{1.3.18}$$

$$f_2(t) = k_2 x \tag{1.3.19}$$

and

$$f(t) = f_1(t) + f_2(t) \tag{1.3.20}$$

It is important to note that both springs, which are in parallel, undergo the same amount of deflection. Substituting Equations 1.3.18 and 1.3.19 into Equation 1.3.20,

$$f(t) = k_{eq} x \tag{1.3.21}$$

where

$$k_{eq} = k_1 + k_2 \tag{1.3.22}$$

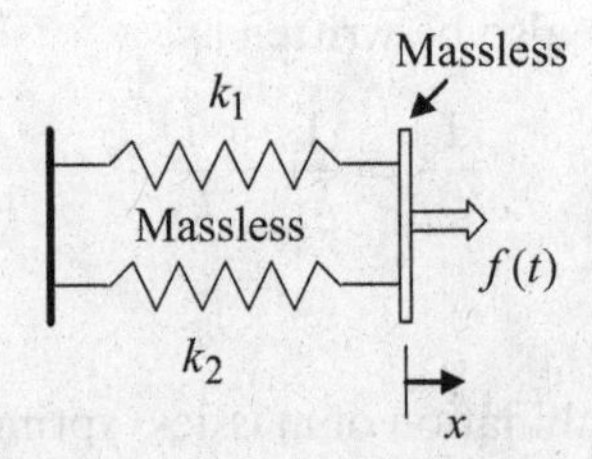

Figure 1.3.9 A parallel combination of springs

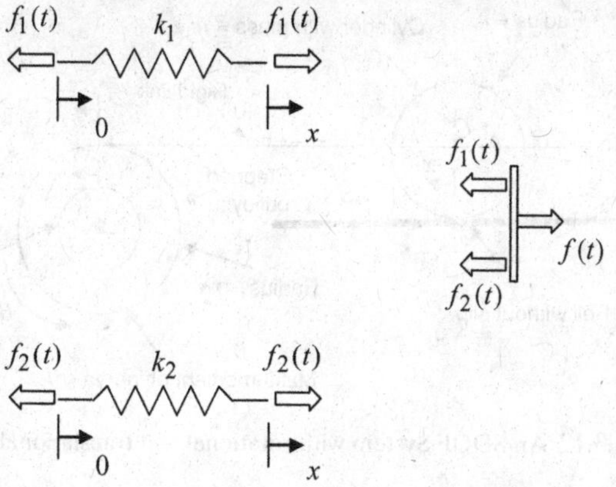

Figure 1.3.10 Free body diagrams for springs in parallel

Here, k_{eq} is the equivalent stiffness and the system in Figure 1.3.9 can be replaced by a system with only one spring with the stiffness k_{eq} as shown in Figure 1.3.11.

1.3.4 An SDOF System with Two Springs and Combined Rotational and Translational Motion

Consider the system shown in Figure 1.3.12, in which a cylinder rolls without slipping. The displacement of the cylinder's center of mass, marked as C, is denoted by the symbol $x(t)$. Because of the displacement $x(t)$, the stepped pulley rotates by an amount $\theta(t)$. For a small displacement $x(t)$, from Figure 1.3.13,

$$\theta(t) = \frac{x(t)}{r_2} \tag{1.3.23}$$

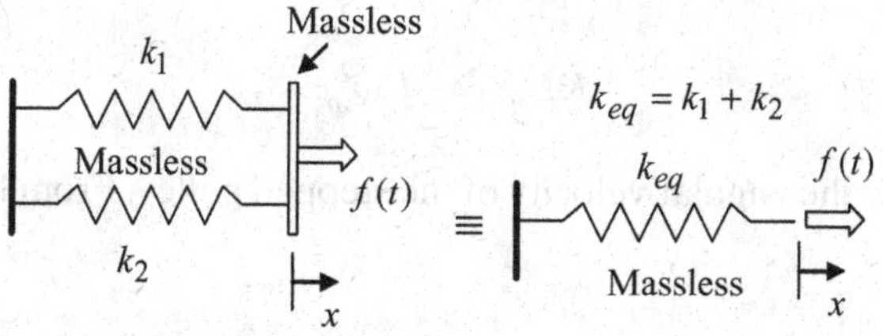

Figure 1.3.11 Equivalent system with only one spring for a parallel combination

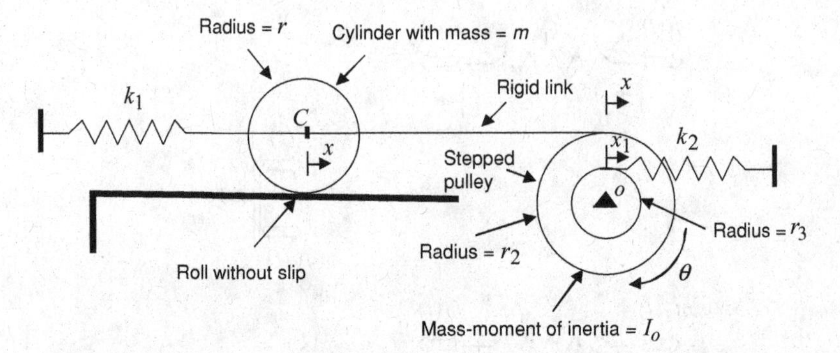

Figure 1.3.12 An SDOF System with rotational and translational motion

Using Equation 1.2.10, the KE of the cylinder is given as

$$KE_{cyl} = \frac{1}{2}m\dot{x}^2 + \frac{1}{2}I_c\omega^2 \tag{1.3.24}$$

where the mass-moment of inertia of the cylinder about C is

$$I_c = \frac{1}{2}mr^2 \tag{1.3.25}$$

and the angular velocity ω of the cylinder for rolling without slipping is

$$\omega = \frac{\dot{x}}{r} \tag{1.3.26}$$

Substituting Equations 1.3.25 and 1.3.26 into Equation 1.3.24,

$$KE_{cyl} = \frac{1}{2}1.5m\dot{x}^2 \tag{1.3.27}$$

The KE of the stepped pulley is

$$KE_{sp} = \frac{1}{2}I_o\omega_{sp}^2 \tag{1.3.28}$$

where ω_{sp} is the angular velocity of the stepped pulley. From Equation 1.3.23,

$$\omega_{sp} = \dot{\theta} = \frac{\dot{x}}{r_2} \tag{1.3.29}$$

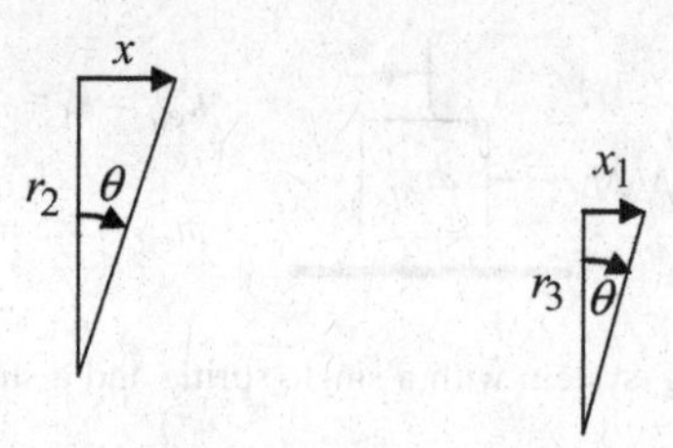

Figure 1.3.13 Displacement and equivalent rotation

Substituting Equation 1.3.29 into Equation 1.3.28,

$$KE_{sp} = \frac{1}{2}\frac{I_o}{r_2^2}\dot{x}^2 \tag{1.3.30}$$

From Equations 1.3.24 and 1.3.30, the total KE in the system is

$$KE_{tot} = KE_{cyl} + KE_{sp} = \frac{1}{2}m_{eq}\dot{x}^2 \tag{1.3.31}$$

where

$$m_{eq} = 1.5m + \frac{I_0}{r_2^2} \tag{1.3.32}$$

From Figure 1.3.12, the total PE in the system is

$$PE_{tot} = \frac{1}{2}k_1x^2 + \frac{1}{2}k_2x_1^2 \tag{1.3.33}$$

where from Figure 1.3.13,

$$x_1 = r_3\theta \tag{1.3.34}$$

Substituting Equation 1.3.34 into Equation 1.3.33 and using Equation 1.3.23,

$$PE_{tot} = \frac{1}{2}k_{eq}x^2 \tag{1.3.35}$$

where

$$k_{eq} = k_1 + k_2\left(\frac{r_3}{r_2}\right)^2 \tag{1.3.36}$$

Here, m_{eq} (Equation 1.3.32) and k_{eq} (Equation 1.3.36) are the equivalent mass and the equivalent stiffness of the SDOF system

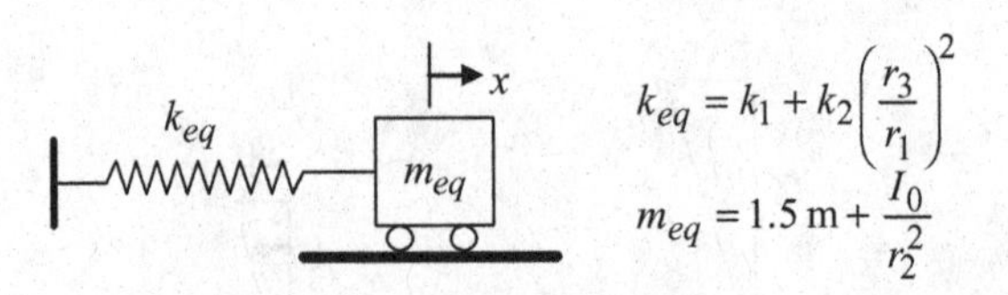

Figure 1.3.14 An SDOF system with a single spring and a single mass equivalent to system in Figure 1.3.12

shown in Figure 1.3.12. And, the system shown in Figure 1.3.12 can also be described as the equivalent SDOF (Figure 1.3.14).

1.3.5 Viscous Dampers in Series and Parallel

Dampers in Series

Consider a series combination of massless viscous dampers with damping coefficients c_1 and c_2 (Figure 1.3.15).

The free body diagram of each damper is shown in Figure 1.3.16 for which the following relationships can be written:

$$f(t) = c_1\dot{x}_1 \tag{1.3.37}$$

$$f(t) = c_2(\dot{x} - \dot{x}_1) \tag{1.3.38}$$

An important point to note here is that both dampers, which are in series, carry the same amount of force. From Equations 1.3.37 and 1.3.38,

$$\dot{x}_1 = \frac{c_2}{c_1 + c_2}\dot{x} \tag{1.3.39}$$

Substituting Equation 1.3.39 into Equation 1.3.38,

$$f(t) = c_{eq}\dot{x} \tag{1.3.40}$$

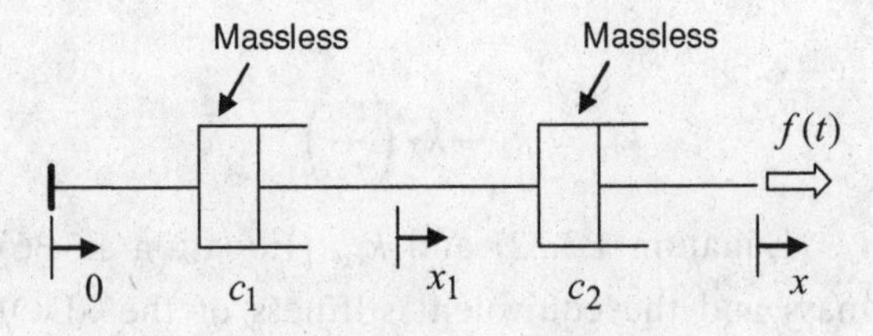

Figure 1.3.15 Dampers in series

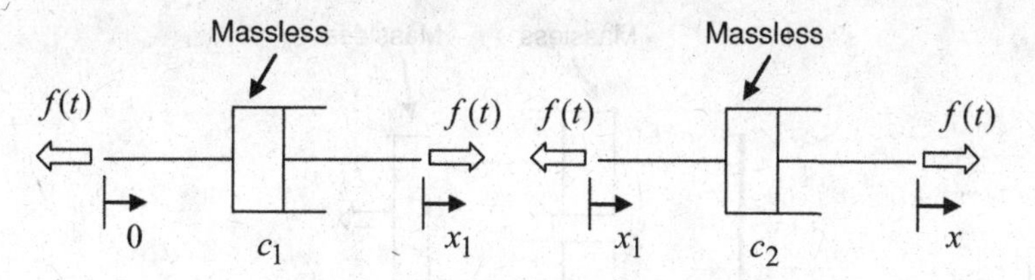

Figure 1.3.16 Free body diagrams for dampers in series

where

$$c_{eq} = \frac{c_1 c_2}{c_1 + c_2} \tag{1.3.41}$$

Here, c_{eq} is the equivalent damping coefficient and the system in Figure 1.3.15 can be replaced by a system with only one damper with the coefficient c_{eq} as shown in Figure 1.3.17. Equation 1.3.41 can also be written as

$$\frac{1}{c_{eq}} = \frac{1}{c_1} + \frac{1}{c_2} \tag{1.3.42}$$

Dampers in Parallel

Consider a parallel combination of massless dampers with coefficients c_1 and c_2 (Figure 1.3.18).

The free body diagram of each damper is shown in Figure 1.3.19 for which the following relationships can be written:

$$f_1(t) = c_1 \dot{x} \tag{1.3.43}$$

$$f_2(t) = c_2 \dot{x} \tag{1.3.44}$$

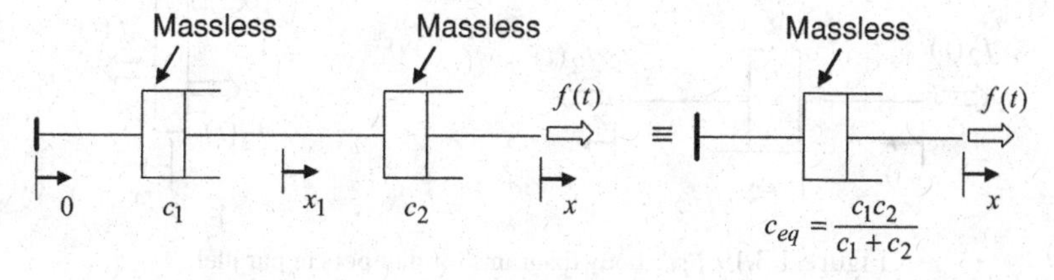

Figure 1.3.17 Equivalent system with only one damper for a series combination

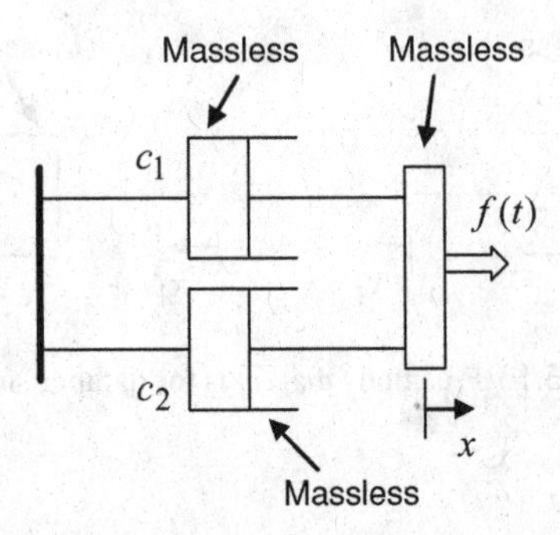

Figure 1.3.18 A parallel combination of dampers

and

$$f(t) = f_1(t) + f_2(t) \tag{1.3.45}$$

An important point to note here is that both dampers, which are in parallel, have the same velocity. Substituting Equations 1.3.43 and 1.3.44 into Equation 1.3.45,

$$f(t) = c_{eq}\dot{x} \tag{1.3.46}$$

where

$$c_{eq} = c_1 + c_2 \tag{1.3.47}$$

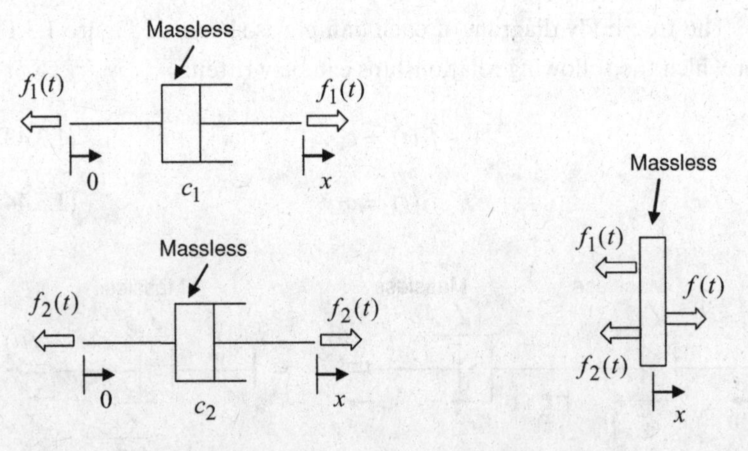

Figure 1.3.19 Free body diagrams for dampers in parallel

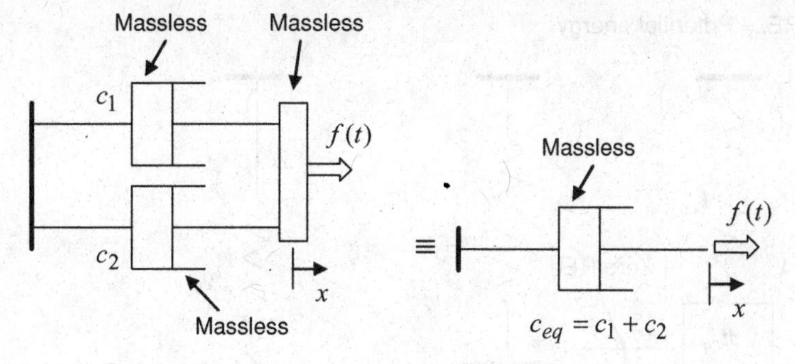

Figure 1.3.20 Equivalent system with only one damper for a parallel combination

Here, c_{eq} is the equivalent damping constant and the system in Figure 1.3.18 can be replaced by a system with only one damper with the coefficient c_{eq} as shown in Figure 1.3.20.

1.4 FREE VIBRATION OF AN UNDAMPED SDOF SYSTEM

This section deals with the derivation and the solution of the differential equation of motion of an undamped SDOF system. The solution of the differential equation of motion is used to characterize the nature of free vibration.

1.4.1 Differential Equation of Motion

Consider an SDOF spring–mass system with equivalent spring stiffness k_{eq} and equivalent mass m_{eq} (Figure 1.4.1). First, consider the system with the unstretched spring (Figure 1.4.1). As we let the mass be under gravity, the spring will deflect due to the weight. There will be a static equilibrium configuration where the net force on the mass will be zero. The free body diagram is shown in Figure 1.4.2, where Δ is the static deflection or the deflection of the spring in the static equilibrium configuration. For the static equilibrium condition,

$$k_{eq}\Delta = m_{eq}g \tag{1.4.1}$$

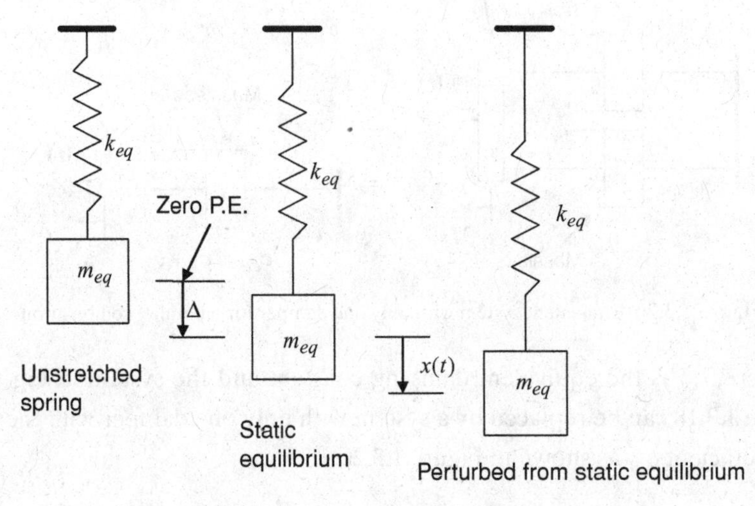

Figure 1.4.1 An undamped SDOF spring–mass system

Let the displacement $x(t)$ be a perturbation from the static equilibrium (Figure 1.4.1). From the free body diagram in Figure 1.4.2 (Perturbed from Static Equilibrium),

$$\text{net force in } x\text{-direction} = -k_{eq}(x + \Delta) + m_{eq}g \qquad (1.4.2)$$

Newton's second law of motion states that

$$\text{Net force in } x\text{-direction} = \text{mass} \times \text{acceleration} \qquad (1.4.3)$$

Therefore,

$$-k_{eq}(x + \Delta) + m_{eq}g = m_{eq}\ddot{x} \qquad (1.4.4)$$

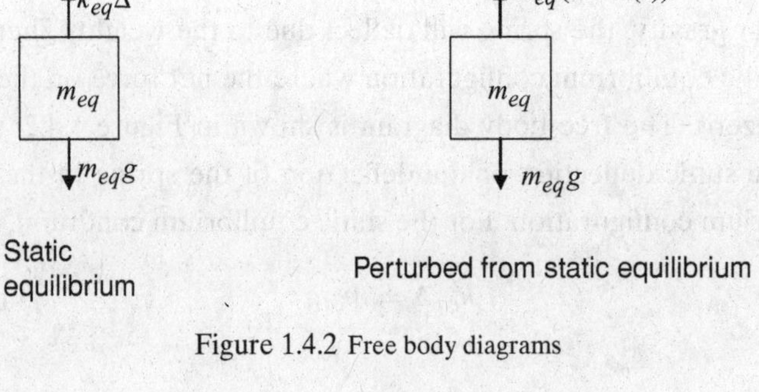

$k_{eq}\Delta$

m_{eq}

$m_{eq}g$

Static equilibrium

$k_{eq}(\Delta + x(t))$

m_{eq}

$m_{eq}g$

Perturbed from static equilibrium

Figure 1.4.2 Free body diagrams

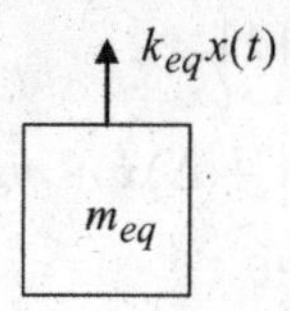

Perturbed from static equilibrium

Figure 1.4.3 Equivalent free body diagram after canceling $k_{eq}\Delta$ and $m_{eq}g$

Using Equation 1.4.1,

$$m_{eq}\ddot{x} + k_{eq}x = 0 \tag{1.4.5}$$

The same differential equation is obtained by neglecting the weight $m_{eq}g$ and the spring force $k_{eq}\Delta$ as well. The resulting free body diagram is shown in Figure 1.4.3, and

$$\text{net force in } x\text{-direction} = -k_{eq}x \tag{1.4.6}$$

Newton's second law of motion yields

$$-k_{eq}x = m_{eq}\ddot{x} \tag{1.4.7}$$

Energy Approach

The KE of the system (perturbed from static equilibrium in Figure 1.4.1 is given by

$$U = \frac{1}{2}m_{eq}\dot{x}^2 \tag{1.4.8}$$

and the PE is given by

$$P = \frac{1}{2}k_{eq}(x + \Delta)^2 - m_{eq}g(x + \Delta) \tag{1.4.9}$$

The total energy is

$$T = U + P = \frac{1}{2}m_{eq}\dot{x}^2 + \frac{1}{2}k_{eq}(x + \Delta)^2 - m_{eq}g(x + \Delta) \tag{1.4.10}$$

Since there is no sink (damping) or source (external force) of energy, the total energy T is a constant. Therefore,

$$\frac{dT}{dt} = 0 \tag{1.4.11}$$

From Equations 1.4.10 and 1.4.11,

$$\frac{1}{2}m_{eq}2\dot{x}\ddot{x} + \frac{1}{2}k_{eq}2(x+\Delta)\dot{x} - m_{eq}g\dot{x} = 0 \tag{1.4.12}$$

or

$$(m_{eq}\ddot{x} + k_{eq}x + k_{eq}\Delta - m_{eq}g)\dot{x} = 0 \tag{1.4.13}$$

Because of the static equilibrium condition in Equation 1.4.1,

$$(m_{eq}\ddot{x} + k_{eq}x)\dot{x} = 0 \tag{1.4.14}$$

As $\dot{x}(t)$ is not zero for all t,

$$m_{eq}\ddot{x} + k_{eq}x = 0 \tag{1.4.15}$$

The same differential equation of motion can be obtained by writing the PE as

$$P = \frac{1}{2}k_{eq}x^2 \tag{1.4.16}$$

Note that the contributions of the weight mg and the corresponding static deflection Δ have been simultaneously neglected.

The total energy is given by

$$T = U + P = \frac{1}{2}m_{eq}\dot{x}^2 + \frac{1}{2}k_{eq}x^2 \tag{1.4.17}$$

The condition in Equation 1.4.11 yields

$$(m_{eq}\ddot{x} + k_{eq}x)\dot{x} = 0 \tag{1.4.18}$$

Therefore,

$$m_{eq}\ddot{x} + k_{eq}x = 0 \tag{1.4.19}$$

Example 1.4.1: A Horizontal Rigid Bar

Consider the system shown in Figure 1.4.4a, in which a rigid bar, pinned at point A, and is supported by springs with stiffnesses k_1 and k_2 which are located at distances a and b from the pin A, respectively. Under the gravity, the bar will rotate to be in the static equilibrium

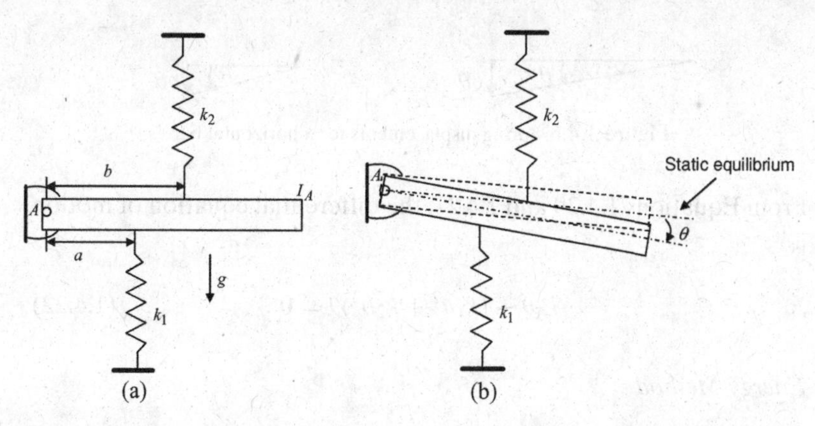

Figure 1.4.4 (a) A horizontal bar with an unstretched spring; (b) Perturbation from static equilibrium configuration

configuration. Let $\theta(t)$ be the **small** angular displacement of the bar from the static equilibrium position (Figure 1.4.4b). The free body diagram of the bar is shown in Figure 1.4.5. Note that the weight of the bar and spring forces due to static deflections (in static equilibrium configuration) have not been included as they cancel out. It should also be noted that the spring deflections with respect to static equilibrium configurations are calculated for a small angular displacement and are found (Figure 1.4.6) to be $b\theta$ and $a\theta$ for springs with stiffnesses k_2 and k_1, respectively.

$$\text{Net torque about point } A \text{ in } \theta \text{ direction} = -k_1 a\theta a - k_2 b\theta b \quad (1.4.20)$$

Applying Newton's second law of motion,

$$\text{net torque about point } A \text{ in } \theta \text{ direction} = I_A \ddot{\theta} \quad (1.4.21)$$

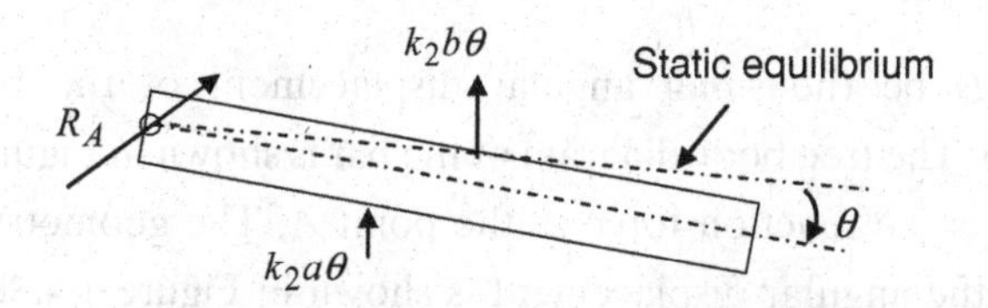

Figure 1.4.5 Free body diagram for a horizontal bar

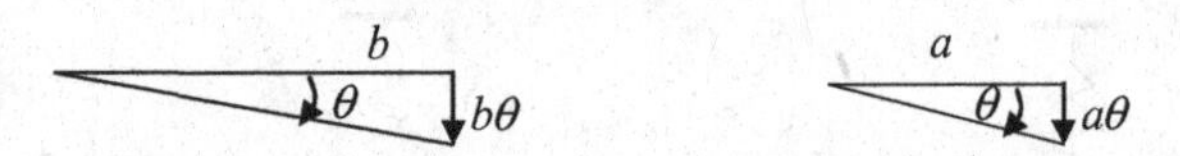

Figure 1.4.6 Spring displacements for a horizontal bar

From Equations 1.4.20 and 1.4.21, the differential equation of motion is

$$I_A\ddot{\theta} + (k_1a^2 + k_2b^2)\theta = 0 \tag{1.4.22}$$

Energy Method

$$\text{KE}, U = \frac{1}{2}I_A\dot{\theta}^2 \tag{1.4.23}$$

$$\text{PE}, P = \frac{1}{2}k_1(a\theta)^2 + \frac{1}{2}k_2(b\theta)^2 \tag{1.4.24}$$

The total energy is

$$T = U + P = \frac{1}{2}I_A\dot{\theta}^2 + \frac{1}{2}k_1(a\theta)^2 + \frac{1}{2}k_2(b\theta)^2 \tag{1.4.25}$$

The condition of a constant value of the total energy yields

$$\frac{dT}{dt} = 0 \Rightarrow I_A\ddot{\theta} + (k_1a^2 + k_2b^2)\theta = 0 \tag{1.4.26}$$

Example 1.4.2: A Vertical Rigid Bar

Consider the rigid bar of mass m in the vertical configuration (Figure 1.4.7a). This bar is pinned at the point A, and is connected to a spring with the stiffness k at a distance b from the point A. The center of the gravity of the bar C, is located at a distance a from the pin A.

Let $\theta(t)$ be the **small** angular displacement of the bar (Figure 1.4.7b). The free body diagram of the bar is shown in Figure 1.4.8a, where R_A is the reaction force at the point A. The geometry associated with the angular displacement is shown in Figure 1.4.8b, where the angle $\theta(t)$ has been magnified for the sake of clarity. The spring

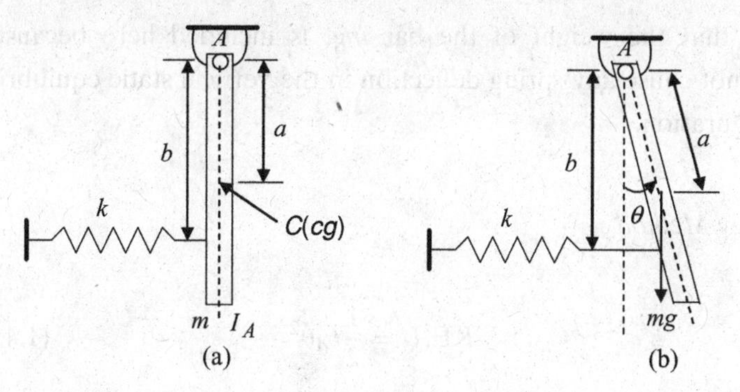

Figure 1.4.7 (a) Vertical bar in static equilibrium; (b) Vertical bar perturbed from static equilibrium

deflection is $b\sin\theta$, which is approximated as $b\theta$ for a small θ. Taking moment about the point A,

$$-kb\theta b\cos\theta - mga\sin\theta = I_A\ddot{\theta} \tag{1.4.27}$$

where I_A is the mass-moment of inertia about A. For a small θ, $\sin\theta \approx \theta$ and $\cos\theta \approx 1$.

Therefore,

$$I_A\ddot{\theta} + (mga + kb^2)\theta = 0 \tag{1.4.28}$$

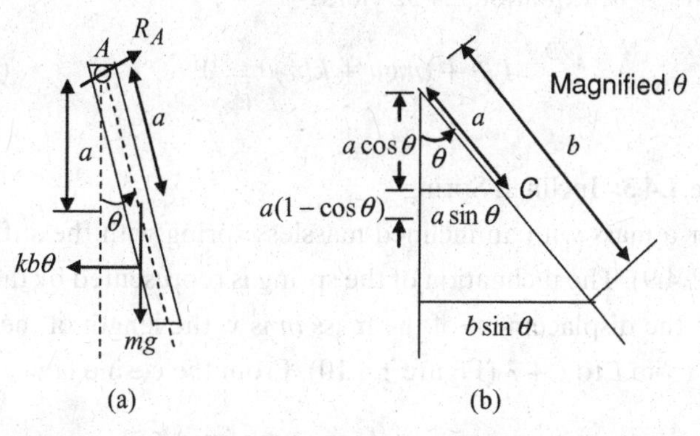

Figure 1.4.8 (a) Free body diagram for vertical rigid bar; (b) Geometry associated with Figure 1.4.8a

Note that the weight of the bar mg, is included here because it does not cause any spring deflection in the vertical static equilibrium configuration.

Energy Method

$$\text{KE}, U = \frac{1}{2} I_A \dot{\theta}^2 \tag{1.4.29}$$

$$\text{PE}, P = mga(1 - \cos\theta) + \frac{1}{2} k(b\theta)^2 \tag{1.4.30}$$

In Equation 1.4.30, $1 - \cos\theta$ is almost equal to zero as $\cos\theta \approx 1$. However, $1 - \cos\theta$ is not negligible in comparison with θ^2, which is also quite small. Therefore, $\cos\theta \approx 1$ should not be used in Equation 1.4.30.

The total energy is given by

$$T = \frac{1}{2} I_A \dot{\theta}^2 + mga(1 - \cos\theta) + \frac{1}{2} k(b\theta)^2 \tag{1.4.31}$$

Because the total energy is a constant,

$$\frac{dT}{dt} = [I_A \ddot{\theta} + mga \sin\theta + kb^2\theta]\dot{\theta} = 0 \tag{1.4.32}$$

Using $\sin\theta \approx \theta$, Equation 1.4.32 yields

$$I_A \ddot{\theta} + (mga + kb^2)\theta = 0 \tag{1.4.33}$$

Example 1.4.3: Inclined Spring

Consider a mass with an inclined massless spring with the stiffness k (Figure 1.4.9). The inclination of the spring is represented by the angle α. When the displacement of the mass m is x, the length of the spring changes from ℓ to $\ell + \delta$ (Figure 1.4.10). From the cosine law,

$$\cos(\pi - \alpha) = \frac{\ell^2 + x^2 - (\ell + \delta)^2}{2\ell x} \tag{1.4.34}$$

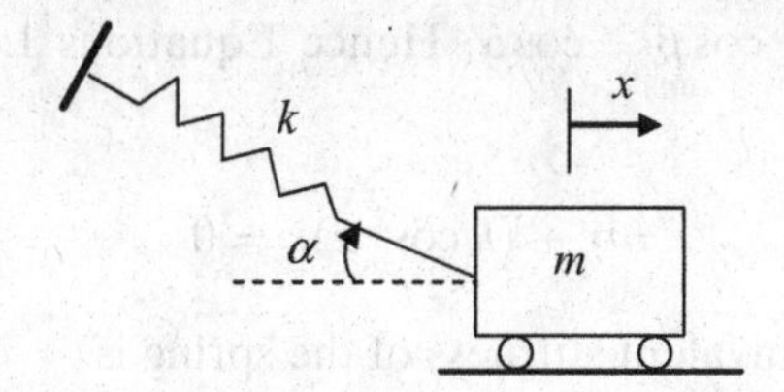

Figure 1.4.9 A mass connected with an inclined spring

After some algebra,

$$(\ell+\delta)^2 = \ell^2\left[1+2\frac{x}{\ell}\cos\alpha+\frac{x^2}{\ell^2}\right] \tag{1.4.35}$$

Assume that $x/\ell \ll 1$. In this case, the term $(x/\ell)^2$ can be neglected and

$$\ell+\delta = \ell\left[1+2\frac{x}{\ell}\cos\alpha\right]^{0.5} \tag{1.4.36}$$

Using the binomial expansion (Appendix B) and neglecting higher-order terms,

$$\ell+\delta \cong \ell\left[1+\frac{x}{\ell}\cos\alpha\right] \tag{1.4.37}$$

Therefore,

$$\delta \cong x\cos\alpha \tag{1.4.38}$$

The spring force $k\delta$ will be directed at an angle β (Figure 1.4.10). Applying Newton's second law in the x-direction,

$$-k\delta\cos\beta = m\ddot{x} \tag{1.4.39}$$

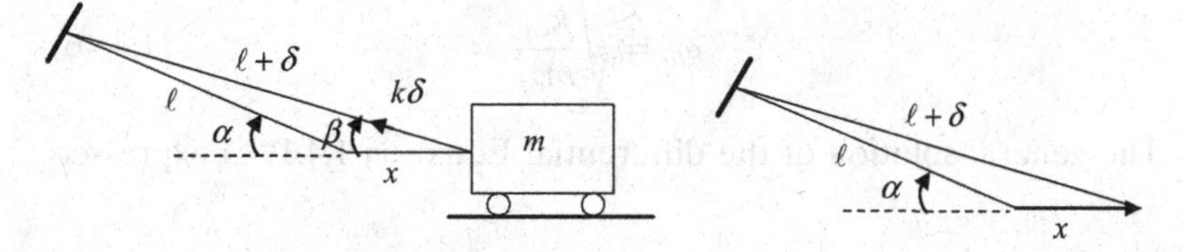

Figure 1.4.10 Free body diagram for system in Figure 1.4.9

Because $x/\ell \ll 1$, $\cos\beta \cong \cos\alpha$. Hence Equations 1.4.38 and 1.4.39 lead to

$$m\ddot{x} + (k\cos^2\alpha)x = 0 \tag{1.4.40}$$

Therefore, the equivalent stiffness of the spring is

$$k_{eq} = k\cos^2\alpha \tag{1.4.41}$$

1.4.2 Solution of the Differential Equation of Motion Governing Free Vibration of an Undamped Spring–Mass System

Assume that (Boyce and DiPrima, 2005)

$$x(t) = De^{st} \tag{1.4.42}$$

where D and s are to be determined. Substituting Equation 1.4.42 into the differential equation of motion in Equation 1.4.19,

$$(m_{eq}s^2 + k_{eq})De^{st} = 0 \tag{1.4.43}$$

Here, D is not zero for a nontrivial solution. Therefore, for Equation 1.4.43 to be true for all time t,

$$m_{eq}s^2 + k_{eq} = 0 \tag{1.4.44}$$

This is called the characteristic equation. The roots of this equation are

$$s_1 = j\omega_n \quad \text{and} \quad s_2 = -j\omega_n \tag{1.4.45a, b}$$

where $j = \sqrt{-1}$ is the imaginary number and

$$\omega_n = \sqrt{\frac{k_{eq}}{m_{eq}}} \tag{1.4.46}$$

The general solution of the differential Equation 1.4.19 is expressed as

$$x(t) = D_1e^{j\omega_nt} + D_2e^{-j\omega_nt} \tag{1.4.47}$$

where D_1 and D_2 are constants. Recall the well-known trigonometric identity,

$$e^{\pm j\omega_n t} = \cos\omega_n t \pm j\sin\omega_n t \tag{1.4.48}$$

Using Equation 1.4.48, Equation 1.4.47 leads to

$$x(t) = (D_1 + D_2)\cos\omega_n t + j(D_1 - D_2)\sin\omega_n t \tag{1.4.49}$$

It can be shown that D_1 and D_2 are complex conjugates. Therefore, both $(D_1 + D_2)$ and $j(D_1 - D_2)$ will be real numbers. Denote

$$A_1 = D_1 + D_2 \quad \text{and} \quad B_1 = j(D_1 - D_2) \qquad \text{(1.4.50a, b)}$$

Equation 1.4.49 is written as

$$x(t) = A_1\cos\omega_n t + B_1\sin\omega_n t \tag{1.4.51}$$

The coefficients A_1 and B_1 depend on initial conditions $x(0)$ and $\dot{x}(0)$. It is easily seen that

$$A_1 = x(0) \tag{1.4.52}$$

Differentiating Equation 1.4.51,

$$\dot{x}(t) = -\omega_n A_1\sin\omega_n t + \omega_n B_1\cos\omega_n t \tag{1.4.53}$$

Substituting $t = 0$ in Equation 1.4.53,

$$B_1 = \frac{\dot{x}(0)}{\omega_n} \tag{1.4.54}$$

Equation 1.4.51 is written as

$$x(t) = x(0)\cos\omega_n t + \frac{\dot{x}(0)}{\omega_n}\sin\omega_n t \tag{1.4.55}$$

Alternatively, Equation 1.4.55 can also be expressed as

$$x(t) = A\sin(\omega_n t + \psi) \tag{1.4.56}$$

where A and ψ are determined as follows:

$$x(t) = A\sin(\omega_n t + \psi) = A\sin\psi\cos\omega_n t + A\cos\psi\sin\omega_n t \tag{1.4.57}$$

Comparing Equations 1.4.55 and 1.4.57,

$$A \sin \psi \cos \omega_n t + A \cos \psi \sin \omega_n t = x(0) \cos \omega_n t + \frac{\dot{x}(0)}{\omega_n} \sin \omega_n t \tag{1.4.58}$$

Equating coefficients of $\cos \omega_n t$ and $\sin \omega_n t$ on both sides,

$$A \sin \psi = x(0) \tag{1.4.59}$$

$$A \cos \psi = \frac{\dot{x}(0)}{\omega_n} \tag{1.4.60}$$

Squaring Equations 1.4.59 and 1.4.60 and then adding them,

$$A^2 \sin^2 \psi + A^2 \cos^2 \psi = (x(0))^2 + \left(\frac{\dot{x}(0)}{\omega_n}\right)^2 \tag{1.4.61}$$

Using the fact that $\sin^2 \psi + \cos^2 \psi = 1$,

$$A = \sqrt{(x(0))^2 + \left(\frac{\dot{x}(0)}{\omega_n}\right)^2} \tag{1.4.62}$$

Dividing Equation 1.4.59 by Equation 1.4.60,

$$\psi = \tan^{-1}\left[\frac{\omega_n x(0)}{\dot{x}(0)}\right] \tag{1.4.63}$$

It should be noted that the value of A is taken to be positive. Furthermore, there is a single solution for the angle ψ. The quadrant in which the angle ψ lies is determined by the signs of the numerator $\omega_n x(0)$ and the denominator $\dot{x}(0)$. For example, if $\omega_n x(0) = 1$ and $\dot{x}(0) = -1$,

$$\psi = \tan^{-1}\left[\frac{+1}{-1}\right] = \frac{3\pi}{4} \text{rad} \tag{1.4.64}$$

In summary, the free vibration of an undamped spring–mass system is given by

$$x(t) = A \sin(\omega_n t + \psi); \; A > 0 \tag{1.4.65}$$

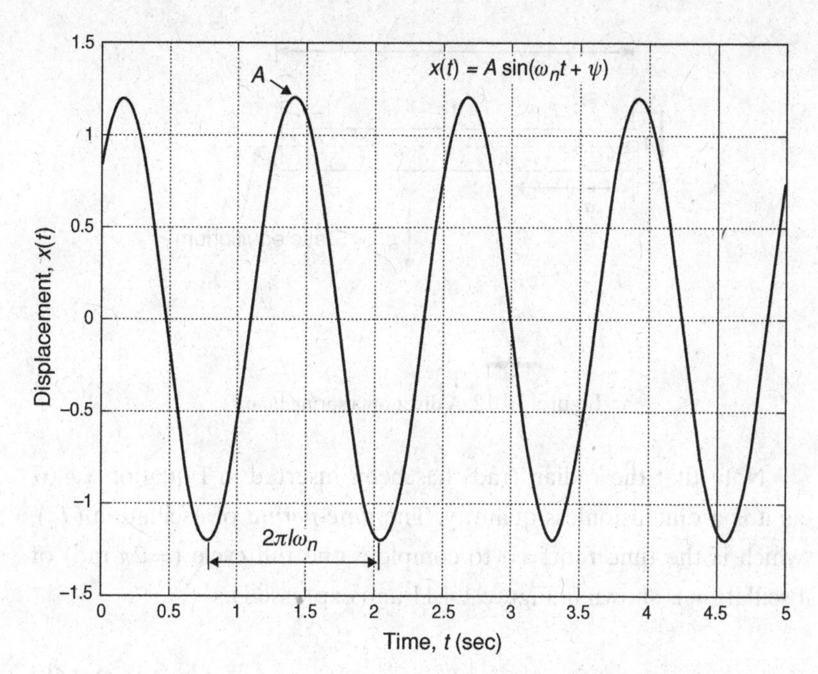

Figure 1.4.11 Free response of an undamped SDOF spring–mass system

where

$$\omega_n = \sqrt{\frac{k_{eq}}{m_{eq}}};\ A = +\sqrt{(x(0))^2 + \left(\frac{\dot{x}(0)}{\omega_n}\right)^2};\quad \text{and}\quad \psi = \tan^{-1}\left[\frac{\omega_n x(0)}{\dot{x}(0)}\right] \tag{1.4.66a, b, c}$$

The free vibration of an undamped SDOF system is purely sinusoidal with the amplitude A and the frequency ω_n (Equation 1.4.65 and Figure 1.4.11). This frequency (ω_n) is called the **natural frequency**, which is an intrinsic property of the SDOF spring–mass system. Denoting Newton as N, meter as m, kilogram as kg, the unit of ω_n is derived as follows:

$$\text{unit of } \omega_n = \sqrt{\frac{\text{unit of } k_{eq}}{\text{unit of } m_{eq}}} = \sqrt{\frac{\text{Nm}^{-1}}{\text{kg}}} = \sqrt{\frac{\text{kg} - \text{m} - \text{sec}^{-2} - \text{m}^{-1}}{\text{kg}}}$$
$$= \text{sec}^{-1} = \text{rad/sec} \tag{1.4.67}$$

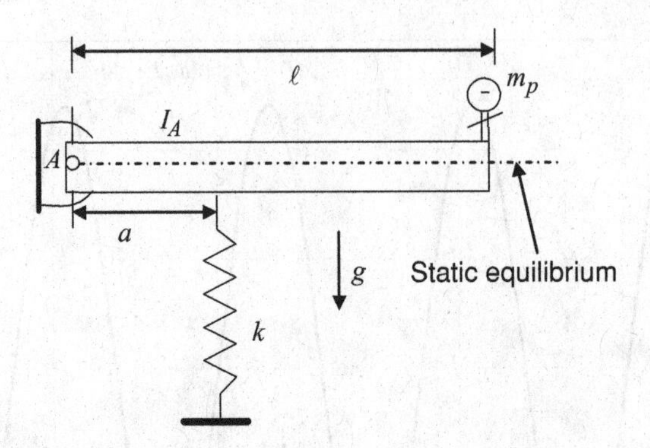

Figure 1.4.12 A diver on a spring board

Note that the radian (rad) has been inserted in Equation 1.4.67 as it is a dimensionless quantity. The *time-period* of oscillation (T_p), which is the time required to complete one full cycle ($= 2\pi$ rad) of oscillation is shown in Figure 1.4.11 and expressed as

$$T_p = \frac{2\pi}{\omega_n} \tag{1.4.68}$$

Using the fact that one cycle $= 2\pi$ rad, the frequency of oscillation (f) can also be expressed in the units of cycles/sec as follows:

$$f = \frac{\omega_n}{2\pi} = \frac{1}{T_p} \text{ cycles/sec} \tag{1.4.69}$$

The unit of cycles/sec is called Hertz (Hz).

Example 1.4.4: A Diving Board

Consider a springboard, which is pinned at the point A and is supported by a spring with stiffness k (Figure 1.4.12). Let the mass-moment of inertia of the board about the point A be I_A. A person of mass m_p is standing at the edge of the board in static equilibrium. Suddenly, this person jumps from the board and the board starts vibrating.

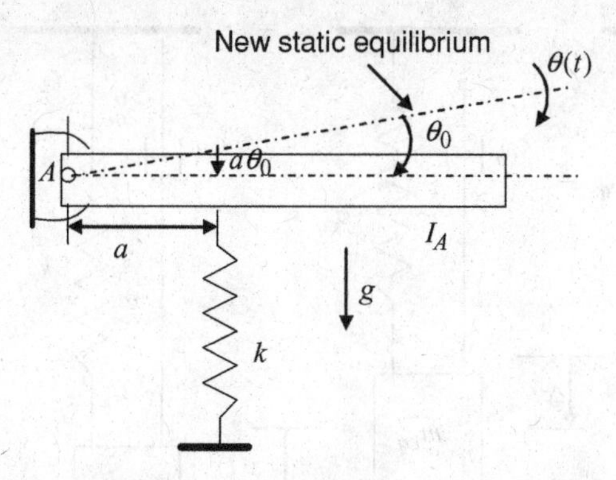

Figure 1.4.13 New static equilibrium for the diving board

When the person leaves the board, there will be a new static equilibrium configuration. Let the new static equilibrium configuration be at an angle θ_0 from the original static equilibrium position (Figure 1.4.13). Then,

$$ka\theta_0 a = m_p g\ell \tag{1.4.70}$$

or,

$$\theta_0 = \frac{m_p g\ell}{ka^2} \tag{1.4.71}$$

From the results in Example 1.4.1, the differential equation of motion (after the person jumps) will be

$$I_A\ddot{\theta} + ka^2\theta = 0 \tag{1.4.72}$$

Therefore, the undamped natural frequency will be

$$\omega_n = \sqrt{\frac{ka^2}{I_A}} \tag{1.4.73}$$

The initial conditions will be

$$\theta(0) = \theta_0 \text{ and } \dot{\theta}(0) = 0 \tag{1.4.74}$$

Hence, from Equation 1.4.65,

$$\theta(t) = \theta_0 \sin\omega_n t \tag{1.4.75}$$

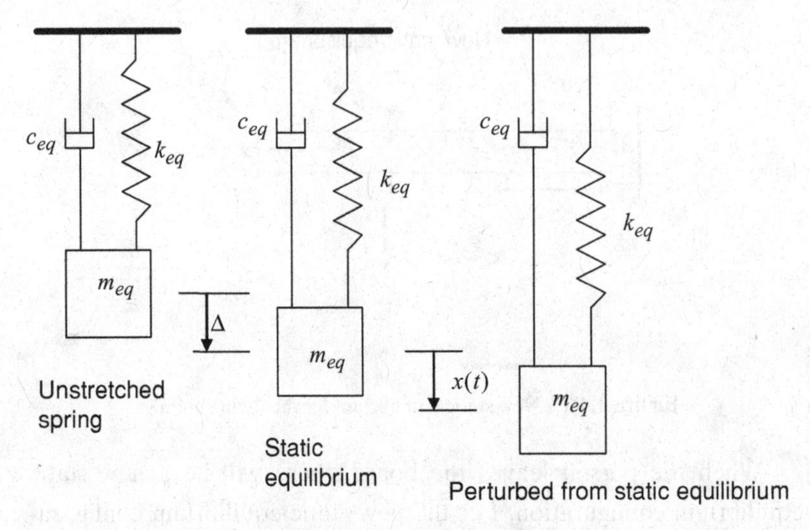

Figure 1.5.1 Spring–mass–damper system

1.5 FREE VIBRATION OF A VISCOUSLY DAMPED SDOF SYSTEM

This section deals with the derivation and the solution of the differential equation of motion of a viscously damped SDOF system. The solution of the differential equation of motion is used to characterize the nature of free vibration for different values of damping.

1.5.1 Differential Equation of Motion

A viscously damped SDOF system is shown in Figure 1.5.1. At the static equilibrium, the velocity of the mass is zero; therefore, the damper does not provide any force, and Equation 1.4.1, $m_{eq}g = k_{eq}\Delta$, still holds. From the free body diagram in Figure 1.5.2 (Perturbed from Static Equilibrium),

$$\text{net force in } x\text{-direction} = -k_{eq}x - c_{eq}\dot{x} \tag{1.5.1}$$

From Newton's second law of motion,

$$-k_{eq}x - c_{eq}\dot{x} = m_{eq}\ddot{x} \tag{1.5.2}$$

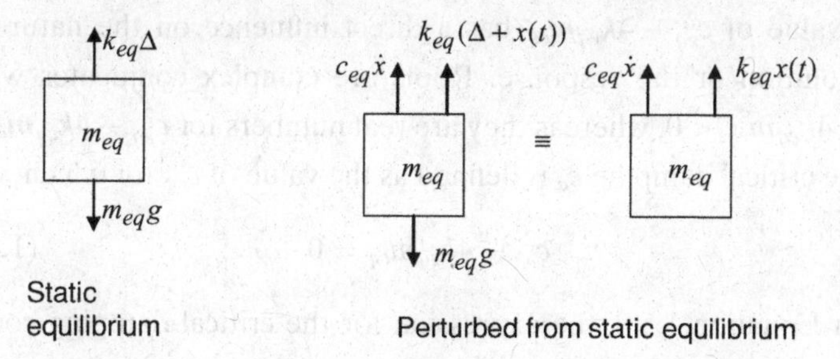

Figure 1.5.2 Free body diagram of the spring–mass–damper system

or

$$m_{eq}\ddot{x} + c_{eq}\dot{x} + k_{eq}x = 0 \tag{1.5.3}$$

Equation 1.5.3 is the governing differential equation of motion.

1.5.2 Solution of the Differential Equation of Motion Governing Free Vibration of a Damped Spring–Mass System

Assume that (Boyce and DiPrima, 2005)

$$x(t) = De^{st} \tag{1.5.4}$$

where D and s are to be determined. Substituting Equation 1.5.4 into Equation 1.5.3,

$$(m_{eq}s^2 + c_{eq}s + k_{eq})De^{st} = 0 \tag{1.5.5}$$

Here, D is not zero for a nontrivial solution. Therefore, for Equation 1.5.5 to be true for all time t,

$$m_{eq}s^2 + c_{eq}s + k_{eq} = 0 \tag{1.5.6}$$

Roots of the quadratic Equation 1.5.6 are given by

$$s_{1,2} = \frac{-c_{eq} \pm \sqrt{c_{eq}^2 - 4k_{eq}m_{eq}}}{2m_{eq}} \tag{1.5.7}$$

The value of $c_{eq}^2 - 4k_{eq}m_{eq}$ has a direct influence on the nature of the solution or the response. Roots are complex conjugates when $c_{eq}^2 - 4k_{eq}m_{eq} < 0$, whereas they are real numbers for $c_{eq}^2 - 4k_{eq}m_{eq} \geq 0$. The critical damping c_c is defined as the value of c_{eq} for which

$$c_{eq}^2 - 4k_{eq}m_{eq} = 0 \tag{1.5.8}$$

From Equation 1.5.8, the expression for the critical damping coefficient c_c is obtained as

$$c_c = 2\sqrt{k_{eq}m_{eq}} = 2m_{eq}\omega_n \tag{1.5.9}$$

Then, the damping ratio ξ is defined as

$$\xi = \frac{c_{eq}}{c_c} \tag{1.5.10}$$

Now, Equation 1.5.7 is written as

$$s_{1,2} = -\frac{c_{eq}}{2m_{eq}} \pm \sqrt{\left(\frac{c_{eq}}{2m_{eq}}\right)^2 - \frac{k_{eq}}{m_{eq}}} \tag{1.5.11}$$

where

$$\frac{c_{eq}}{2m_{eq}} = \frac{c_{eq}}{c_c}\frac{c_c}{2m_{eq}} = \xi\omega_n \tag{1.5.12}$$

Equation 1.5.11 is written as

$$s_{1,2} = -\xi\omega_n \pm \sqrt{\xi^2\omega_n^2 - \omega_n^2} = -\xi\omega_n \pm \omega_n\sqrt{\xi^2 - 1} \tag{1.5.13}$$

Three cases of damping are defined as follows:

a. Underdamped ($0 < \xi < 1$ or $0 < c_{eq} < c_c$)
b. Critically damped ($\xi = 1$ or $c_{eq} = c_c$)
c. Overdamped ($\xi > 1$ or $c_{eq} > c_c$)

Case I: Underdamped ($0 < \xi < 1$ *or* $0 < c_{eq} < c_c$)
In this case, the roots (s_1 and s_2) are complex conjugates. From Equation 1.5.13,

$$s_{1,2} = -\xi\omega_n \pm j\omega_n\sqrt{1 - \xi^2} \tag{1.5.14}$$

or,

$$s_1 = -\xi\omega_n + j\omega_d \text{ and } s_2 = -\xi\omega_n - j\omega_d \qquad (1.5.15\text{a, b})$$

where

$$\omega_d = \omega_n\sqrt{1-\xi^2} \qquad (1.5.16)$$

The general solution is expressed as

$$x(t) = D_1 e^{s_1 t} + D_2 e^{s_2 t} = e^{-\xi\omega_n t}(D_1 e^{j\omega_d t} + D_2 e^{-j\omega_d t}) \qquad (1.5.17)$$

where D_1 and D_2 are constants. Similar to the solution procedure for an undamped spring–mass system, the following well-known trigonometric identity is again used:

$$e^{\pm j\omega_d t} = \cos\omega_d t \pm j\sin\omega_d t \qquad (1.5.18)$$

Therefore, Equation 1.5.17 leads to

$$x(t) = e^{-\xi\omega_n t}(A_1\cos\omega_d t + B_1\sin\omega_d t) \qquad (1.5.19)$$

where A_1 and B_1 are real numbers defined by Equations 1.4.50a,b. To determine A_1 and B_1, initial conditions $x(0)$ and $\dot{x}(0)$ are used. First,

$$x(0) = A_1 \qquad (1.5.20)$$

Differentiating Equation 1.5.19,

$$\begin{aligned}\dot{x}(t) &= e^{-\xi\omega_n t}(-\omega_d A_1\sin\omega_d t + \omega_d B_1\cos\omega_d t)\\ &\quad -\xi\omega_n e^{-\xi\omega_n t}(A_1\cos\omega_d t + B_1\sin\omega_d t)\end{aligned} \qquad (1.5.21)$$

At $t = 0$,

$$\dot{x}(0) = \omega_d B_1 - \xi\omega_n A_1 \qquad (1.5.22)$$

Using Equation 1.5.20,

$$B_1 = \frac{\dot{x}(0) + \xi\omega_n x(0)}{\omega_d} \qquad (1.5.23)$$

Alternatively, Equation 1.5.19 can be written as

$$x(t) = e^{-\xi\omega_n t} A \sin(\omega_d t + \psi) \tag{1.5.24}$$

To determine the amplitude A and the phase ψ in terms of A_1 and B_1, Equations 1.5.19 and 1.5.24 are compared to yield

$$A \sin(\omega_d t + \psi) = A_1 \cos \omega_d t + B_1 \sin \omega_d t \tag{1.5.25}$$

or

$$A \sin \psi \cos \omega_d t + A \cos \psi \sin \omega_d t = A_1 \cos \omega_d t + B_1 \sin \omega_d t \tag{1.5.26}$$

Equating the coefficients of $\cos \omega_d t$ and $\sin \omega_d t$ on both sides,

$$A \sin \psi = A_1 \tag{1.5.27}$$

$$A \cos \psi = B_1 \tag{1.5.28}$$

Therefore, following derivations in Section 1.4,

$$A = \sqrt{[x(0)]^2 + \left[\frac{\dot{x}(0) + \xi\omega_n x(0)}{\omega_d}\right]^2} \tag{1.5.29}$$

$$\psi = \tan^{-1}\left[\frac{\omega_d x(0)}{\dot{x}(0) + \xi\omega_n x(0)}\right] \tag{1.5.30}$$

Summary: The free vibration of an underdamped ($0 < \xi < 1$ or $0 < c_{eq} < c_c$) spring–mass–damper system is given by

$$x(t) = Ae^{-\xi\omega_n t} \sin(\omega_d t + \psi); A > 0 \tag{1.5.31}$$

where ω_d, A, and ψ are given by Equations 1.5.16, 1.5.29, and 1.5.30.

The free response of an underdamped system is shown in Figure 1.5.3. Compared to free vibration of an undamped system in Figure 1.4.11, the following observations are made:

a. Amplitude is exponentially decaying and $\lim x(t) \to 0$ as $t \to \infty$.
b. The natural frequency of the damped system is ω_d, which is smaller than the *undamped natural frequency* ω_n. The time period of the free response of an underdamped system is $2\pi/\omega_d$, which will be larger than that of the undamped system.

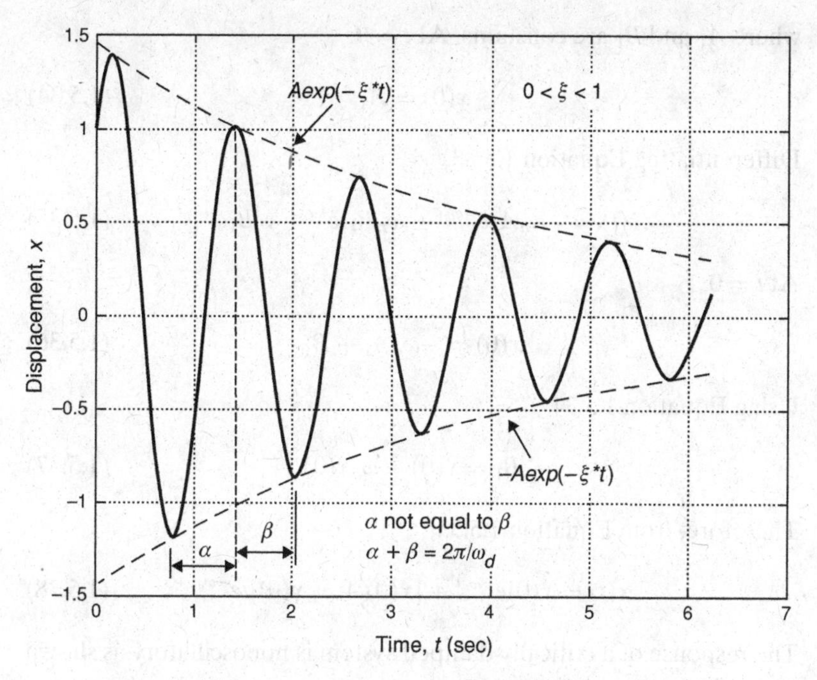

Figure 1.5.3 Free response of an underdamped SDOF system

c. Unlike a pure sinusoidal function, the half-way point in a period of underdamped free response does not correspond to zero velocity, when the starting point of a period is taken to be zero velocity condition. This point is expressed in Figure 1.5.3 by the following relationship: $\alpha \neq \beta$.

Case II: Critically Damped ($\xi = 1$ *or* $c_{eq} = c_c$)
In this case, the roots (s_1 and s_2) are negative real numbers and equal. From Equation 1.5.13,

$$s_1 = s_2 = -\omega_n \tag{1.5.32}$$

Therefore, the general solution (Boyce and DiPrima, 2005) of the differential Equation 1.5.3 will be

$$x(t) = A_1 e^{s_1 t} + B_1 t e^{s_1 t} = A_1 e^{-\omega_n t} + B_1 t e^{-\omega_n t} \tag{1.5.33}$$

where A_1 and B_1 are constants. At $t = 0$,

$$x(0) = A_1 \tag{1.5.34}$$

Differentiating Equation 1.5.33,

$$\dot{x}(t) = -\omega_n A_1 e^{-\omega_n t} - \omega_n B_1 t e^{-\omega_n t} + B_1 e^{-\omega_n t} \tag{1.5.35}$$

At $t = 0$,

$$\dot{x}(0) = -\omega_n A_1 + B_1 \tag{1.5.36}$$

Using Equation 1.5.34,

$$B_1 = \dot{x}(0) + \omega_n x(0) \tag{1.5.37}$$

Therefore, from Equation 1.5.33,

$$x(t) = x(0)e^{-\omega_n t} + [\dot{x}(0) + \omega_n x(0)]te^{-\omega_n t} \tag{1.5.38}$$

The response of a critically damped system is nonoscillatory as shown in Figure 1.5.4. Furthermore, $\lim x(t) \to 0$ as $t \to \infty$.

Case III: Overdamped ($\xi > 1$ *or* $c_{eq} > c_c$)
From Equation 1.5.13, both roots s_1 and s_2 are negative real numbers.

$$s_1 = -\xi\omega_n + \omega_n\sqrt{\xi^2 - 1} < 0 \tag{1.5.39}$$

$$s_2 = -\xi\omega_n - \omega_n\sqrt{\xi^2 - 1} < 0 \tag{1.5.40}$$

And, the general solution of the differential Equation 1.5.3 will be

$$x(t) = A_1 e^{s_1 t} + B_1 e^{s_2 t} \tag{1.5.41}$$

where A_1 and B_1 are constants. At $t = 0$,

$$x(0) = A_1 + B_1 \tag{1.5.42}$$

Differentiating Equation 1.5.41,

$$\dot{x}(t) = A_1 s_1 e^{s_1 t} + B_1 s_2 e^{s_2 t} \tag{1.5.43}$$

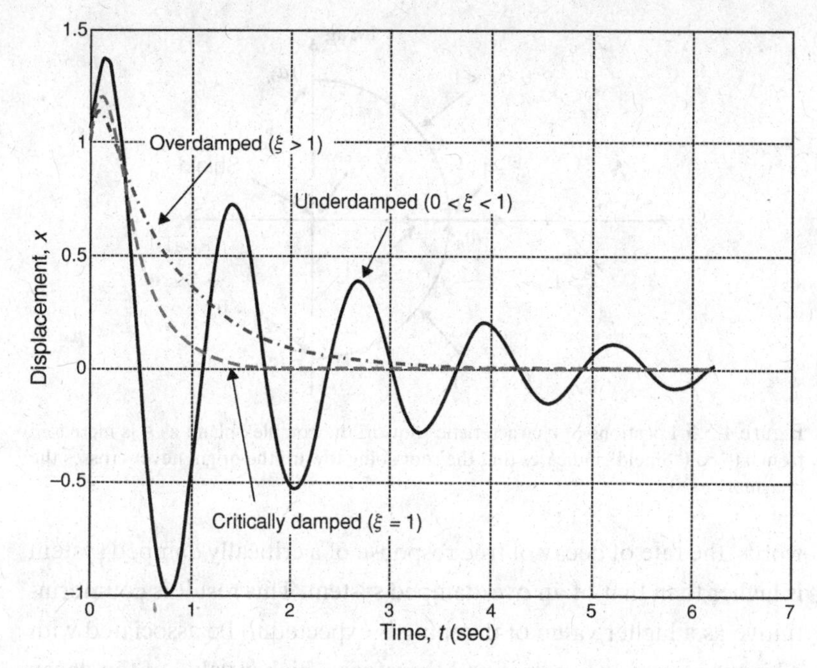

Figure 1.5.4 Free responses of underdamped, critically damped, and overdamped SDOF systems

At $t = 0$,

$$\dot{x}(0) = A_1 s_1 + B_1 s_2 \tag{1.5.44}$$

Solving Equations 1.5.42 and 1.5.44,

$$A_1 = \frac{s_2 x(0) - \dot{x}(0)}{s_2 - s_1} \tag{1.5.45}$$

and

$$B_1 = \frac{-s_1 x(0) + \dot{x}(0)}{s_2 - s_1} \tag{1.5.46}$$

The free response of an overdamped system is also nonoscillatory as shown in Figure 1.5.4. Furthermore, $\lim x(t) \to 0$ as $t \to \infty$.

Comparing responses of underdamped, critically damped, and overdamped systems, it is seen in Figure 1.5.4 that ***the free response of a critically damped system decays at the fastest rate.*** In other

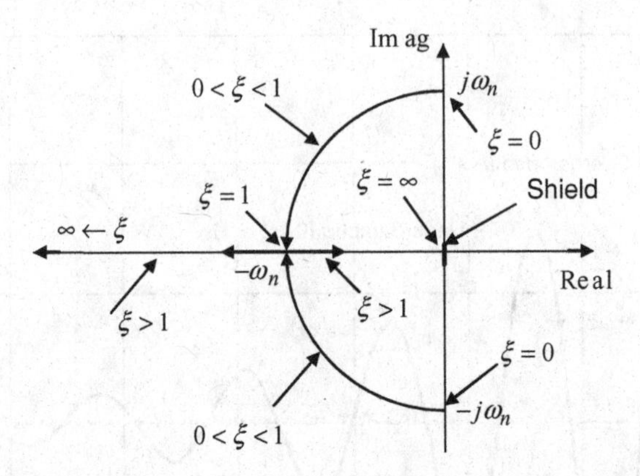

Figure 1.5.5 Locations of characteristic roots in the complex plane as ξ is increased from 0 to ∞ ("Shield" indicates that the root going toward the origin never crosses the imaginary axis)

words, the rate of decay of free response of a critically damped system is higher than that of an overdamped system. This result is counterintuitive as a higher value of damping is expected to be associated with a higher rate of energy loss and therefore a higher value of the decay rate of free response. To understand this result, locations of characteristic roots s_1 and s_2 are plotted in Figure 1.5.5 as ξ is increased from 0 to ∞. The following information is used for the construction of Figure 1.5.5:

a. For $\xi = 0$,

$$s_1 = +j\omega_n \quad \text{and} \quad s_2 = -j\omega_n \tag{1.5.47}$$

b. For $0 < \xi < 1$,

$$s_1 = -\xi\omega_n + j\omega_n\sqrt{1-\xi^2} \quad \text{and} \quad s_2 = -\xi\omega_n - j\omega_n\sqrt{1-\xi^2} \tag{1.5.48}$$

The real part (RP) and the imaginary part (IP) of s_1 and s_2 are as follows:

$$RP = -\xi\omega_n \quad \text{and} \quad IP = \pm\omega_n\sqrt{1-\xi^2} \tag{1.5.49}$$

Therefore,

$$RP^2 + IP^2 = \xi^2\omega_n^2 + \omega_n^2(1-\xi^2) = \omega_n^2 \tag{1.5.50}$$

This is the equation of a circle with the radius ω_n and the center at the origin of the complex plane. Since *RP* is negative, the roots move along the semicircle in the left half of the complex plane as ξ varies from 0 to 1 (Figure 1.5.5).

c. For $\xi = 1$,

$$s_1 = s_2 = -\omega_n \tag{1.5.51}$$

d. For $\xi > 1$,

$$s_1 = -\xi\omega_n + \omega_n\sqrt{\xi^2 - 1} < 0 \tag{1.5.52}$$

$$s_2 = -\xi\omega_n - \omega_n\sqrt{\xi^2 - 1} < 0 \tag{1.5.53}$$

Note that

$$|s_1| < \omega_n\,, \; |s_2| > \omega_n \tag{1.5.54}$$

Equation 1.5.41 is rewritten as

$$x(t) = A_1 e^{-|s_1|t} + B_1 e^{-|s_2|t} \tag{1.5.55}$$

Because $|s_2| > |s_1|$ for an overdamped system, the term with $e^{-|s_2|t}$ dies at a rate faster than the term with $e^{-|s_1|t}$. As a result, the dominant term in Equation 1.5.55 is the one with $e^{-|s_1|t}$. The critically damped system decays at the rate of $e^{-\omega_n t}$. Since $|s_1| < \omega_n$, the decay rate of an overdamped system is slower than that of a critically damped system.

Example 1.5.1: A Rigid Bar Supported by a Spring and a Damper
Consider a rigid bar of length ℓ, which is pinned at the point *A* and is supported by a spring with the stiffness *k* and a damper with the coefficient *c* (Figure 1.5.6). The mass of the bar is *m* and is concentrated at its right end as shown in Figure 1.5.6.

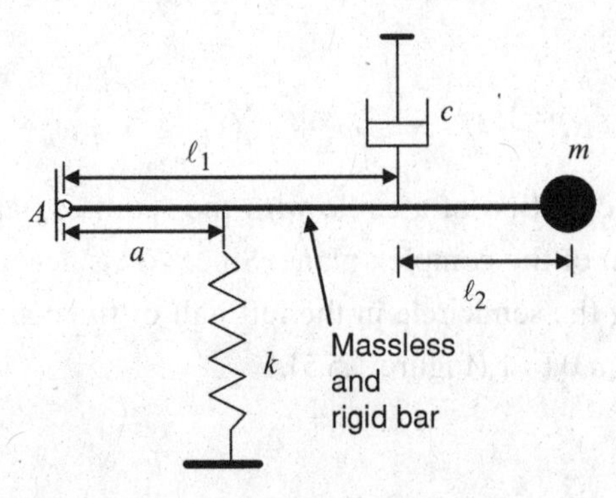

Figure 1.5.6 Rigid bar with a spring and a damper

The free body diagram is shown in Figure 1.5.7 for a small angular displacement θ from its static equilibrium position, where R_A is the unknown reaction force at the point A. Taking moment about the point A,

$$-ka\theta a - c\ell_1\dot{\theta}\ell_1 = I_A\ddot{\theta} \tag{1.5.56}$$

where I_A is the mass-moment of inertia of the bar about A. Here,

$$I_A = m\ell^2 \tag{1.5.57}$$

Substituting Equation 1.5.57 into Equation 1.5.56,

$$m\ell^2\ddot{\theta} + c\ell_1^2\dot{\theta} + ka^2\theta = 0 \tag{1.5.58}$$

Therefore,

$$m_{eq} = m\ell^2; \;\; k_{eq} = ka^2; \quad \text{and} \quad c_{eq} = c\ell_1^2 \tag{1.5.59}$$

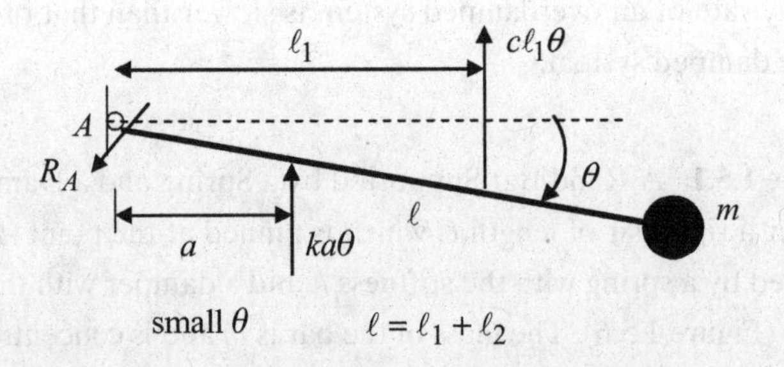

Figure 1.5.7 Free body diagram of a rigid bar in Figure 1.5.6

The undamped natural frequency is

$$\omega_n = \sqrt{\frac{k_{eq}}{m_{eq}}} = \frac{a}{\ell}\sqrt{\frac{k}{m}} \tag{1.5.60}$$

The critical damping for this system is

$$c_c = 2\sqrt{k_{eq}m_{eq}} = 2\ell a\sqrt{km} \tag{1.5.61}$$

Consider the following numerical values:

$m = 1\,\text{kg}$, $c = 20\,\text{N} - \text{sec/m}$, $a = 0.4\,\text{m}$, $\ell_1 = 0.5\,\text{m}$, and $\ell = 1\,\text{m}$

Case I: If the stiffness $k = 100$ N/m,

$$c_c = 2\ell a\sqrt{km} = 8\,\text{N} - \text{sec/m}$$

As a result, $c > c_c$ and the system is overdamped. The damping ratio $\xi = 2.5$ and the damped natural frequency is not defined.

Case II: If the stiffness $k = 900$ N/m,

$$c_c = 2\ell a\sqrt{km} = 24\,\text{N} - \text{sec/m}$$

As a result, $c < c_c$ and the system is underdamped. The damping ratio $\xi = 0.833$ and the damped natural frequency $\omega_d = \omega_n\sqrt{1-\xi^2} = 5.5277$ rad/sec.

1.5.3 Logarithmic Decrement: Identification of Damping Ratio from Free Response of an Underdamped System ($0 < \xi < 1$)

Let us assume that two successive peak displacements in the response of an underdamped system are known. If the first known peak displacement occurs at $t = t_1$ (Figure 1.5.8), Equation 1.5.31 yields

$$x(t_1) = x_1 = Ae^{-\xi\omega_n t_1}\sin(\omega_d t_1 + \psi) \tag{1.5.62}$$

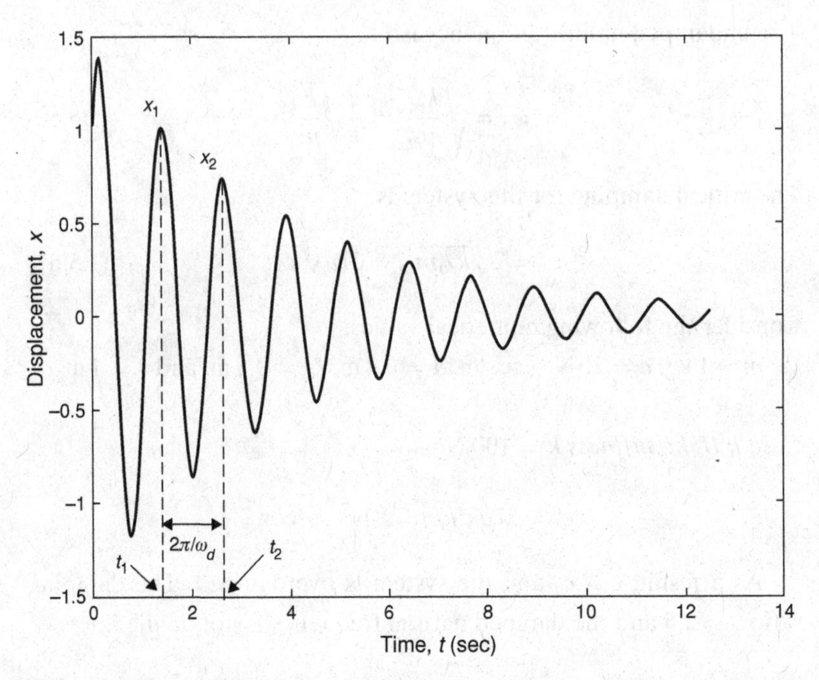

Figure 1.5.8 Free vibration of an underdamped system (illustration of log decrement)

Let the second peak displacement be at $t = t_2$ (Figure 1.5.8). From Equation 1.5.31,

$$x(t_2) = x_2 = Ae^{-\xi\omega_n t_2}\sin(\omega_d t_2 + \psi) \tag{1.5.63}$$

From the characteristic of free underdamped response,

$$t_2 = t_1 + \frac{2\pi}{\omega_d} \tag{1.5.64}$$

As a result,

$$\sin(\omega_d t_2 + \psi) = \sin(\omega_d t_1 + \psi + 2\pi) = \sin(\omega_d t_1 + \psi) \tag{1.5.65}$$

Dividing Equation 1.5.62 by Equation 1.5.63 and using Equations 1.5.64–1.5.65,

$$\frac{x_1}{x_2} = e^{\xi\omega_n(t_2 - t_1)} = e^{\frac{2\pi\xi\omega_n}{\omega_d}} \tag{1.5.66}$$

Using the fact that $\omega_d = \omega_n\sqrt{1-\xi^2}$ (Equation 1.5.16),

$$\frac{x_1}{x_2} = e^{\frac{2\pi\xi}{\sqrt{1-\xi^2}}} \tag{1.5.67}$$

The ratio of the two successive amplitudes only depends on the damping ratio ξ, that is, it is independent of the undamped natural frequency ω_n. Taking natural logarithm of Equation 1.5.67 with respect to the base e,

$$\ln\frac{x_1}{x_2} = \frac{2\pi\xi}{\sqrt{1-\xi^2}} \tag{1.5.68}$$

The natural logarithm of two successive amplitudes or peak displacements is known as the *logarithmic decrement* δ, that is,

$$\delta = \ln\frac{x_1}{x_2} \tag{1.5.69}$$

Therefore, from Equations 1.5.68 and 1.5.69,

$$\delta = \frac{2\pi\xi}{\sqrt{1-\xi^2}} \tag{1.5.70}$$

For a small ξ, $\sqrt{1-\xi^2} \approx 1$ and from Equation 1.5.70,

$$\xi = \frac{\delta}{2\pi} \tag{1.5.71}$$

In general, Equation 1.5.70 is directly solved to yield

$$\xi = \frac{\delta}{\sqrt{(2\pi)^2+\delta^2}} \tag{1.5.72}$$

The approximation Equation 1.5.71 is found to be valid for $\xi \leq 0.2$ or equivalently $\delta \leq 0.4\pi$.

Since ratio of two successive amplitudes only depends on the damping ratio,

$$\frac{x_1}{x_2} = \frac{x_2}{x_3} = \cdots = \frac{x_{m-1}}{x_m} = \frac{x_m}{x_{m+1}} \tag{1.5.73}$$

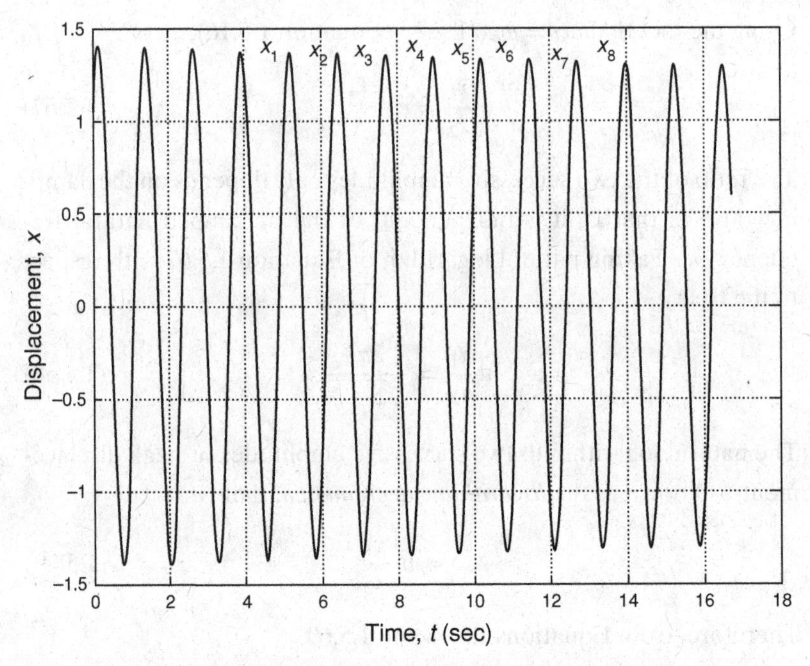

Figure 1.5.9 Free response of an underdamped system for a small damping ratio

where x_{i+1} is the amplitude after i cycles of oscillation, $i = 1, 2, \ldots, m-1, m$ (Figure 1.5.9). Therefore,

$$\frac{x_1}{x_{m+1}} = \frac{x_1}{x_2}\frac{x_2}{x_3} \cdots \frac{x_{m-1}}{x_m}\frac{x_m}{x_{m+1}} \tag{1.5.74}$$

Using Equation 1.5.73,

$$\frac{x_1}{x_{m+1}} = \left(\frac{x_1}{x_2}\right)^m \tag{1.5.75}$$

Taking natural logarithms of both sides of Equation 1.5.75 and using the definition of logarithmic decrement Equation 1.5.69,

$$\ln\frac{x_1}{x_{m+1}} = \ln\left(\frac{x_1}{x_2}\right)^m = m\ln\frac{x_1}{x_2} = m\delta \tag{1.5.76}$$

Therefore, the logarithmic decrement δ can also be computed by the following relationship:

$$\delta = \frac{1}{m}\ln\frac{x_1}{x_{m+1}} \tag{1.5.77}$$

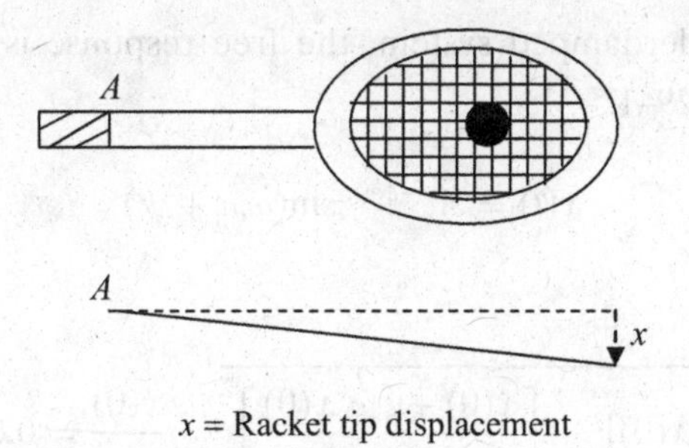

Figure 1.5.10 A tennis racket and a ball

where x_{m+1} is the amplitude after m cycles of oscillation. The expression in Equation 1.5.77 is useful for low values of the damping ratio ξ for which the difference between the two successive amplitudes can be so small that they may not be accurately differentiable by a measuring instrument, that is, the expression in Equation 1.5.69 may not accurately predict the value of the logarithmic decrement. But, if one considers the amplitude after a certain number of cycles of oscillation, for example, $m = 7$ in Figure 1.5.9, the difference between x_1 and x_{m+1} can be quite significant. As a result, the ratio of x_1 and x_{m+1} can be estimated quite accurately via a measuring instrument and Equation 1.5.77 can lead to an accurate value of δ and the damping ratio ξ using Equation 1.5.72.

Example 1.5.2: Vibration of a Tennis Racket

A tennis ball hits the tennis racket (Figure 1.5.10) and imparts a velocity of 1.5 m/sec to the racket tip. The natural frequency and the damping ratio of the tennis racket (Oh and Yum, 1986) are given to be 31.45 Hz and 0.0297, respectively. Determine the maximum displacement of the racket tip.

Solution

Given: $\omega_n = 31.45\,\text{Hz} = 197.606\,\text{rad/sec}$, $\xi = 0.0297$, $x(0) = 0$, and $\dot{x}(0) = 1.5\,\text{m/sec}$

Therefore, $\omega_d = \omega_n\sqrt{1-\xi^2} = 197.519$ rad/sec

For an underdamped system, the free response is described by Equations 1.5.29–1.5.31:

$$x(t) = Ae^{-\xi\omega_n t}\sin(\omega_d t + \psi)$$

where

$$A = \sqrt{[x(0)]^2 + \left[\frac{\dot{x}(0) + \xi\omega_n x(0)}{\omega_d}\right]^2} = \frac{\dot{x}(0)}{\omega_d} = 0.0076\,\mathrm{m}$$

and

$$\psi = \tan^{-1}\left[\frac{\omega_d x(0)}{\dot{x}(0) + \xi\omega_n x(0)}\right] = 0$$

Therefore,

$$x(t) = Ae^{-\xi\omega_n t}\sin(\omega_d t)$$

For the maximum displacement,

$$\dot{x}(t) = Ae^{-\xi\omega_n t}\omega_d\cos(\omega_d t) + A(-\xi\omega_n)e^{-\xi\omega_n t}\sin(\omega_d t) = 0$$

Let t^* be the time corresponding to the maximum displacement. Then,

$$\tan(\omega_d t^*) = \frac{\omega_d}{\xi\omega_n} = \frac{\sqrt{1-\xi^2}}{\xi} = 33.6652$$

Therefore,

$$\omega_d t^* = 1.541\,\mathrm{rad}$$

or

$$t^* = \frac{1.541}{\omega_d} = 0.0078\ \mathrm{sec}$$

The maximum displacement is computed as

$$x(t^*) = 0.0076\,e^{-0.0297\times 197.606\times 0.0078}\sin(1.541) = 0.0073\,\mathrm{m}$$

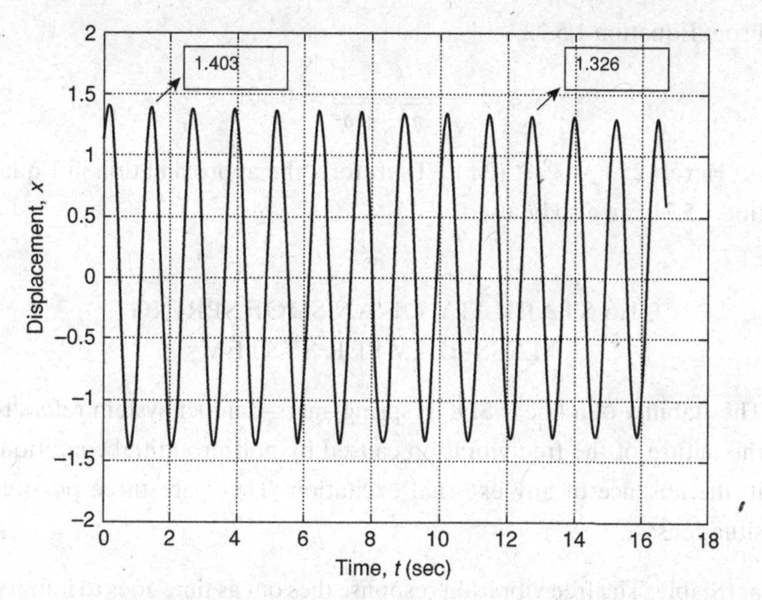

Figure 1.5.11 Measured free vibration of a damped spring–mass system (displacement in mm)

Example 1.5.3: Damping Ratio and Undamped Natural Frequency from Free Response

For the free vibration of an SDOF system (Figure 1.5.11), the amplitudes 1.403 mm and 1.326 mm are measured at 1.242 sec and 12.734 sec, respectively. Find the undamped natural frequency ω_n and the damping ratio ξ.

The number of cycles of oscillation between two measured amplitudes is 9. Hence, the time period of damped oscillation is

$$T_d = \frac{12.734 - 1.242}{9} = 1.2769 \text{ sec}$$

Therefore,

$$\omega_d = \frac{2\pi}{T_d} = 4.9207 \text{ rad/sec}$$

From Equation 1.5.77, the logarithmic decrement δ is

$$\delta = \frac{1}{9} \ell n \frac{1.403}{1.326} = 0.0063$$

From Equation 1.5.72,

$$\xi = \frac{\delta}{\sqrt{(2\pi)^2 + \delta^2}} = 0.001$$

Here, $(2\pi)^2 + \delta^2 \cong (2\pi)^2$. Therefore, the approximation in Equation 1.5.71 can also be used.

1.6 STABILITY OF AN SDOF SPRING–MASS–DAMPER SYSTEM

The stability of a linear SDOF spring–mass–damper system refers to the nature of the free vibration caused by nonzero initial conditions in the absence of any external excitation. There are three possible situations:

a. Stable: The free vibration response dies out as time goes to infinity, that is, $(x(t) \to 0 \text{ as } t \to \infty)$
b. Marginally Stable/Unstable: The response remains bounded but nonzero as time goes to infinity.
c. Unstable: The response becomes unbounded as the time goes to infinity, that is, $(x(t) \to \infty \text{ as } t \to \infty)$

As shown in Section 1.5.2, the characteristic of the free response is governed by e^{st} term, where s is a root of the characteristic Equation 1.5.6:

$$m_{eq}s^2 + c_{eq}s + k_{eq}s = 0 \tag{1.6.1}$$

In general, a root s is represented as a complex number:

$$s = s_R + js_I;\ j = \sqrt{-1} \tag{1.6.2}$$

where s_R and s_I are real and imaginary parts of s. For a purely real root, $s_I = 0$. Similarly, for a purely imaginary root, $s_R = 0$. From Equation 1.6.2,

$$e^{st} = e^{(s_R + js_I)t} = e^{s_R t}(\cos(s_I t) + j\sin(s_I t)) \tag{1.6.3}$$

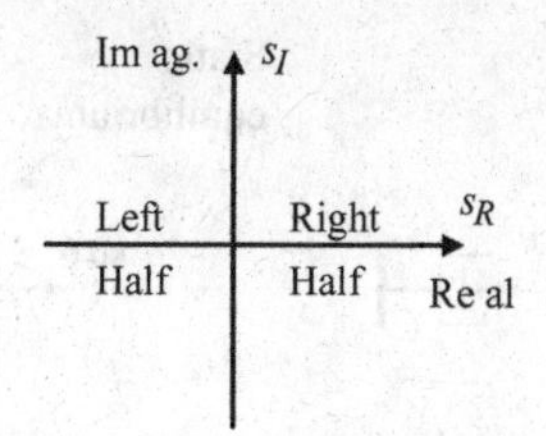

Figure 1.6.1 Complex s–plane

Because cosine and sine terms are bounded between −1 and +1, the sign of the real part s_R will determine whether the response will die out, remain nonzero and bounded, or grow to be unbounded as $t \to \infty$. The system will be stable, marginally stable/unstable, and unstable when $s_R < 0$ (left half of the complex plane, Figure 1.6.1), $s_R = 0$ (imaginary axis of the complex plane, Figure 1.6.1), and $s_R > 0$ (right half of the complex plane, Figure 1.6.1), respectively. This fact can be summarized as follows:

a. The spring–mass–damper system will be stable, provided both roots are located in the left half (excluding imaginary axis) of the complex plane.
b. The spring–mass–damper system will be marginally stable/unstable, provided at least one root is on the imaginary axis, and no root is in the right half of the complex plane.
c. The spring–mass–damper system will be unstable, provided one root is in the right half of the complex plane.

Equation 1.6.1 is a second-order polynomial or a quadratic equation. The roots of this equation can be easily calculated, and therefore their locations in the complex plane can be easily determined to evaluate the stability of the system. The necessary and sufficient conditions for the stability in the context of the quadratic Equation 1.6.1 are as follows:

a. None of the coefficients vanishes.
b. All coefficients must have the same sign.

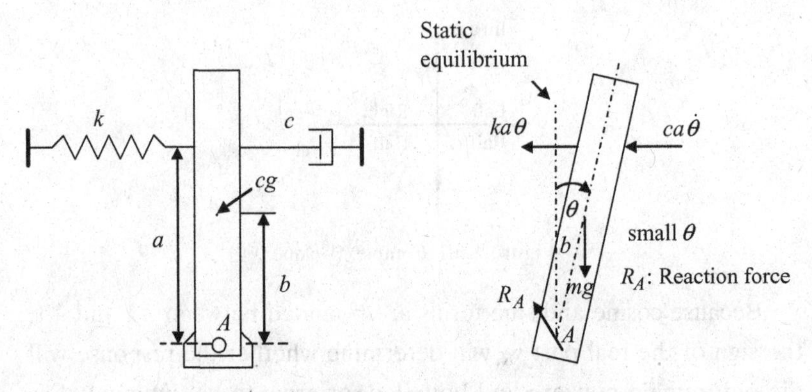

Figure 1.6.2 Inverted pendulum and free body diagram

Since $m_{eq} > 0$, necessary and sufficient conditions for stability are

$$c_{eq} > 0 \tag{1.6.4}$$

and

$$k_{eq} > 0 \tag{1.6.5}$$

In other words, a spring–mass–damper system is guaranteed to be stable, provided both equivalent damping and equivalent stiffness are positive.

Example 1.6.1: Inverted Pendulum (Negative Stiffness)
Consider an inverted pendulum which is supported by a spring with the stiffness k, located at a distance a from the pivot point A (Figure 1.6.2). The mass of the pendulum is m and the center of gravity is located at a distance b from the pivot point A.

Let θ be the clockwise small rotation of the bar from its static equilibrium configuration. Taking moment about the point A,

$$-ka\theta a + mgb\theta - ca\dot{\theta}a = I_A\ddot{\theta} \tag{1.6.6}$$

where I_A is the mass-moment of inertia of bar about the point A. Rearranging Equation 1.6.6,

$$I_A\ddot{\theta} + ca^2\dot{\theta} + (ka^2 - mgb)\theta = 0 \tag{1.6.7}$$

v_r : Relative wind velocity
f_a : Aerodynamic force

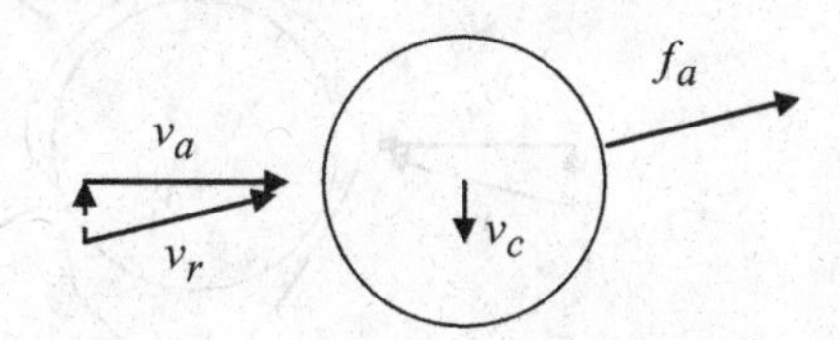

Figure 1.6.3 Wind across a cable with a circular cross section

The equivalent stiffness k_{eq} for the system is

$$k_{eq} = ka^2 - mgb \tag{1.6.8}$$

The equivalent stiffness k_{eq} is negative if

$$k < \frac{mgb}{a^2} \tag{1.6.9}$$

When the condition in Equation 1.6.9 is satisfied, the vertical static equilibrium configuration is unstable.

Example 1.6.2: Galloping of Transmission Lines During Winter (Negative Damping)
There are long electric transmission cables (Den Hartog, 1956), which have circular cross sections. Assume that there is also a cross wind with the velocity v_a and the cable is undergoing a small transverse oscillation. Let v_c be the instantaneous cable velocity, which is directed downwards as shown in Figure 1.6.3. The cable will experience the relative wind velocity v_r, and the effective aerodynamic force f_a on the cable will be along the direction of the relative velocity v_r. The vertical component of this aerodynamic force is in the direction opposite to the downward velocity of the cable. In other words, the aerodynamic force opposes the cable motion, and as a result, aerodynamic forces dissipate energy and the equivalent damping c_{eq} is positive.

During the winter season in certain regions, there is an ice formation and effective cross section of the cable becomes noncircular

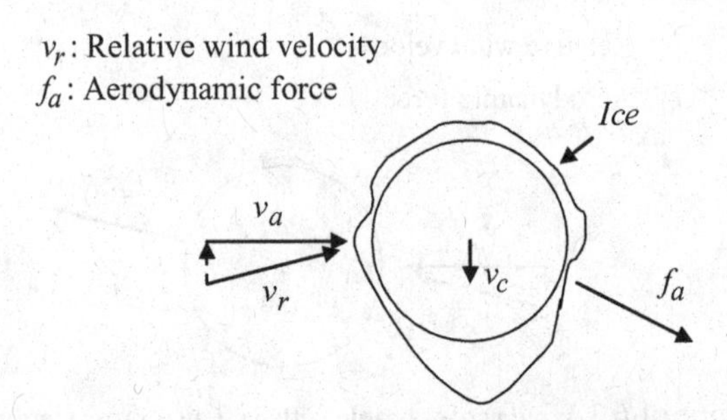

Figure 1.6.4 Wind across an ice-coated cable

as shown in Figure 1.6.4. In this case, the aerodynamic force can also be downwards when the velocity of the cable is downward. In other words, the wind will further strengthen the downward motion of the cable, and as a result, aerodynamic forces add energy to the system and the equivalent damping c_{eq} is negative.

Example 1.6.3: Galloping of a Square Prism (Negative Damping)
Consider a square prism with the mass m, which is subjected to a cross wind with the velocity v_a (Figure 1.6.5). The width and the height of the prism are w and h, respectively. The displacement of the prism with respect to the static equilibrium position is denoted by $x(t)$. For a small velocity $\dot{x}$, the vertical component of the aerodynamic force $f_{av}(t)$ is given by (Thompson, 1982)

$$f_{av}(t) = \frac{1}{2}\rho v_a w h \beta \dot{x} \tag{1.6.10}$$

where ρ is the air density. The constant β has been experimentally found to be a positive number. The differential equation of motion is easily derived to be

$$m\ddot{x} + c\dot{x} + kx = f_{av}(t) \tag{1.6.11}$$

Substituting Equation 1.6.10 into Equation 1.6.11,

$$m\ddot{x} + (c - c_n)\dot{x} + kx = 0 \tag{1.6.12}$$

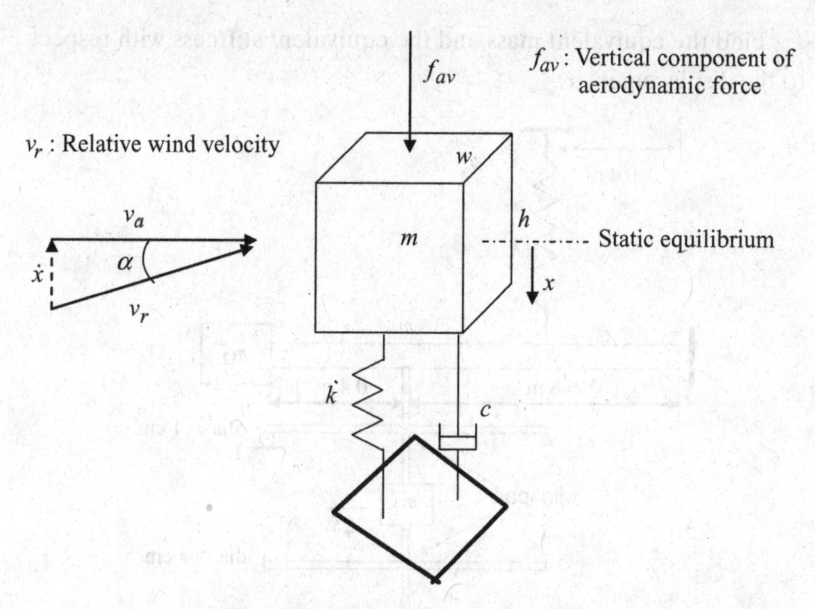

Figure 1.6.5 Wind across a square prism

where

$$c_n = \frac{1}{2}\rho v_a w h \beta \tag{1.6.13}$$

When $c < c_n$, the equivalent damping is negative and system becomes unstable.

EXERCISE PROBLEMS

P1.1 Consider the SDOF system shown in Figure P1.1. All the shafts and the connections among them are massless. The material of the shaft is steel. Also,

$$m_1 = 1.1\,\text{kg},\ m_2 = 1.4\,\text{kg},\ m_b = 2.1\,\text{kg}$$

The stiffness k is half of the cantilever beam stiffness.

Find the equivalent mass and the equivalent stiffness with respect to the displacement x.

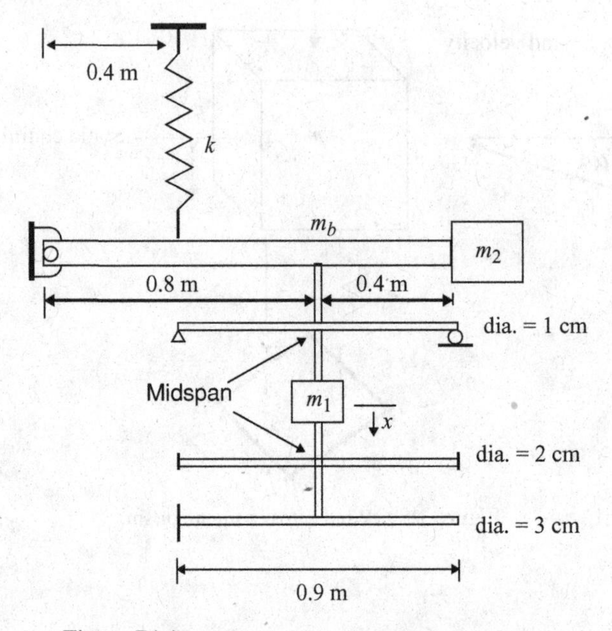

Figure P1.1 A combination of beams and a rigid bar

P1.2 Find the equivalent mass of a spring under the assumption that the velocity distribution along the length of the spring is parabolic.

P1.3 Consider the cantilever beam with mass m and length ℓ (Figure P1.3). Obtain the equivalent mass of the cantilever beam under the assumption that the beam deflection is $y(z) = \frac{x}{2}[1 - \cos(\frac{\pi z}{\ell})]$.

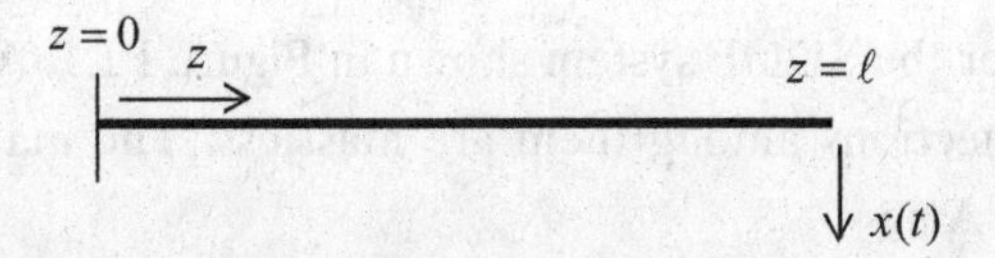

Figure P1.3 A cantilever beam

P1.4 An object with mass m and rectangular cross section A is floating in a liquid with mass density ρ (Figure P1.4). Derive the governing differential equation of motion, and obtain the natural frequency of the system.

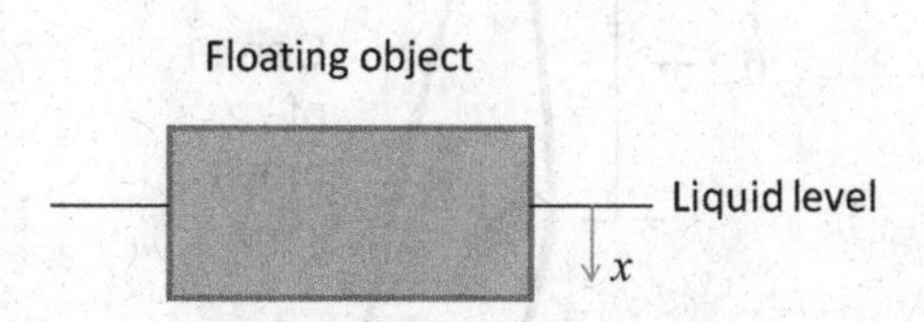

Figure P1.4 A floating object

P1.5 An L-shaped bracket hinged at point A is supported by two springs with stiffnesses k_1 and k_2 (Figure P1.5). The mass of the bracket is m and is uniformly distributed.

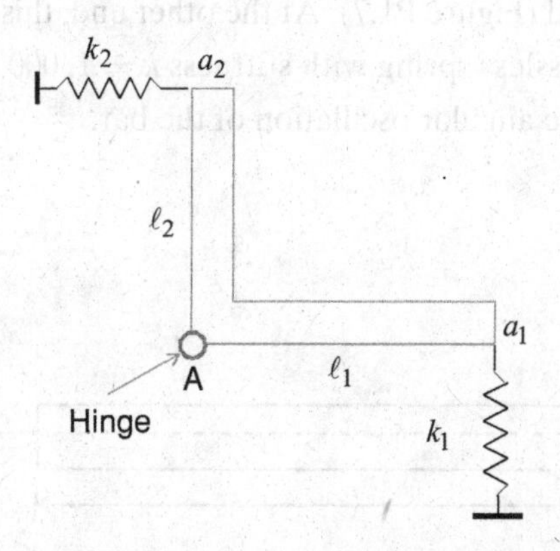

Figure P1.5 L-shaped bracket supported by springs

Find the equivalent mass, the equivalent stiffness and the natural frequency of the system.

P1.6 The mass of a complex-shaped object is 3 kg. When this object is suspended like a pendulum (Figure P1.6), its frequency of oscillation is 30 cycles/min. The center of mass is at a distance of 0.2 m from the pivot point A.

Find the mass-moment of inertia of the object about its center of mass.

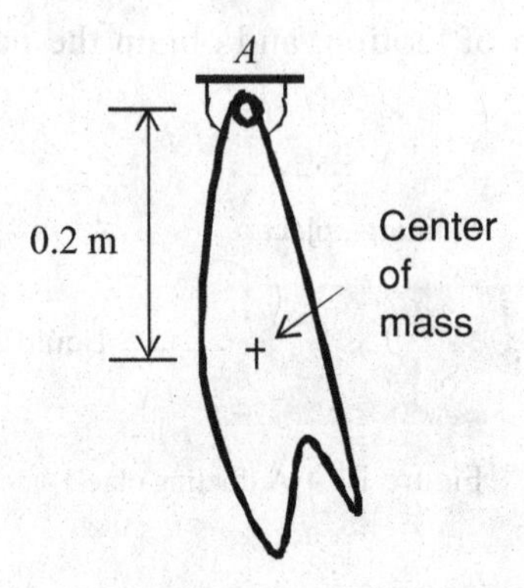

Figure P1.6 Complex-shaped object as a pendulum

P1.7 An uniform rigid bar of length $\ell = 50\,\text{cm}$ and mass $m = 7\,\text{kg}$ is hinged at one end (Figure P1.7). At the other end, this bar is suddenly attached to a massless spring with stiffness $k = 1,000\,\text{N/m}$. Derive an expression for the angular oscillation of the bar.

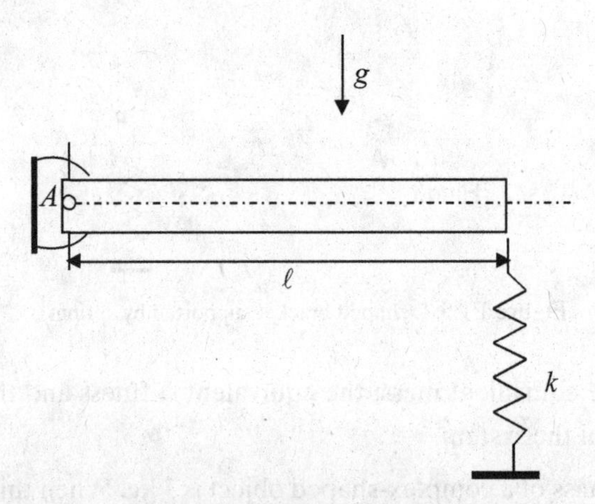

Figure P1.7 A uniform rigid bar

P1.8 A cylinder of mass m_2 rolls without slipping inside the box with mass m_1 (Figure P1.8). Derive the equivalent mass and the stiffness of the system.

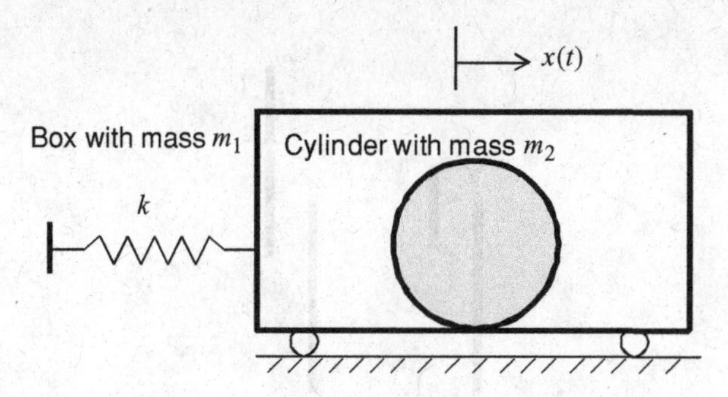

Figure P1.8 A cylinder inside a box

P1.9 Consider the gear shaft system in Figure P1.9. The length and the diameter of shaft A are 50 cm and 4 cm, respectively. Similarly, the length and the diameter of shaft B are 60 cm and 3 cm, respectively. Masses of gears A and B are 1.5 kg and 0.5 kg, respectively. The gear ratio is 2.

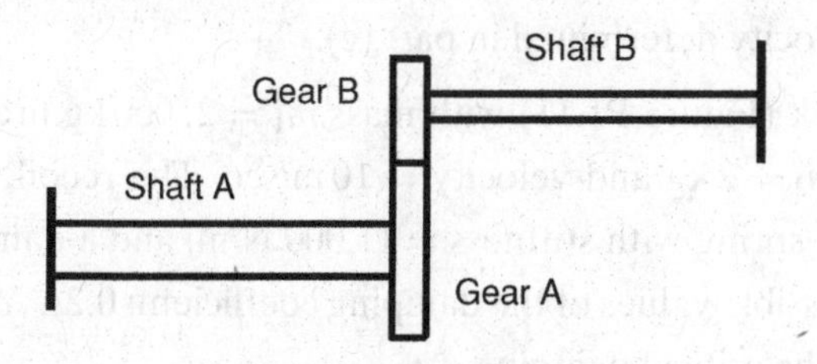

Figure P1.9 Gear shaft system

Assuming that the shaft material is steel, determine the equivalent mass and the equivalent stiffness with respect to the angular displacement of gear A. What is the natural frequency?

P1.10 An object with mass = 500 kg is attached to a table with four steel legs of diameter = 0.015 m and length = 0.1 m (Figure P1.10).

a. Derive the differential equation for vibration of the object in vertical direction.
b. Determine the natural frequency of vibration.
c. What is the maximum possible amplitude of vibration so that the maximum vibratory stress in each table leg is less than 70% of the

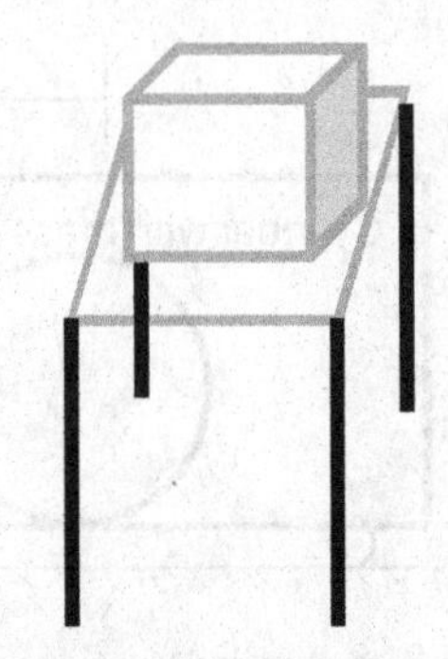

Figure P1.10 An object on the table

yield point stress? For this maximum amplitude, plot the allowable region of initial displacement and the initial velocity of the object.

d. Develop a MATLAB program to solve the governing differential equation. Compare the solution from your program to that obtained analytically for an allowable initial displacement and the initial velocity determined in part (c).

P1.11 A tank (Figure P1.11) with mass $m_1 = 2,000$ kg fires a cannon with mass $m_2 = 2$ kg and velocity $= 10$ m/sec. The recoil mechanism consists of a spring with stiffness = 11,000 N/m, and a damper. There are three possible values of the damping coefficient: $0.2c_c$, c_c, and $1.5c_c$ where c_c is the critical damping.

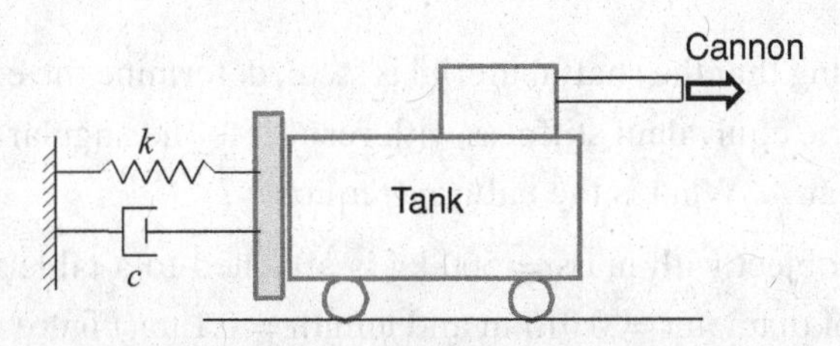

Figure P1.11 A tank with recoil mechanism

a. For each value of the damping constant, determine the time required to come back to the original firing position. Validate your analytical results via comparison with results from numerical integration of the solution of the differential equation via MATLAB.

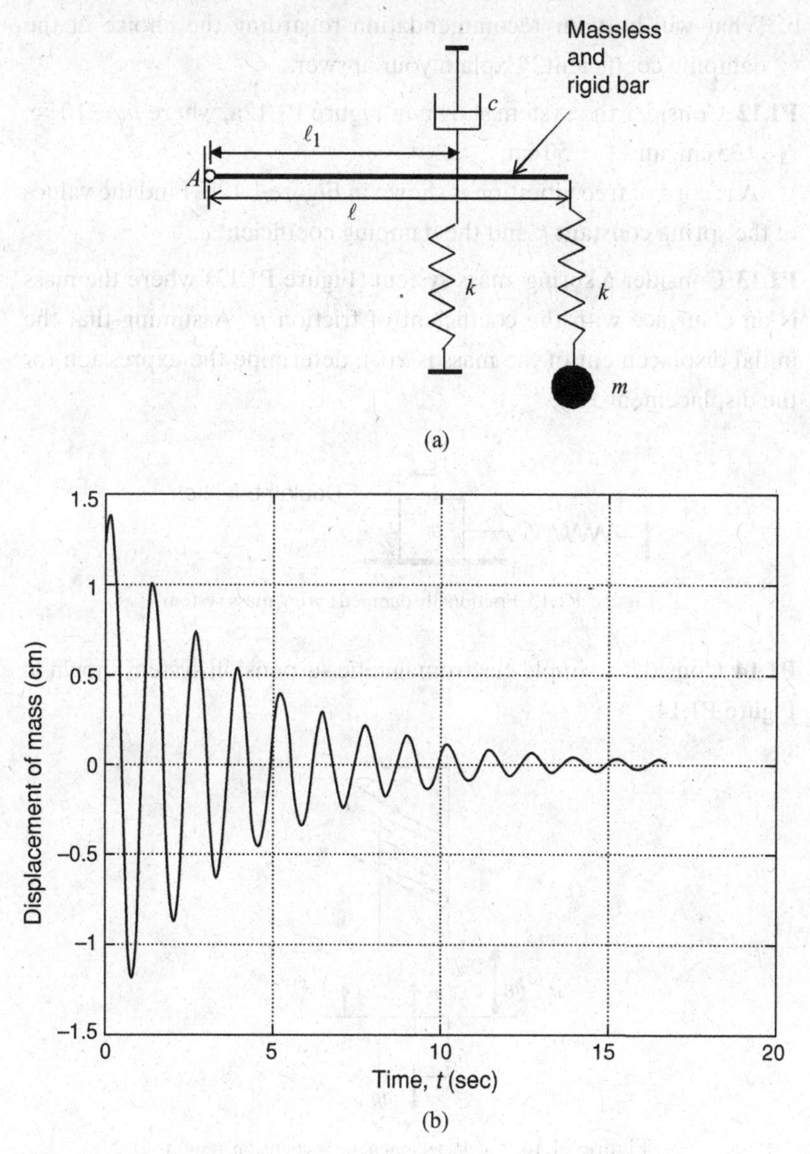

Figure P1.12 (a) Mass supported via a spring and a massless rigid bar; (b) Free vibration of mass m

b. What will be your recommendation regarding the choice of the damping coefficient? Explain your answer.

P1.12 Consider the system shown in Figure P1.12a, where $m = 10\,\text{kg}$, $\ell_1 = 35\,\text{cm}$, and $\ell = 50\,\text{cm}$.

A record of free vibration is shown in Figure 1.12b. Find the values of the spring constant k and the damping coefficient c.

P1.13 Consider a spring–mass system (Figure P1.13) where the mass is on a surface with the coefficient of friction μ. Assuming that the initial displacement of the mass is $x(0)$, determine the expression for the displacement $x(t)$.

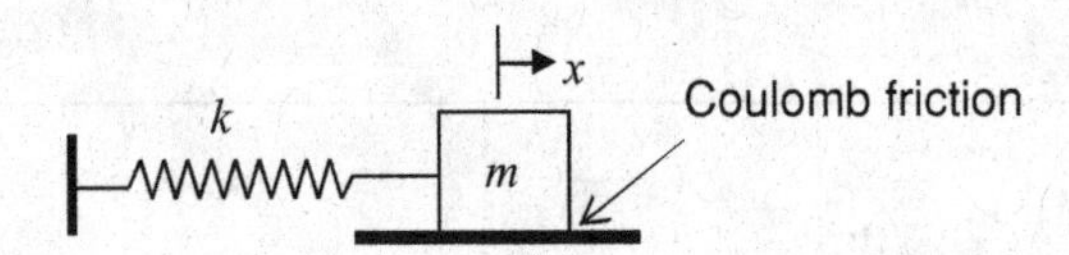

Figure P1.13 Frictionally damped spring mass system

P1.14 Consider a simple electromagnetic suspension system shown in Figure P1.14.

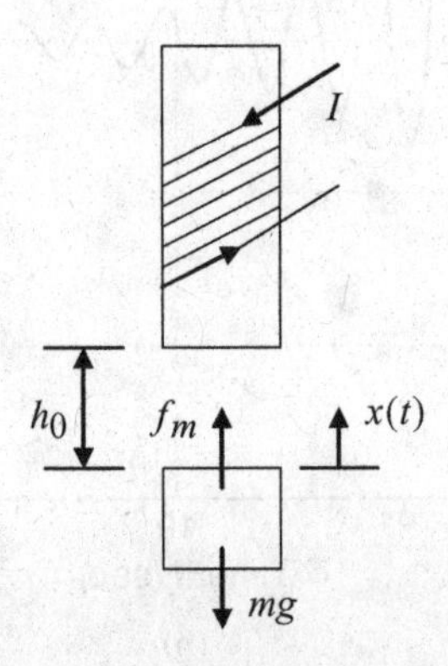

Figure P1.14 An electromagnetic suspension system

The electromagnetic force f_m is given by

$$f_m = \alpha \frac{I^2}{h^2}$$

where I and h are the coil current and the air gap, respectively. The constant $\alpha = \mu_0 N^2 A_p$ where μ_0, N, and A_p are the air permeability, the number of coil turns, and the face area per single pole of the magnet, respectively. Let h_0 be the desired air gap. Then, the current I_0 is calculated from the following static equilibrium condition:

$$\alpha \frac{I_0^2}{h_0^2} = mg$$

Let $x(t)$ be the dynamic displacement of the mass with respect to the static equilibrium position. Derive the differential equation of motion and determine the stability of system. Show that the dynamic characteristic of this system is equivalent to that of an inverted pendulum.

P1.15 Consider the system in Figure P1.15. Determine the natural frequencies when the mass m is constrained to move along x and y directions, respectively.

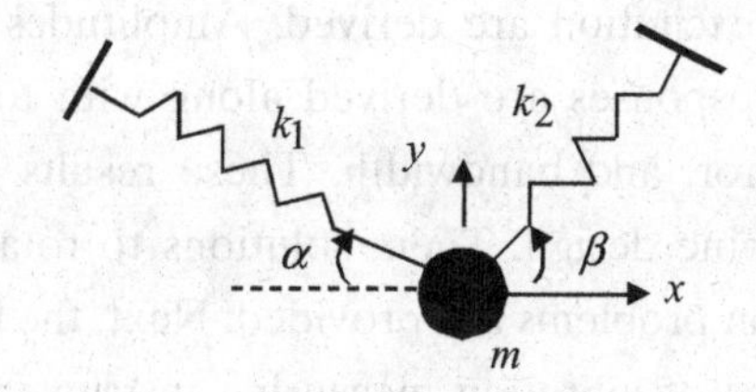

Figure P1.15 Mass supported by two inclined springs

2

VIBRATION OF A SINGLE-DEGREE-OF-FREEDOM SYSTEM UNDER CONSTANT AND PURELY HARMONIC EXCITATION

First, responses of undamped and damped single-degree-of-freedom (SDOF) spring–mass systems are presented in the presence of a constant external force. An important example of input shaping is shown. Using the input shaping procedure, the system settles to a steady state in a finite time in spite of a low level of damping. Next, complete solutions of both undamped and damped spring–mass systems under sinusoidal excitation are derived. Amplitudes and phases of the steady-state responses are derived along with force transmissibility, quality factor, and bandwidth. These results are fundamental tools for machine design. Then, solutions to rotating unbalance and base excitation problems are provided. Next, the basic principles behind the designs of vibration measuring instruments (vibrometer and accelerometer) are presented. Last, the concept of equivalent viscous damping is presented for nonviscous energy dissipation.

2.1 RESPONSES OF UNDAMPED AND DAMPED SDOF SYSTEMS TO A CONSTANT FORCE

Consider a damped SDOF system subjected to a force $f(t)$ (Figure 2.1.1). Using the free body diagram in Figure 2.1.1,

$$\text{net force in } x\text{-direction} = -k_{eq}x - c_{eq}\dot{x} + f(t) \qquad (2.1.1)$$

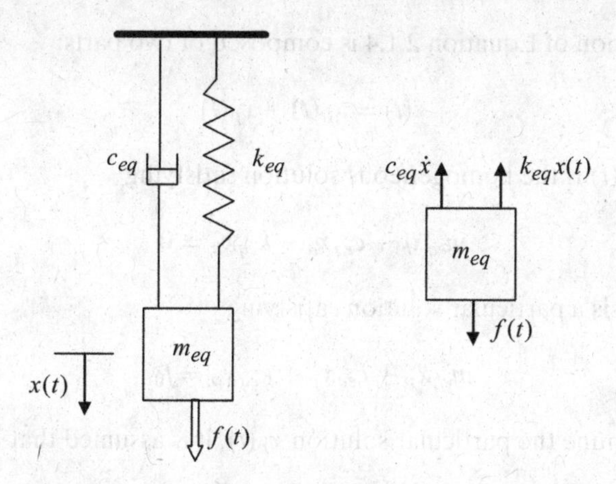

Figure 2.1.1 An SDOF spring–mass–damper system subjected to external excitation

Applying Newton's second law of motion,

$$-k_{eq}x - c_{eq}\dot{x} + f(t) = m_{eq}\ddot{x} \tag{2.1.2}$$

Therefore, the differential equation of motion is

$$m_{eq}\ddot{x} + c_{eq}\dot{x} + k_{eq}x = f(t) \tag{2.1.3}$$

Let the force $f(t)$ be a step function as shown in Figure 2.1.2.

From Equation 2.1.3, the differential equation of motion for a step forcing function is

$$m_{eq}\ddot{x} + c_{eq}\dot{x} + k_{eq}x = f_0;\ t \geq 0 \tag{2.1.4}$$

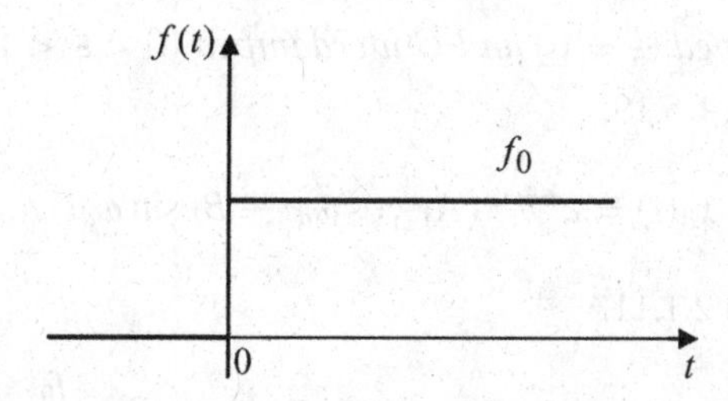

Figure 2.1.2 A step forcing function

The solution of Equation 2.1.4 is composed of two parts:

$$x(t) = x_h(t) + x_p(t) \tag{2.1.5}$$

where $x_h(t)$ is the homogeneous solution satisfying

$$m_{eq}\ddot{x}_h + c_{eq}\dot{x}_h + k_{eq}x_h = 0 \tag{2.1.6}$$

and $x_p(t)$ is a particular solution satisfying

$$m_{eq}\ddot{x}_p + c_{eq}\dot{x}_p + k_{eq}x_p = f_0 \tag{2.1.7}$$

To determine the particular solution $x_p(t)$, it is assumed that

$$x_p(t) = x_0, \text{ a constant} \tag{2.1.8}$$

Substituting Equation 2.1.8 into Equation 2.1.7,

$$m_{eq}\ddot{x}_0 + c_{eq}\dot{x}_0 + k_{eq}x_0 = f_0 \tag{2.1.9}$$

Since x_0 is a constant, $\dot{x}_0 = 0$ and $\ddot{x}_0 = 0$. Therefore,

$$x_0 = \frac{f_0}{k_{eq}} \tag{2.1.10}$$

Therefore, from Equation 2.1.5,

$$x(t) = x_h(t) + \frac{f_0}{k_{eq}} \tag{2.1.11}$$

where the homogeneous part $x_h(t)$ depends on the amount of damping as discussed in Sections 1.4 and 1.5.

Case I: Undamped ($\xi = 0$) *and Underdamped* ($0 < \xi < 1$)
From Equation 1.5.19,

$$x_h(t) = e^{-\xi\omega_n t}(A_1 \cos\omega_d t + B_1 \sin\omega_d t) \tag{2.1.12}$$

From Equation 2.1.11,

$$x(t) = e^{-\xi\omega_n t}(A_1 \cos\omega_d t + B_1 \sin\omega_d t) + \frac{f_0}{k_{eq}} \tag{2.1.13}$$

At $t = 0$,

$$x(0) = A_1 + \frac{f_0}{k_{eq}} \Rightarrow A_1 = x(0) - \frac{f_0}{k_{eq}} \tag{2.1.14}$$

Differentiating Equation 2.1.13,

$$\begin{aligned} \dot{x}(t) = \; & e^{-\xi\omega_n t}(-\omega_d A_1 \sin\omega_d t + \omega_d B_1 \cos\omega_d t) \\ & - \xi\omega_n e^{-\xi\omega_n t}(A_1 \cos\omega_d t + B_1 \sin\omega_d t) \end{aligned} \tag{2.1.15}$$

At $t = 0$,

$$\dot{x}(0) = \omega_d B_1 - \xi\omega_n A_1 \tag{2.1.16}$$

Using Equation 2.1.14,

$$B_1 = \frac{\dot{x}(0) + \xi\omega_n(x(0) - f_0/k_{eq})}{\omega_d} \tag{2.1.17}$$

With zero initial conditions ($x(0) = 0$ and $\dot{x}(0) = 0$), Equations 2.1.14 and 2.1.17 yield

$$A_1 = -\frac{f_0}{k_{eq}} \tag{2.1.18}$$

$$B_1 = -\frac{\xi\omega_n}{\omega_d}\frac{f_0}{k_{eq}} = -\chi\frac{f_0}{k_{eq}} \tag{2.1.19}$$

where

$$\chi = \frac{\xi}{\sqrt{1-\xi^2}} \tag{2.1.20}$$

For zero initial conditions,

$$x(t) = \frac{f_0}{k_{eq}}[1 - e^{-\xi\omega_n t}\cos\omega_d t - e^{-\xi\omega_n t}\chi\sin\omega_d t]; \quad t \geq 0 \tag{2.1.21}$$

Case II: Critically Damped ($\xi = 1$ *or* $c_{eq} = c_c$)
From Equation 1.5.33,

$$x_h(t) = A_1 e^{-\omega_n t} + B_1 t e^{-\omega_n t} \tag{2.1.22}$$

From Equation 2.1.11,

$$x(t) = A_1 e^{-\omega_n t} + B_1 t e^{-\omega_n t} + \frac{f_0}{k_{eq}} \tag{2.1.23}$$

At $t = 0$,

$$x(0) = A_1 + \frac{f_0}{k_{eq}} \Rightarrow A_1 = x(0) - \frac{f_0}{k_{eq}} \tag{2.1.24}$$

Differentiating Equation 2.1.23,

$$\dot{x}(t) = -\omega_n A_1 e^{-\omega_n t} - \omega_n B_1 t e^{-\omega_n t} + B_1 e^{-\omega_n t} \tag{2.1.25}$$

At $t = 0$,

$$\dot{x}(0) = -\omega_n A_1 + B_1 \tag{2.1.26}$$

Using Equation 2.1.24,

$$B_1 = \dot{x}(0) + \omega_n \left(x(0) - \frac{f_0}{k_{eq}} \right) \tag{2.1.27}$$

For zero initial conditions ($x(0) = 0$ and $\dot{x}(0) = 0$), substitution of Equations 2.1.24 and 2.1.27 into Equation 2.1.23 yields

$$x(t) = \frac{f_0}{k_{eq}} [1 - e^{-\omega_n t} - \omega_n t e^{-\omega_n t}]; \; t \geq 0 \tag{2.1.28}$$

Case III: Overdamped ($\xi > 1$ *or* $c_{eq} > c_c$)
From Equations 1.5.39 and 1.5.40, both characteristic roots s_1 and s_2 are negative real numbers,

$$s_1 = -\xi\omega_n + \omega_n\sqrt{\xi^2 - 1} < 0 \tag{2.1.29}$$

$$s_2 = -\xi\omega_n - \omega_n\sqrt{\xi^2 - 1} < 0 \tag{2.1.30}$$

From Equation 1.5.41,

$$x_h(t) = A_1 e^{s_1 t} + B_1 e^{s_2 t} \tag{2.1.31}$$

From Equation 2.1.11,

$$x(t) = A_1 e^{s_1 t} + B_1 e^{s_2 t} + \frac{f_0}{k_{eq}} \tag{2.1.32}$$

At $t = 0$,

$$x(0) = A_1 + B_1 + \frac{f_0}{k_{eq}} \tag{2.1.33}$$

Differentiating Equation 2.1.32,

$$\dot{x}(t) = A_1 s_1 e^{s_1 t} + B_1 s_2 e^{s_2 t} \tag{2.1.34}$$

At $t = 0$,

$$\dot{x}(0) = A_1 s_1 + B_1 s_2 \tag{2.1.35}$$

Solving Equations 2.1.33 and 2.1.35,

$$A_1 = \frac{s_2(x(0) - f_0/k_{eq}) - \dot{x}(0)}{s_2 - s_1} \tag{2.1.36}$$

and

$$B_1 = \frac{-s_1(x(0) - f_0/k_{eq}) + \dot{x}(0)}{s_2 - s_1} \tag{2.1.37}$$

For zero initial conditions ($x(0) = 0$ and $\dot{x}(0) = 0$), substitution of Equations 2.1.36 and 2.1.37 into Equation 2.1.32 yields

$$x(t) = \frac{f_0}{k_{eq}} \left[1 - \frac{s_2}{s_2 - s_1} e^{s_1 t} + \frac{s_1}{s_2 - s_1} e^{s_2 t} \right]; \quad t \geq 0 \tag{2.1.38}$$

where s_1 and s_2 are given by Equations 2.1.29 and 2.1.30.

Finally, it should be noted that

$$x_h(t) \to 0 \quad \text{as} \quad t \to \infty \tag{2.1.39}$$

for any damping $c_{eq} > 0$ or $\xi > 0$. Therefore, in the steady state ($t \to \infty$), from Equation 2.1.11

$$x_{ss} = \frac{f_0}{k_{eq}} \tag{2.1.40}$$

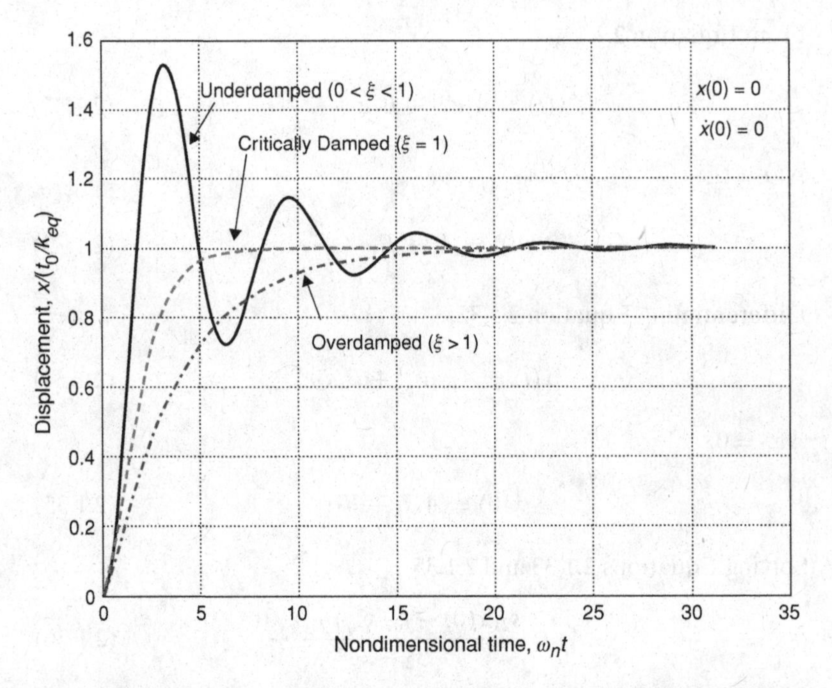

Figure 2.1.3 Response to the unit step forcing function with zero initial conditions

The step responses (Equations 2.1.21, 2.1.28, and 2.1.38) are plotted in Figure 2.1.3 for zero initial conditions.

Example 2.1.1: Robot Vibration

Consider a single link robot manipulator with a rigid link but with a flexible revolute joint (Figure 2.1.4a). The length of the link and the torsional stiffness of the joint are ℓ and k_t, respectively.

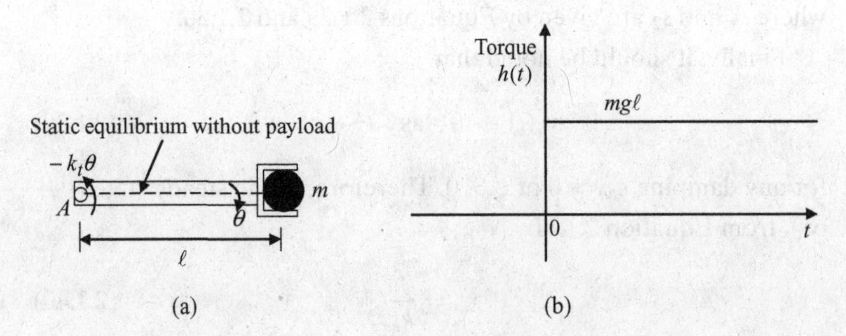

Figure 2.1.4 (a) A robot with a payload; and (b) Torque due to sudden payload

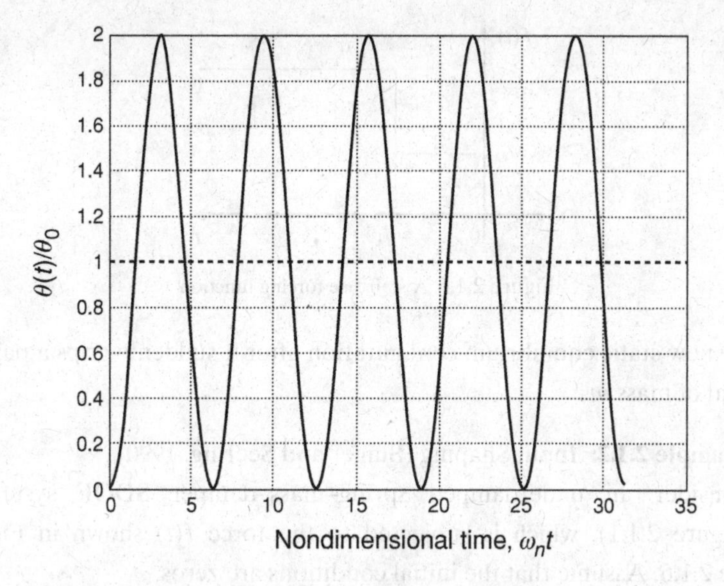

Figure 2.1.5 Angular oscillation of a robot arm

When the gripper suddenly gets a payload of mass m, the robot link undergoes a torque that is a step function $h(t)$ (Figure 2.1.4). The differential equation of motion is

$$I_A \ddot{\theta} + k_t \theta = h(t) \tag{2.1.41}$$

where $\theta(t)$ is measured from the static equilibrium configuration without the payload, and I_A is the mass-moment of inertia of the link with the payload about the joint axis A. Initial conditions are $\theta(0) = 0$ and $\dot{\theta}(0) = 0$. Using Equation 2.1.21 with $\xi = 0$,

$$\theta(t) = \theta_0(1 - \cos \omega_n t); \quad t \geq 0 \tag{2.1.42}$$

where

$$\theta_0 = \frac{mg\ell}{k_t} \tag{2.1.43}$$

It should be noted that θ_0 represents the new static equilibrium configuration. The response (Figure 2.1.5) clearly indicates that the robot arm will sustain a nondecaying oscillation of magnitude θ_0 about

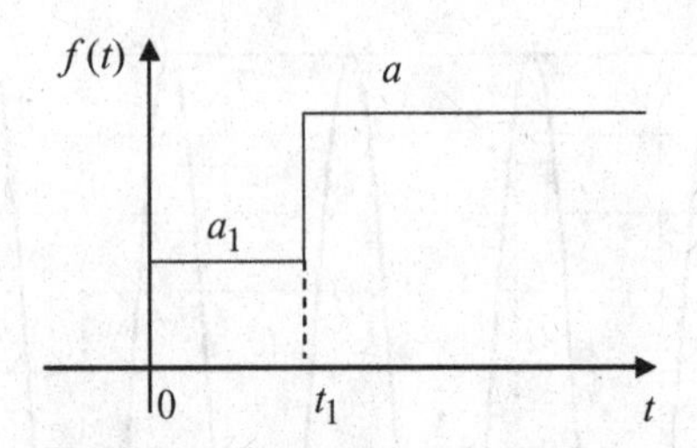

Figure 2.1.6 A staircase forcing function

the new static equilibrium configuration after it suddenly grips a payload of mass m.

Example 2.1.2: Input Shaping (Singer and Seering, 1990)
Consider an underdamped spring–mass–damper SDOF system (Figure 2.1.1), which is subjected to the force $f(t)$ shown in Figure 2.1.6. Assume that the initial conditions are zeros.

The differential equation of motion is

$$m_{eq}\ddot{x} + c_{eq}\dot{x} + k_{eq}x = f(t) \tag{2.1.44}$$

To determine the response $x(t)$, the force $f(t)$ is expressed as

$$f(t) = f_1(t) + f_2(t) \tag{2.1.45}$$

where $f_1(t)$ and $f_2(t)$ are shown in Figure 2.1.7.

Using Equation 2.1.21, the response due to $f_1(t)$ will be

$$x_1(t) = \frac{a_1}{k_{eq}}[1 - e^{-\xi\omega_n t}\cos\omega_d t - e^{-\xi\omega_n t}\chi\sin\omega_d t]; \quad t \geq 0 \tag{2.1.46}$$

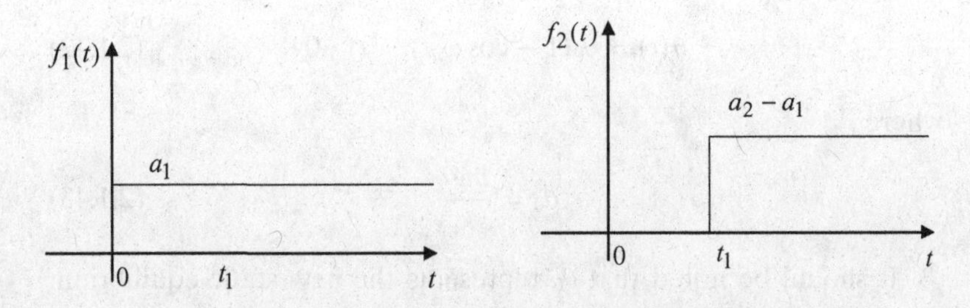

Figure 2.1.7 Components of staircase forcing function in Figure 2.1.6

Similarly, using Equation 2.1.21, the response due to $f_2(t)$ will be

$$x_2(t) = \frac{a_2 - a_1}{k_{eq}}[1 - e^{-\xi\omega_n(t-t_1)} \cos\omega_d(t - t_1)$$

$$- e^{-\xi\omega_n(t-t_1)} \chi \sin\omega_d(t - t_1)]; \quad t \geq t_1 \qquad (2.1.47)$$

It is obvious that $x_2(t) = 0$ for $t \leq t_1$. Using the principle of superposition,

$$x(t) = x_1(t) + x_2(t) \qquad (2.1.48)$$

Substituting Equations 2.1.46 and 2.1.47 into Equation 2.1.48 and after some algebra,

$$x(t) = \frac{a_2}{k_{eq}} - \frac{e^{-\xi\omega_n t}}{k_{eq}}(\alpha \sin\omega_d t + \beta \cos\omega_d t); \quad t \geq t_1 \qquad (2.1.49)$$

where

$$\alpha = a_1\chi + e^{\xi\omega_n t_1}(a_2 - a_1)(\sin\omega_d t_1 + \chi \cos\omega_d t_1) \qquad (2.1.50)$$

$$\beta = a_1 + e^{\xi\omega_n t_1}(a_2 - a_1)(\cos\omega_d t_1 - \chi \sin\omega_d t_1) \qquad (2.1.51)$$

It should be noted that $\alpha = 0$ and $\beta = 0$ when

$$t_1 = \frac{\pi}{\omega_d} \quad \text{and} \quad a_1 = \frac{a_2}{1 + q} \qquad (2.1.52\text{a,b})$$

where

$$q = e^{-\frac{\xi\pi}{\sqrt{1-\xi^2}}} \qquad (2.1.53)$$

When t_1 and a_1 are chosen according to Equation 2.1.52,

$$x(t) = \frac{a_2}{k_{eq}} \quad \text{for} \quad t > t_1 \qquad (2.1.54)$$

In other words, the system reaches its steady state in a finite time t_1 without any oscillation, even when the damping is zero or almost zero. Note that the input command $a_2, t > 0$, will yield a sustained oscillation as shown in Figure 2.1.5. Hence, by shaping the input command as shown in Figure 2.1.6, the system reaches the steady state without any vibration.

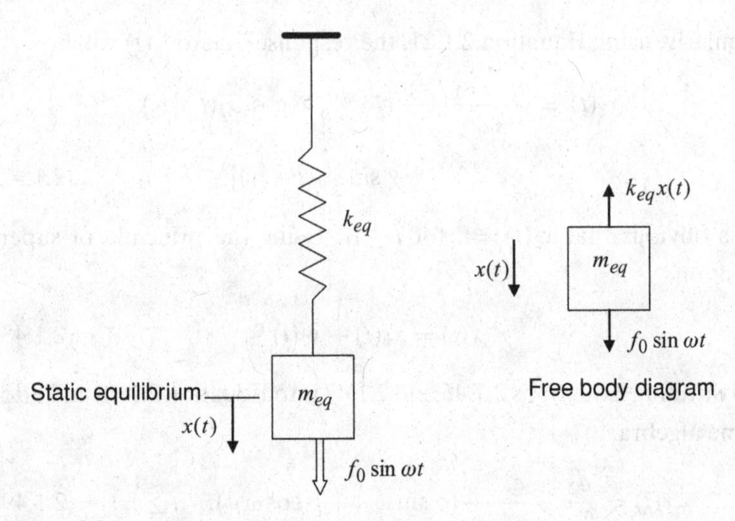

Figure 2.2.1 An undamped spring–mass system subjected to sinusoidal excitation

2.2 RESPONSE OF AN UNDAMPED SDOF SYSTEM TO A HARMONIC EXCITATION

Consider an undamped SDOF system subjected to a sinusoidal excitation with the amplitude f_0 and the frequency ω_0. Using the free body diagram in Figure 2.2.1,

$$\text{net force in } x\text{-direction} = -k_{eq}x + f_0 \sin \omega t \tag{2.2.1}$$

Applying Newton's second law of motion,

$$-k_{eq}x + f_0 \sin \omega t = m_{eq}\ddot{x} \tag{2.2.2}$$

Therefore, the differential equation of motion is

$$m_{eq}\ddot{x} + k_{eq}x = f_0 \sin \omega t \tag{2.2.3}$$

The solution of Equation 2.2.3 is composed of two parts:

$$x(t) = x_h(t) + x_p(t) \tag{2.2.4}$$

where $x_h(t)$ is the homogeneous solution satisfying

$$m_{eq}\ddot{x}_h + k_{eq}x_h = 0 \tag{2.2.5}$$

and $x_p(t)$ is a particular solution satisfying

$$m_{eq}\ddot{x}_p + k_{eq}x_p = f_0 \sin \omega t \tag{2.2.6}$$

The homogeneous solution in Equation 2.2.5 is

$$x_h(t) = A_1 \cos \omega_n t + B_1 \sin \omega_n t \tag{2.2.7}$$

where A_1 and B_1 are constants. And,

$$\omega_n = \sqrt{\frac{k_{eq}}{m_{eq}}} \tag{2.2.8}$$

The form of the particular solution is dependent on whether the excitation frequency ω equals the natural frequency ω_n.

Case I: $\omega \neq \omega_n$

$$x_p(t) = A \sin \omega t \tag{2.2.9}$$

where A can be positive or negative. To determine A, Equation 2.2.9 is substituted into Equation 2.2.6,

$$(-\omega^2 m_{eq} + k_{eq})A \sin \omega t = f_0 \sin \omega t \tag{2.2.10}$$

Equating coefficients of $\sin \omega t$ on both sides,

$$(-\omega^2 m_{eq} + k_{eq})A = f_0 \tag{2.2.11}$$

or

$$A = \frac{f_0}{k_{eq} - \omega^2 m_{eq}} \tag{2.2.12}$$

or

$$\frac{A}{f_0/k_{eq}} = \frac{1}{1 - r^2} \tag{2.2.13}$$

where r is the frequency ratio defined as

$$r = \frac{\omega}{\omega_n} \tag{2.2.14}$$

Note that f_0/k_{eq} is the steady-state deflection when a constant ($\omega = 0$ or $r = 0$) force f_0 is applied. A constant force corresponds to $\omega = 0$ or $r = 0$.

From Equations 2.2.4, 2.2.7, and 2.2.9, the total solution will be

$$x(t) = A_1 \cos \omega_n t + B_1 \sin \omega_n t + A \sin \omega t \qquad (2.2.15)$$

At $t = 0$,

$$x(0) = A_1 \qquad (2.2.16)$$

Differentiating Equation 2.2.15,

$$\dot{x}(t) = -A_1 \omega_n \sin \omega_n t + B_1 \omega_n \cos \omega_n t + A\omega \cos \omega t \qquad (2.2.17)$$

At $t = 0$,

$$\dot{x}(0) = B_1 \omega_n + A\omega \qquad (2.2.18)$$

Solving Equation 2.2.18,

$$B_1 = \frac{\dot{x}(0) - A\omega}{\omega_n} \qquad (2.2.19)$$

Substituting Equations 2.2.16 and 2.2.19 into Equation 2.2.15,

$$x(t) = x(0) \cos \omega_n t + \frac{\dot{x}(0) - A\omega}{\omega_n} \sin \omega_n t + A \sin \omega t \qquad (2.2.20)$$

where the amplitude A is given by Equation 2.2.13.

Case II: $\omega = \omega_n$ (Resonance)

In this case, the form of the particular integral Equation 2.2.9 is not valid. Note that the expression in Equation 2.2.9 is already represented by the homogeneous solution in Equation 2.2.7 when $\omega = \omega_n$. If the form given in Equation 2.2.9 was valid, Equation 2.2.13 would yield $A = \infty$ at $\omega = \omega_n$; that is, the displacement would become infinite from any finite initial displacement as soon as the external force is applied, which is physically impossible.

The particular integral for $\omega = \omega_n$ is

$$x_p(t) = A_r t \cos \omega t \tag{2.2.21}$$

Differentiating Equation 2.2.21 twice,

$$\ddot{x}_p(t) = -2A_r \omega \sin \omega t - A_r t \omega^2 \cos \omega t \tag{2.2.22}$$

Substituting Equations 2.2.21 and 2.2.22 into Equation 2.2.6,

$$-2A_r m_{eq} \omega \sin \omega t - A_r t m_{eq} \omega^2 \cos \omega t + A_r t k_{eq} \cos \omega t = f_0 \sin \omega t \tag{2.2.23}$$

Because $k_{eq} = m_{eq}\omega_n^2 = m_{eq}\omega^2$,

$$-2A_r m_{eq} \omega \sin \omega t = f_0 \sin \omega t \tag{2.2.24}$$

Equating coefficients of $\sin \omega t$ on both sides,

$$-2A_r m_{eq} \omega_n = f_0 \tag{2.2.25}$$

or

$$A_r = -\frac{f_0}{2m_{eq}\omega_n} = -\frac{f_0 \omega_n}{2k_{eq}} \tag{2.2.26}$$

From, 2.2.7, and 2.2.21, the total solution will be

$$x(t) = A_1 \cos \omega_n t + B_1 \sin \omega_n t + A_r t \cos \omega_n t \tag{2.2.27}$$

At $t = 0$,

$$x(0) = A_1 \tag{2.2.28}$$

Differentiating Equation 2.2.27,

$$\dot{x}(t) = -A_1 \omega_n \sin \omega_n t + B_1 \omega_n \cos \omega_n t - A_r t \omega_n \sin \omega_n t + A_r \cos \omega_n t \tag{2.2.29}$$

At $t = 0$,

$$\dot{x}(0) = B_1 \omega_n + A_r \tag{2.2.30}$$

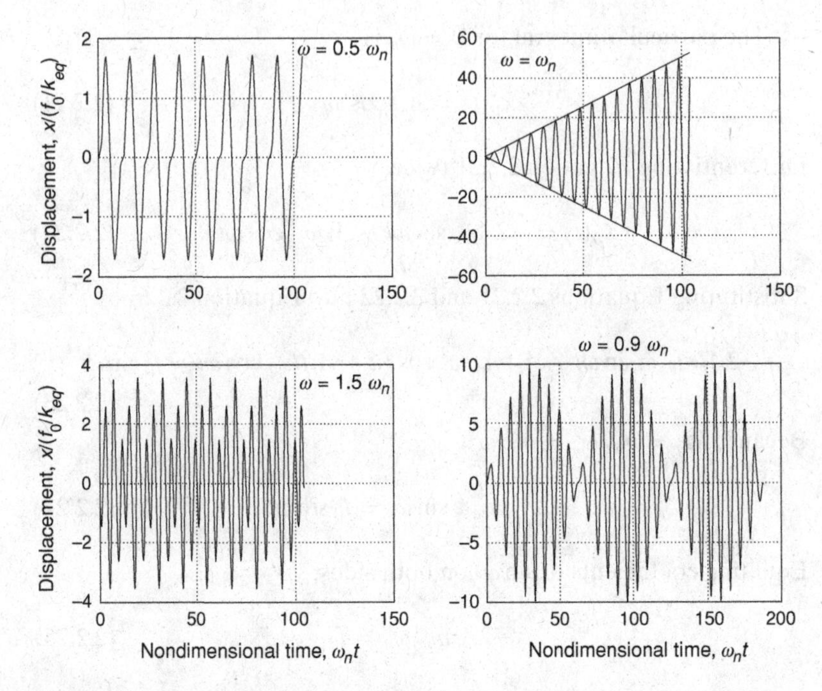

Figure 2.2.2 Responses at excitation frequencies $\omega = 0.5\omega_n, \omega_n, 1.5\omega_n$, and $0.9\omega_n$ with zero initial conditions

or

$$B_1 = \frac{\dot{x}(0) - A_r}{\omega_n} \tag{2.2.31}$$

Substituting Equations 2.2.28 and 2.2.31 into Equation 2.2.27,

$$x(t) = x(0)\cos\omega_n t + \frac{\dot{x}(0) - A_r}{\omega_n}\sin\omega_n t + A_r t\cos\omega t \tag{2.2.32}$$

where A_r is given by Equation 2.2.26.

Example 2.2.1: Responses at Different Excitation Frequencies with Zero Initial Conditions

In Figure 2.2.2, responses are plotted for different values of excitation frequencies ω with zero initial conditions ($x(0) = 0$ and $\dot{x}(0) = 0$).

Case I: $\omega \neq \omega_n$
With zero initial conditions, Equation 2.2.20 yields

$$x(t) = -\frac{A\omega}{\omega_n}\sin\omega_n t + A\sin\omega t \tag{2.2.33}$$

where A is given by Equation 2.2.13. When $\omega = 0.5\omega_n$ and $\omega = 1.5\omega_n$, responses contain both the frequencies ω and ω_n (Figure 2.2.2). When ω is close to ω_n, for example, $\omega = 0.9\omega_n$, the response exhibits a beating phenomenon (Figure 2.2.2) which can be explained by expressing Equation 2.2.33 as

$$x(t) = A(\sin\omega t - \sin\omega_n t) - A\frac{\omega - \omega_n}{\omega_n}\sin\omega_n t \tag{2.2.34}$$

or

$$x(t) = 2A\sin(0.5(\omega - \omega_n)t)\cos(0.5(\omega + \omega_n)t) - A\frac{\omega - \omega_n}{\omega_n}\sin\omega_n t \tag{2.2.35}$$

The second term on the right-hand side of Equation 2.2.35 is small. Ignoring this term,

$$x(t) = 2A\sin(0.5(\omega - \omega_n)t)\cos(0.5(\omega + \omega_n)t) \tag{2.2.36}$$

Equation 2.2.36 can be interpreted as the response with the frequency $0.5(\omega + \omega_n)$ with the time-varying amplitude. The frequency of the amplitude variation is $0.5(\omega - \omega_n)$, as found in Figure 2.2.2, and is much less than $0.5(\omega + \omega_n)$.

Case II: $\omega = \omega_n$
With zero initial conditions, Equation 2.2.32 yields

$$x(t) = -\frac{A_r}{\omega_n}\sin\omega_n t + A_r t\cos\omega t \tag{2.2.37}$$

Using Equation 2.2.26,

$$x(t) = \frac{f_0}{2k_{eq}}\sin\omega_n t - \frac{f_0\omega_n}{2k_{eq}}t\cos\omega t \tag{2.2.38}$$

It is obvious that the second term on the right-hand side of Equation 2.2.38 will be dominant after some time t. This will be true for a

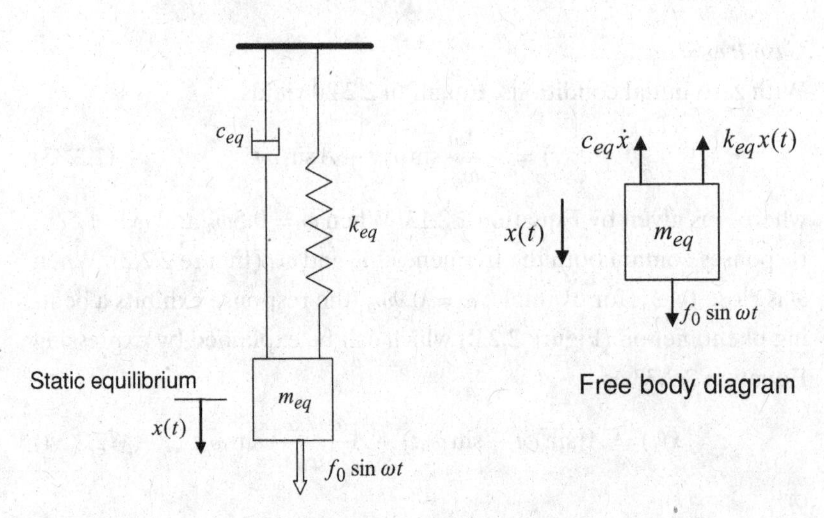

Figure 2.3.1 A damped SDOF spring–mass system subjected to sinusoidal excitation

large ω_n even for a small time t. In other words, for a large ω_n and/or after some time t,

$$x(t) \approx -\frac{f_0 \omega_n}{2k_{eq}} t \cos \omega t \tag{2.2.39}$$

The response in Figure 2.2.2 is described by Equation 2.2.39, in which the amplitude of vibration grows to infinity in a linear manner with respect to time. From the structural integrity point of view, the bad news is that the amplitude of vibration grows to infinity. But, the good news is that it takes time for the amplitude to build up to infinity, and there is time to save the structure from catastrophic failure. This fact is used in determining the rate at which a rotor shaft must be accelerated past its resonant speed (or critical speed) when the desired rotor speed is greater than the critical speed.

2.3 RESPONSE OF A DAMPED SDOF SYSTEM TO A HARMONIC EXCITATION

Consider a damped SDOF system subjected to a sinusoidal excitation with the amplitude f_0 and the frequency ω (Figure 2.3.1). Using the free body diagram in Figure 2.3.1,

$$\text{net force in } x\text{-direction} = -k_{eq}x - c_{eq}\dot{x} + f_0 \sin \omega t \tag{2.3.1}$$

Applying Newton's second law of motion,

$$-k_{eq}x - c_{eq}\dot{x} + f_0 \sin \omega t = m_{eq}\ddot{x} \tag{2.3.2}$$

Therefore, the differential equation of motion is

$$m_{eq}\ddot{x} + c_{eq}\dot{x} + k_{eq}x = f_0 \sin \omega t \tag{2.3.3}$$

The solution of Equation 2.3.3 is composed of two parts:

$$x(t) = x_h(t) + x_p(t) \tag{2.3.4}$$

where $x_h(t)$ is the homogeneous solution satisfying

$$m_{eq}\ddot{x}_h + c_{eq}\dot{x}_h + k_{eq}x_h = 0 \tag{2.3.5}$$

and $x_p(t)$ is a particular solution satisfying

$$m_{eq}\ddot{x}_p + c_{eq}\dot{x}_p + k_{eq}x_p = f_0 \sin \omega t \tag{2.3.6}$$

Particular Solution.
Assume that

$$x_p(t) = A \sin(\omega t - \phi) \tag{2.3.7}$$

Substituting Equation 2.3.7 into Equation 2.3.6,

$$\begin{aligned} -m_{eq}\omega^2 A \sin(\omega t - \phi) + c_{eq}\omega A \cos(\omega t - \phi) \\ + k_{eq}A \sin(\omega t - \phi) = f_0 \sin \omega t \end{aligned} \tag{2.3.8}$$

or

$$(k_{eq} - m_{eq}\omega^2)A \sin(\omega t - \phi) + c_{eq}\omega A \cos(\omega t - \phi) = f_0 \sin \omega t \tag{2.3.9}$$

or

$$\begin{aligned} (k_{eq} - m_{eq}\omega^2)A[\sin \omega t \cos \phi - \cos \omega t \sin \phi] \\ + c_{eq}\omega A[\cos \omega t \cos \phi + \sin \omega t \sin \phi] = f_0 \sin \omega t \end{aligned} \tag{2.3.10}$$

or

$$[(k_{eq} - m_{eq}\omega^2)A\cos\phi + c_{eq}\omega A\sin\phi]\sin\omega t$$
$$+ [c_{eq}\omega A\cos\phi - (k_{eq} - m_{eq}\omega^2)A\sin\phi]\cos\omega t = f_0\sin\omega t \quad (2.3.11)$$

Equating coefficients of $\sin\omega t$ and $\cos\omega t$ on both sides,

$$(k_{eq} - m_{eq}\omega^2)A\cos\phi + c_{eq}\omega A\sin\phi = f_0 \quad (2.3.12a)$$

and

$$c_{eq}\omega A\cos\phi - (k_{eq} - m_{eq}\omega^2)A\sin\phi = 0 \quad (2.13.12b)$$

Representing Equations 2.3.12a and 2.3.12b in matrix form,

$$\begin{bmatrix} k_{eq} - \omega^2 m_{eq} & c_{eq}\omega \\ c_{eq}\omega & -(k_{eq} - \omega^2 m_{eq}) \end{bmatrix} \begin{bmatrix} A\cos\phi \\ A\sin\phi \end{bmatrix} = \begin{bmatrix} f_0 \\ 0 \end{bmatrix} \quad (2.3.13)$$

or

$$\begin{bmatrix} A\cos\phi \\ A\sin\phi \end{bmatrix} = \begin{bmatrix} k_{eq} - \omega^2 m_{eq} & c_{eq}\omega \\ c_{eq}\omega & -(k_{eq} - \omega^2 m_{eq}) \end{bmatrix}^{-1} \begin{bmatrix} f_0 \\ 0 \end{bmatrix} \quad (2.3.14)$$

or

$$\begin{bmatrix} A\cos\phi \\ A\sin\phi \end{bmatrix} = \frac{1}{\Delta} \begin{bmatrix} -(k_{eq} - \omega^2 m_{eq}) & -c_{eq}\omega \\ -c_{eq}\omega & +(k_{eq} - \omega^2 m_{eq}) \end{bmatrix} \begin{bmatrix} f_0 \\ 0 \end{bmatrix} \quad (2.3.15)$$

where

$$\Delta = -(k_{eq} - \omega^2 m_{eq})^2 - (c_{eq}\omega)^2 \quad (2.3.16)$$

From Equation 2.3.15,

$$A\cos\phi = \frac{(k_{eq} - \omega^2 m_{eq})f_0}{(k_{eq} - \omega^2 m_{eq})^2 + (c_{eq}\omega)^2} \quad (2.3.17)$$

and

$$A\sin\phi = \frac{c_{eq}\omega f_0}{(k_{eq} - \omega^2 m_{eq})^2 + (c_{eq}\omega)^2} \quad (2.3.18)$$

Using Equations 2.3.17 and 2.3.18,

$$A = \sqrt{(A\cos\phi)^2 + (A\sin\phi)^2} = +\frac{f_0}{\sqrt{(k_{eq} - \omega^2 m_{eq})^2 + (c_{eq}\omega)^2}} \tag{2.3.19}$$

and

$$\tan\phi = \frac{A\sin\phi}{A\cos\phi} = \frac{c_{eq}\omega}{(k_{eq} - \omega^2 m_{eq})} \tag{2.3.20}$$

Dividing the numerators and denominators of Equations 2.3.19 and 2.3.20 by k_{eq},

$$A = \frac{\frac{f_0}{k_{eq}}}{\sqrt{\left(1 - \omega^2 \frac{m_{eq}}{k_{eq}}\right)^2 + \left(\frac{c_{eq}\omega}{k_{eq}}\right)^2}} \tag{2.3.21}$$

and

$$\tan\phi = \frac{\frac{c_{eq}\omega}{k_{eq}}}{1 - \omega^2 \frac{m_{eq}}{k_{eq}}} \tag{2.3.22}$$

Now,

$$\omega^2 \frac{m_{eq}}{k_{eq}} = \frac{\omega^2}{\omega_n^2} = r^2 \tag{2.3.23}$$

$$\frac{c_{eq}\omega}{k_{eq}} = \frac{c_c \xi \omega}{k_{eq}} = \frac{2 m_{eq} \omega_n \xi \omega}{k_{eq}} = 2\xi \frac{\omega}{\omega_n} = 2\xi r \tag{2.3.24}$$

Substituting Equations 2.3.23 and 2.3.24 into Equations 2.3.21 and 2.3.22,

$$\frac{A}{f_0/k_{eq}} = \frac{1}{\sqrt{(1 - r^2)^2 + (2\xi r)^2}} \tag{2.3.25}$$

and

$$\tan\phi = \frac{2\xi r}{1 - r^2} \tag{2.3.26}$$

where r is the frequency ratio defined as

$$r = \frac{\text{excitation frequency}}{\text{natural frequency}} = \frac{\omega}{\omega_n} \tag{2.3.27}$$

Case I: Underdamped ($0 < \xi < 1$ *or* $0 < c_{eq} < c_c$)
From Equation 1.5.19,

$$x_h(t) = e^{-\xi\omega_n t}(A_1 \cos\omega_d t + B_1 \sin\omega_d t) \tag{2.3.28}$$

From Equations 2.3.4, 2.3.28, and 2.3.7,

$$x(t) = e^{-\xi\omega_n t}(A_1 \cos\omega_d t + B_1 \sin\omega_d t) + A\sin(\omega t - \phi) \tag{2.3.29}$$

At $t = 0$,

$$x(0) = A_1 - A\sin\phi \Rightarrow A_1 = x(0) + A\sin\phi \tag{2.3.30}$$

Differentiating Equation 2.3.29,

$$\begin{aligned}\dot{x}(t) &= e^{-\xi\omega_n t}(-\omega_d A_1 \sin\omega_d t + \omega_d B_1 \cos\omega_d t) \\ &\quad - \xi\omega_n e^{-\xi\omega_n t}(A_1 \cos\omega_d t + B_1 \sin\omega_d t) + \omega A\cos(\omega t - \phi)\end{aligned} \tag{2.3.31}$$

At $t = 0$,

$$\dot{x}(0) = \omega_d B_1 - \xi\omega_n A_1 + \omega A\cos\phi \tag{2.3.32}$$

Using Equation 2.3.30,

$$B_1 = \frac{\dot{x}(0) + \xi\omega_n(x(0) + A\sin\phi) - \omega A\cos\phi}{\omega_d} \tag{2.3.33}$$

The total response $x(t) = x_h(t) + x_p(t)$ is shown in Figure 2.3.2 for $\xi = 0.04$ and $\xi = 0.5$.

Case II: Critically Damped ($\xi = 1$ *or* $c_{eq} = c_c$)
From Equation 1.5.33,

$$x_h(t) = A_1 e^{-\omega_n t} + B_1 t e^{-\omega_n t} \tag{2.3.34}$$

From Equations 2.3.4, 2.3.34, and 2.3.7,

$$x(t) = A_1 e^{-\omega_n t} + B_1 t e^{-\omega_n t} + A\sin(\omega t - \phi) \tag{2.3.35}$$

At $t = 0$,

$$x(0) = A_1 - A\sin\phi \Rightarrow A_1 = x(0) + A\sin\phi \tag{2.3.36}$$

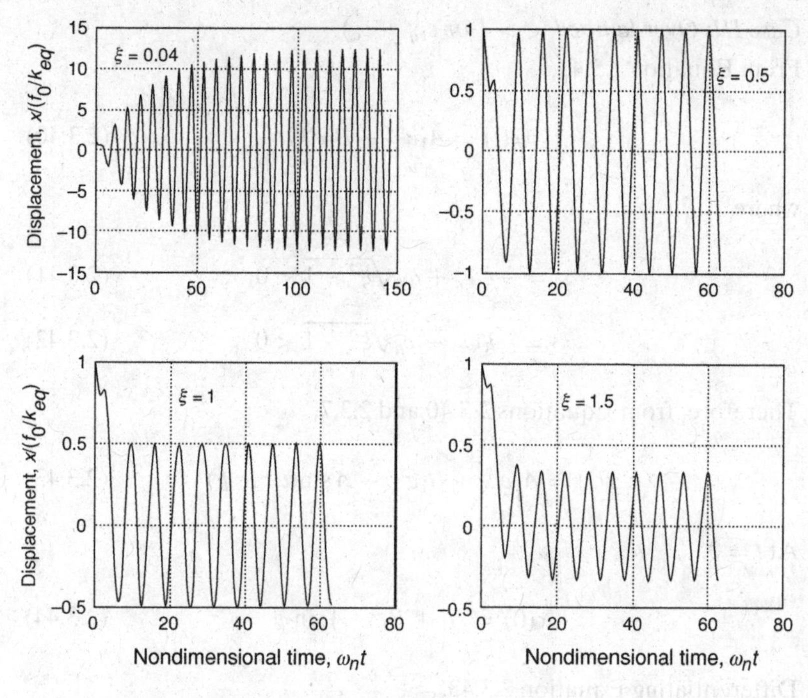

Figure 2.3.2 Response ($x(0) = f_0/k_{eq}$ and $\dot{x}(0) = 0$) for $\xi = 0.04$, 0.5, 1, and 1.5

Differentiating Equation 2.3.35,

$$\dot{x}(t) = -\omega_n A_1 e^{-\omega_n t} - \omega_n B_1 t e^{-\omega_n t} + B_1 e^{-\omega_n t} + \omega A \cos(\omega t - \phi) \tag{2.3.37}$$

At $t = 0$,

$$\dot{x}(0) = -\omega_n A_1 + B_1 + \omega A \cos\phi \tag{2.3.38}$$

Using Equation 2.3.36,

$$B_1 = \dot{x}(0) + \omega_n (x(0) + A \sin\phi) - \omega A \cos\phi \tag{2.3.39}$$

The total response $x(t) = x_h(t) + x_p(t)$ is shown in Figure 2.3.2 for $\xi = 1$.

Case III: Overdamped ($\xi > 1$ *or* $c_{eq} > c_c$)
From Equation 1.5.41,

$$x_h(t) = A_1 e^{s_1 t} + B_1 e^{s_2 t} \tag{2.3.40}$$

where

$$s_1 = -\xi\omega_n + \omega_n\sqrt{\xi^2 - 1} < 0 \tag{2.3.41}$$

$$s_2 = -\xi\omega_n - \omega_n\sqrt{\xi^2 - 1} < 0 \tag{2.3.42}$$

Therefore, from Equations 2.3.40 and 2.3.7,

$$x(t) = A_1 e^{s_1 t} + B_1 e^{s_2 t} + A\sin(\omega t - \phi) \tag{2.3.43}$$

At $t = 0$,

$$x(0) = A_1 + B_1 - A\sin\phi \tag{2.3.44}$$

Differentiating Equation 2.3.43,

$$\dot{x}(t) = A_1 s_1 e^{s_1 t} + B_1 s_2 e^{s_2 t} + \omega A\cos(\omega t - \phi) \tag{2.3.45}$$

At $t = 0$,

$$\dot{x}(0) = A_1 s_1 + B_1 s_2 + \omega A\cos\phi \tag{2.3.46}$$

Solving Equations 2.3.44 and 2.3.46,

$$A_1 = \frac{s_2(x(0) + A\sin\phi) - (\dot{x}(0) - \omega A\cos\phi)}{s_2 - s_1} \tag{2.3.47}$$

and

$$B_1 = \frac{-s_1(x(0) + A\sin\phi) + (\dot{x}(0) - \omega A\cos\phi)}{s_2 - s_1} \tag{2.3.48}$$

The total response $x(t) = x_h(t) + x_p(t)$ is shown in Figure 2.3.2 for $\xi > 1$.

2.3.1 Steady State Response

For all cases of damping $c_{eq} > 0$ or $\xi > 0$ (Equations 2.3.28, 2.3.34, and 2.3.40)

$$x_h(t) \to 0 \quad \text{as} \quad t \to \infty \tag{2.3.49}$$

Therefore, the homogeneous part $x_h(t)$ is also called "transient response." In the steady state ($t \to \infty$),

$$x(t) = x_h(t) + x_p(t) \to x_p(t) \quad \text{as} \quad t \to \infty \tag{2.3.50}$$

Therefore, the particular integral $x_p(t)$ is also called the steady state response $x_{ss}(t)$:

$$x_{ss}(t) = A\sin(\omega t - \phi) \tag{2.3.51}$$

where (from Equations 2.3.25 and 2.3.26),

$$\frac{A}{f_0/k_{eq}} = \frac{1}{\sqrt{(1-r^2)^2 + (2\xi r)^2}} \tag{2.3.52}$$

and

$$\tan\phi = \frac{2\xi r}{1-r^2} \tag{2.3.53}$$

Note that A is the amplitude of the steady state response. And ϕ is the phase of the steady state response which is the angle by which the steady state response **lags** behind the forcing function.

Amplitude (A) and phase (ϕ) are plotted in Figures 2.3.3 and 2.3.4 as functions of the frequency ratio r for different values of the damping ratio ξ. For all values of the damping ratio ξ, $A \to 0$ as $r \to \infty$. Near resonance condition, the amplitude can be very large, particularly for a small value of the damping.

To find the peak amplitude, Equation 2.3.52 is differentiated with respect to r,

$$\frac{k_{eq}}{f_0}\frac{dA}{dr} = -\frac{1}{2}[(1-r^2)^2 + (2\xi r)^2]^{-\frac{3}{2}}[2(1-r^2)(-2r) + (2\xi)^2 2r] \tag{2.3.54}$$

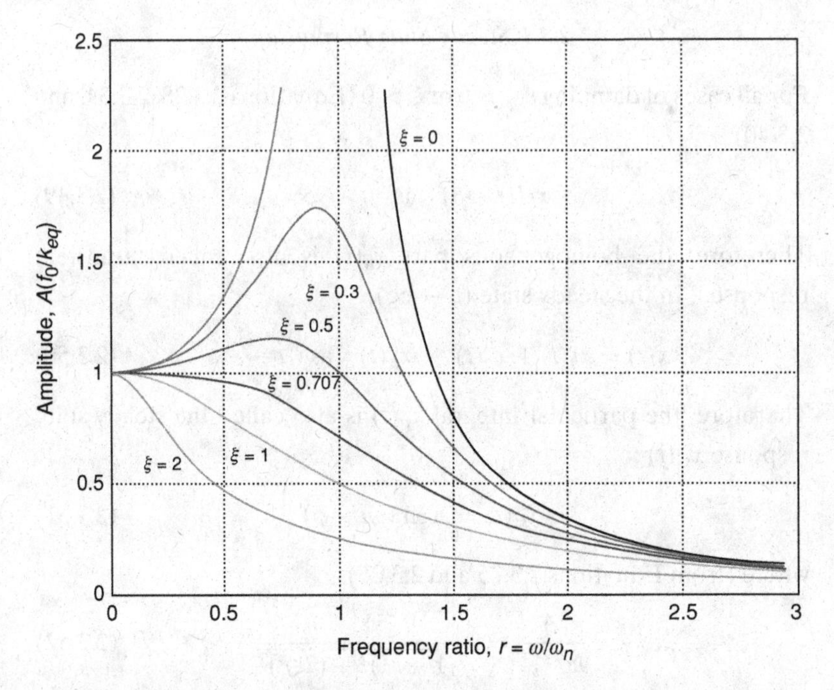

Figure 2.3.3 Amplitude versus frequency ratio (direct excitation on mass)

The condition for the maximum value of the amplitude A is

$$\frac{dA}{dr} = 0 \tag{2.3.55}$$

Therefore, from Equations 2.3.54 and 2.3.55,

$$-(1 - r^2) + 2\xi^2 = 0 \tag{2.3.56}$$

or

$$r = \sqrt{1 - 2\xi^2} \quad \text{provided} \quad \xi \leq \frac{1}{\sqrt{2}} = 0.707 \tag{2.3.57}$$

Equation 2.3.57 yields the frequency ratio at which the amplitude is maximum.

The corresponding peak amplitude (A_p) is given by

$$\frac{A_p}{f_0/k_{eq}} = \frac{1}{2\xi\sqrt{1 - \xi^2}}; \quad \xi \leq 0.707 \tag{2.3.58}$$

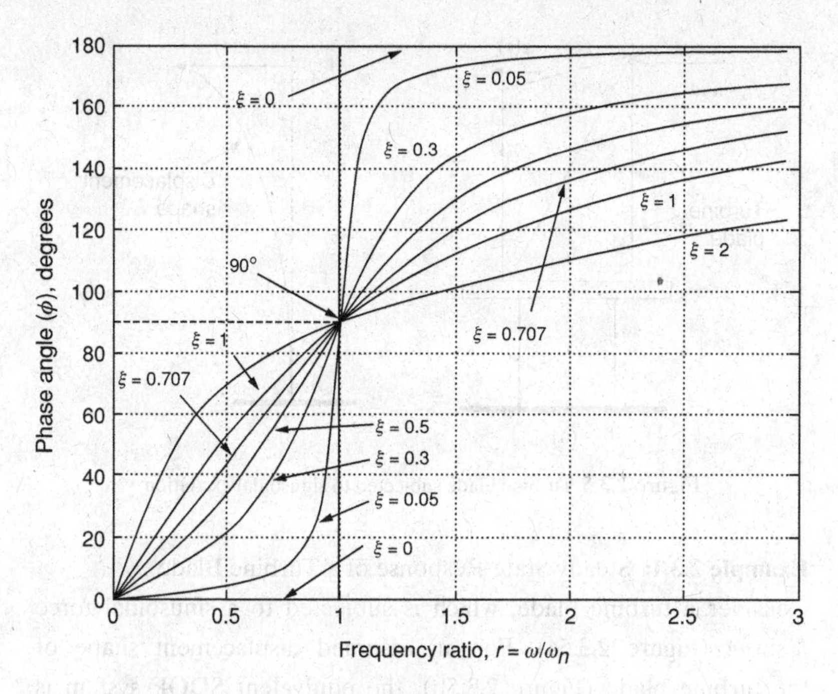

Figure 2.3.4 Phase versus frequency ratio (direct excitation on mass)

It should be noted that the maximum value of the amplitude occurs at $r = 0$ for $\xi > 0.707$.

For the phase plot (Figure 2.3.4), two important points should be noted:

a. The phase angle $\phi = 90°$ at $\omega = \omega_n$ for all values of damping ratios greater than zero. In other words, the phase angle ϕ of a damped ($\xi > 0$) system is always 90° when the excitation frequency ω equals the undamped natural frequency ω_n of the system.
b. The phase angles ϕ of an undamped ($\xi = 0$) system are 0° and 180° for $\omega < \omega_n$ and $\omega > \omega_n$, respectively. In other words, the phase angle ϕ changes abruptly from 0° to 180° across $\omega = \omega_n$. It should be noted that the phase angle ϕ of an undamped system is not defined for $\omega = \omega_n$, as $A \sin(\omega t - \phi)$ is not the particular integral in this case.

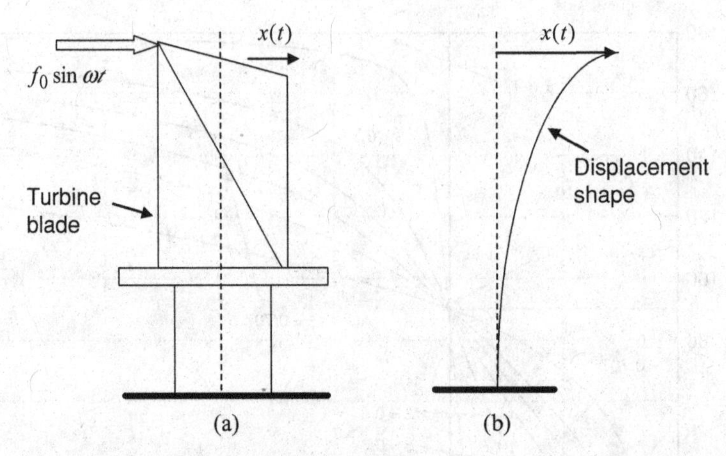

Figure 2.3.5 Turbine blade subjected to sinusoidal excitation

Example 2.3.1: Steady State Response of a Turbine Blade
Consider a turbine blade, which is subjected to a sinusoidal force $f_0 \sin \omega t$ (Figure 2.3.5a). For an assumed displacement shape of the turbine blade (Figure 2.3.5b), the equivalent SDOF system is shown in Figure 2.3.6, where $m_{eq} = 0.0114$ kg and $k_{eq} = 430{,}000$ N/m (Griffin and Sinha, 1985). Assuming that the damping ratio $\xi = 0.01$ and the force amplitude $f_0 = 1$ N, determine the steady state amplitudes and the phases for the following values of frequency ratio: 0.4, 0.95, 1, and 2. Also, plot the response with zero initial conditions for each frequency ratio.

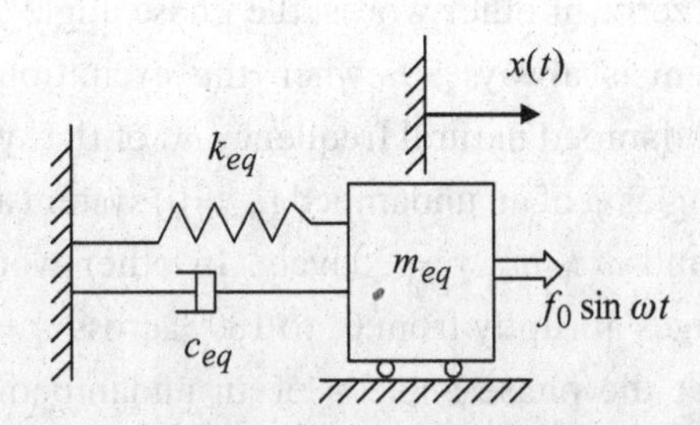

Figure 2.3.6 Equivalent SDOF model for the turbine blade

Solution

$$\frac{f_0}{k_{eq}} = 2.3256 \times 10^{-6} \text{ m}$$

$$\omega_n = \sqrt{\frac{k_{eq}}{m_{eq}}} = 6{,}141.6 \text{ rad/sec}$$

a. $r = 0.4$

$$A = \frac{f_0/k_{eq}}{\sqrt{(1-r^2)^2 + (2\xi r)^2}} = 2.7684 \times 10^{-6} \text{ m}$$

$$\phi = \tan^{-1}\frac{2\xi r}{1-r^2} = \tan^{-1}\frac{0.008}{0.84} = 0.0095 \text{ rad}$$

b. $r = 0.95$

$$A = \frac{f_0/k_{eq}}{\sqrt{(1-r^2)^2 + (2\xi r)^2}} = 2.3412 \times 10^{-5} \text{ m}$$

$$\phi = \tan^{-1}\frac{2\xi r}{1-r^2} = \tan^{-1}\frac{0.019}{0.0975} = 0.1925 \text{ rad}$$

c. $r = 1$

$$A = \frac{f_0/k_{eq}}{\sqrt{(1-r^2)^2 + (2\xi r)^2}} = 1.1628 \times 10^{-4} \text{ m}$$

$$\phi = \tan^{-1}\frac{2\xi r}{1-r^2} = \tan^{-1}\frac{0.02}{0} = \frac{\pi}{2} \text{ rad}$$

d. $r = 2$

$$A = \frac{f_0/k_{eq}}{\sqrt{(1-r^2)^2 + (2\xi r)^2}} = 7.7512 \times 10^{-7} \text{ m}$$

$$\phi = \tan^{-1}\frac{2\xi r}{1-r^2} = \tan^{-1}\frac{0.04}{-3} = 3.1283 \text{ rad}$$

For each frequency ratio, the response is computed with zero initial conditions (Equation 2.3.29). The steady state response and the excitation force are plotted together in Figure 2.3.7, primarily to

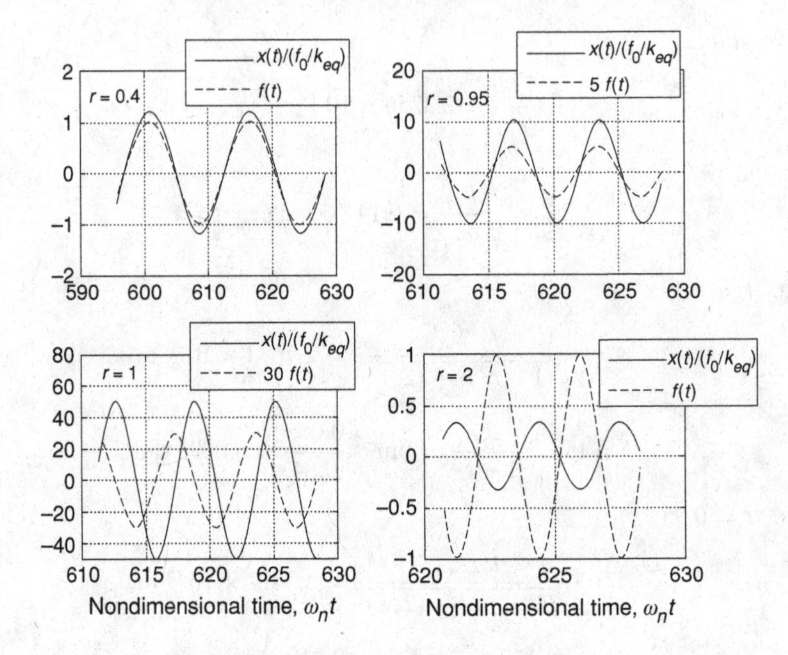

Figure 2.3.7 Steady state vibration of a turbine blade

see the phase relationships. For $r = 1$, the excitation force is maximum when $x(t) = 0$ and $\dot{x}(t) > 0$. This confirms that the steady state response lags behind the forcing function by 90°.

The subplot for $r = 0.95$ in Figure 2.3.7 is re-plotted in Figure 2.3.8, where it is estimated that

$$\omega_n \Delta t \cong 0.20 \text{ rad}$$

where Δt is the time difference for peaks of excitation and steady state response to occur. Therefore,

$$\omega \Delta t \cong \frac{\omega}{\omega_n} 0.20 \text{ rad} = 0.19 \text{ rad}$$

This is the phase difference between the excitation and the steady state response. Also, the excitation peak occurs before the response peak. Therefore, the steady state response lags behind the excitation

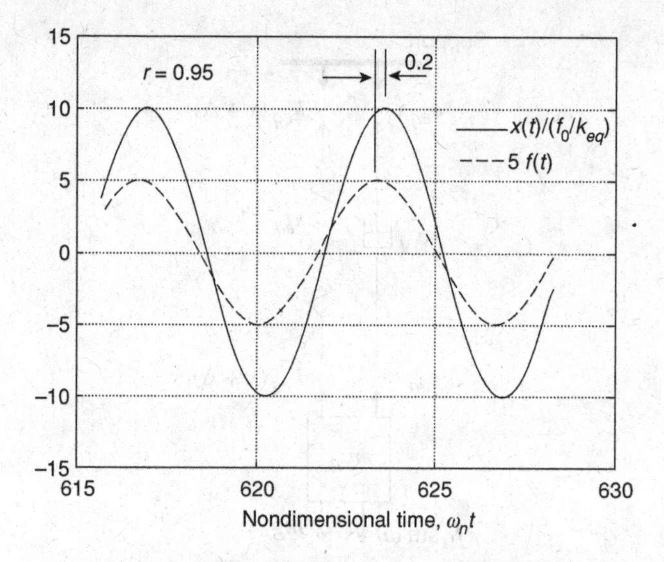

Figure 2.3.8 A zoomed subplot in Figure 2.3.7

by 0.19 rad, which matches well with the theoretical result. The slight difference between the numerical and the theoretical result is due to approximations involved in obtaining $\omega_n \Delta t$ from the plot.

2.3.2 *Force Transmissibility*

When a force is applied to the mass, it is important to determine the force transmitted to the support. Therefore, free body diagrams are constructed for mass, spring, and damper in Figure 2.3.1. From these free body diagrams in Figure 2.3.9,

$$\begin{aligned}\text{force transmitted to the support} &= k_{eq}(x + \Delta) + c_{eq}\dot{x} \\ &= mg + k_{eq}x + c_{eq}\dot{x}\end{aligned} \qquad (2.3.59)$$

The time-varying part of the force transmitted to the support is given by

$$f_T(t) = k_{eq}x + c_{eq}\dot{x} \qquad (2.3.60)$$

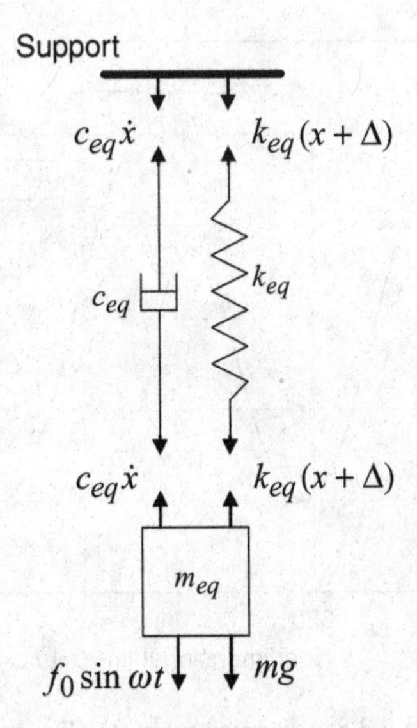

Figure 2.3.9 Free body diagram of each element in Figure 2.3.1

In the steady state, the force transmitted to the support is obtained by substituting Equation 2.3.51 into Equation 2.3.60,

$$f_T(t) = k_{eq}A\sin(\omega t - \phi) + c_{eq}\omega A\cos(\omega t - \phi) = f_{T0}\sin(\omega t - \psi) \tag{2.3.61}$$

where

$$\psi = \phi - \theta \tag{2.3.62}$$

$$f_{T0} = A\sqrt{k_{eq}^2 + c_{eq}^2\omega^2} \tag{2.3.63a}$$

and

$$\tan\theta = \frac{c_{eq}\omega}{k_{eq}} \tag{2.3.63b}$$

Substituting Equation 2.3.52 into Equation 2.3.63a, and using Equation 2.3.24,

$$\frac{f_{T0}}{f_0} = \frac{\sqrt{1+(2\xi r)^2}}{\sqrt{(1-r^2)^2+(2\xi r)^2}} \tag{2.3.64}$$

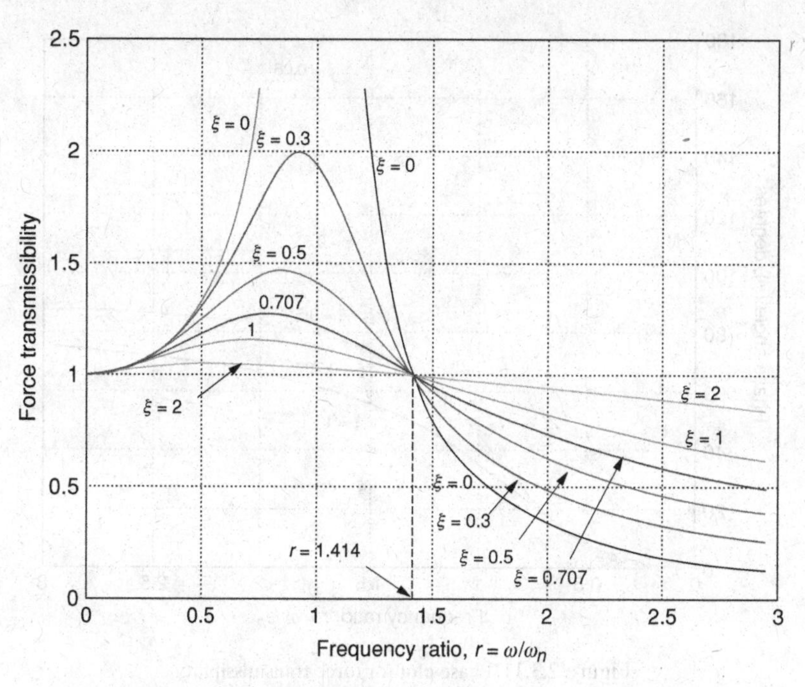

Figure 2.3.10 Force transmissibility as a function of frequency ratio

Using Equations 2.3.62, 2.3.24, and 2.3.26,

$$\psi = \phi - \theta = \tan^{-1}\frac{2\xi r}{1-r^2} - \tan^{-1} 2\xi r = \tan^{-1}\frac{2\xi r^3}{1+r^2(4\xi^2-1)} \tag{2.3.65}$$

Since f_0 and f_{T0} are, respectively, the amplitudes of the applied sinusoidal force and the time-varying part of the force transmitted to the support, the ratio f_{T0}/f_0 is known as the **force transmissibility**, which is plotted in Figure 2.3.10 as a function of the frequency ratio r for many values of the damping ratio ξ. As the applied force and the time-varying part of the force transmitted to the support are $f_0 \sin \omega t$ and $f_{T0} \sin(\omega t - \psi)$ respectively, ψ is the phase angle by which the transmitted force lags behind the applied force. The phase angle ψ has been plotted in Figure 2.3.11 as a function of the frequency ratio r for many values of the damping ratio ξ.

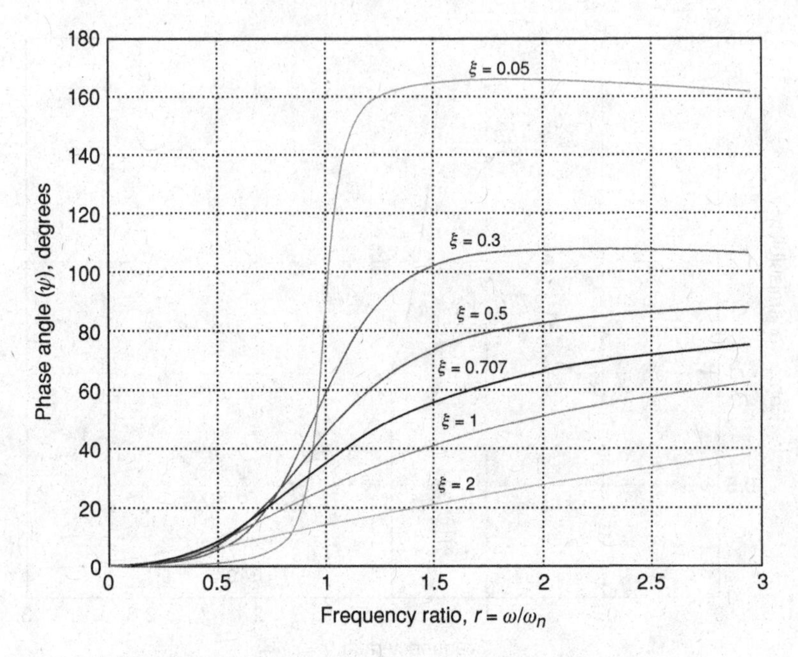

Figure 2.3.11 Phase plot for force transmissibility

It is often desired to select the spring stiffness and the damping coefficient such that the force transmissibility is as small as possible, but never greater than one. To determine the frequency ratio at which f_{T0}/f_0 is one, from Equation 2.3.64,

$$\frac{\sqrt{1+(2\xi r)^2}}{\sqrt{(1-r^2)^2+(2\xi r)^2}} = 1 \tag{2.3.66}$$

Squaring both sides of Equation 2.3.66,

$$\frac{1+(2\xi r)^2}{(1-r^2)^2+(2\xi r)^2} = 1 \tag{2.3.67}$$

After some algebra, Equation 2.3.67 yields

$$r^2(r^2-2) = 0 \tag{2.3.68}$$

Solving Equation 2.3.68,

$$r = 0 \quad \text{or} \quad r = \sqrt{2} \tag{2.3.69}$$

Therefore, the force transmissibility equals one at $r = 0$ and $r = \sqrt{2}$ for all values of the damping ratio. In Figure 2.3.10, it can also be seen that the force transmissibility is greater than one for $0 < r < \sqrt{2}$, whereas it is less than one for $r > \sqrt{2}$. Further, for $r > \sqrt{2}$, the force transmissibility increases as the damping ratio increases. Therefore, a support system, that is, the spring stiffness and the damping constant, should be designed such that $r > \sqrt{2}$ and the damping constant c is as small as possible. For $r > \sqrt{2}$,

$$\frac{\omega}{\omega_n} > \sqrt{2} \Rightarrow \omega_n^2 < \frac{\omega^2}{2} \Rightarrow k_{eq} < \frac{m_{eq}\omega^2}{2} \tag{2.3.70}$$

In general, a design guideline is to choose small values for the spring stiffness and the damping constant. Rubber pads are often used to minimize the force transmitted to the support because rubber material has small values for the stiffness and the damping constant.

Example 2.3.2: Force Transmitted to Turbine Blade Support
For the frequency ratio = 1, find the steady state force transmitted to the support of the turbine blade in Example 2.3.1.

Solution
Here, $r = 1$ and $\xi = 0.01$.

$$f_{T0} = f_0 \frac{\sqrt{1 + (2\xi r)^2}}{\sqrt{(1 - r^2)^2 + (2\xi r)^2}} = f_0 \sqrt{1 + \frac{1}{(2\xi)^2}} = 50.01 \text{ N}$$

$$\psi = \tan^{-1} \frac{2\xi r^3}{1 + r^2(4\xi^2 - 1)} = \tan^{-1} \frac{1}{2\xi} = 1.5508 \text{ rad}$$

The steady state force transmitted to the support is

$$f_T(t) = f_{T0} \sin(\omega t - \psi) = 50.01 \sin(6141.6t - 1.5508) \text{ N}$$

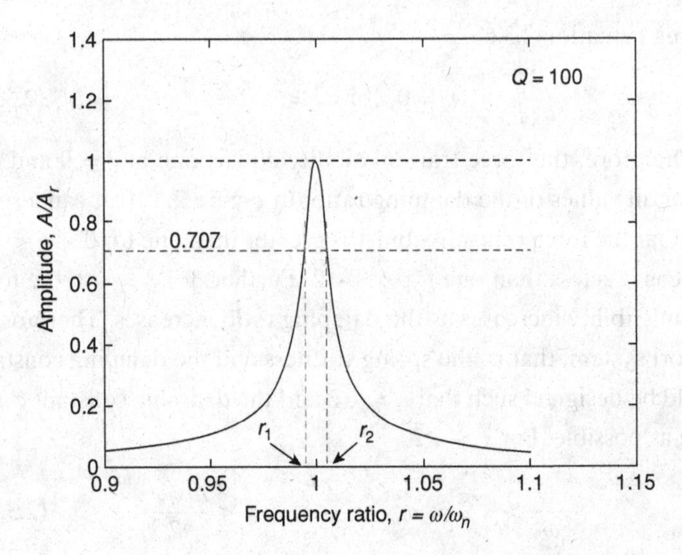

Figure 2.3.12 Frequency ratios for $A/A_r = 0.707$

2.3.3 Quality Factor and Bandwidth

Here, two widely used terms, quality factor (Q) and bandwidth, are defined.

Quality Factor

The amplitude A_r at $r = 1$ is obtained from Equation 2.3.52,

$$\frac{A_r}{f_0/k_{eq}} = \frac{1}{2\xi} \tag{2.3.71}$$

For the definition of quality factor, it is important to determine the frequency ratios where the values of A are $1/\sqrt{2}$ times the amplitude A_r (Figure 2.3.12). Using Equations 2.3.52 and 2.3.71, the amplitude ratio (A/A_r) is determined and set equal to $1/\sqrt{2}$ for this purpose:

$$\frac{A}{A_r} = \frac{2\xi}{\sqrt{(1-r^2)^2 + (2\xi r)^2}} = \frac{1}{\sqrt{2}} \tag{2.3.72}$$

Squaring both sides of Equation 2.3.72,

$$(r^2)^2 - (2 - 4\xi^2)r^2 + (1 - 8\xi^2) = 0 \tag{2.3.73}$$

Solving this quadratic equation in r^2,

$$r^2 = (1 - 2\xi^2) \pm 2\xi\sqrt{1 - \xi^2} \tag{2.3.74}$$

For a small ξ, $\sqrt{1-\xi^2} \approx 1$ and $1 - 2\xi^2 \approx 1$, that is,

$$r^2 \approx 1 \pm 2\xi \tag{2.3.75}$$

Using binomial expansion (Appendix B),

$$r \approx (1 \pm 2\xi)^{\frac{1}{2}} = 1 \pm \xi \pm \text{higher power terms of } \xi \tag{2.3.76}$$

Neglecting higher powers of ξ, frequencies ω_1 and ω_2 where the values of $A/(f_0/k_{eq})$ are $1/\sqrt{2}$ times the value at $r = 1$ can be approximated for small ξ as

$$r_1 = \frac{\omega_1}{\omega_n} = 1 - \xi \tag{2.3.77}$$

$$r_2 = \frac{\omega_2}{\omega_n} = 1 + \xi \tag{2.3.78}$$

The quality factor Q is defined as

$$Q = \frac{\omega_n}{\omega_2 - \omega_1} \tag{2.3.79}$$

From Equations 2.3.77–2.3.79,

$$Q = \frac{1}{2\xi} \tag{2.3.80}$$

A higher value of Q implies a lower value of the damping ratio ξ and vice versa.

Bandwidth

The bandwidth of a system is defined as the frequency below which the steady state amplitude is above $1/\sqrt{2}$ $(= 0.707)$ times the steady state amplitude at zero frequency or the steady state amplitude under constant force, for example (Figure 2.3.13).

The bandwidth is a measure of the frequency range for which the system responds strongly to the forcing function.

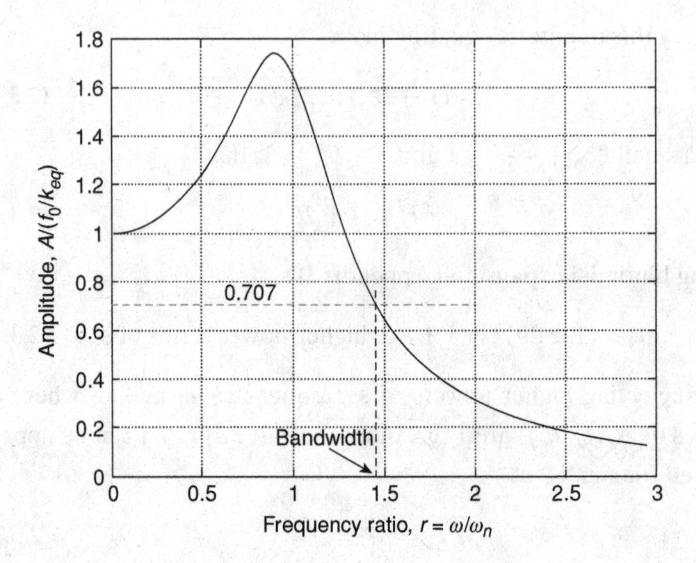

Figure 2.3.13 Definition of bandwidth

The bandwidth (frequency) is obtained by setting

$$\frac{A}{f_0/k_{eq}} = \frac{1}{\sqrt{(1-r^2)^2 + (2\xi r)^2}} = \frac{1}{\sqrt{2}} \tag{2.3.81}$$

Squaring both sides of Equation 2.3.81, the following quadratic equation in r^2 is obtained:

$$(r^2)^2 - (2 - 4\xi^2)r^2 - 1 = 0 \tag{2.3.82}$$

Solving Equation 2.3.82,

$$r^2 = \frac{(2-4\xi^2) \pm \sqrt{(2-4\xi^2)^2 + 4}}{2} \tag{2.3.83}$$

The negative sign in front of the square root sign in Equation 2.3.83 will lead to a negative value of r^2, which is meaningless. Therefore, the positive sign in front of the square root sign is chosen to obtain

$$r^2 = 1 - 2\xi^2 + \sqrt{4\xi^4 - 4\xi^2 + 2} \tag{2.3.84}$$

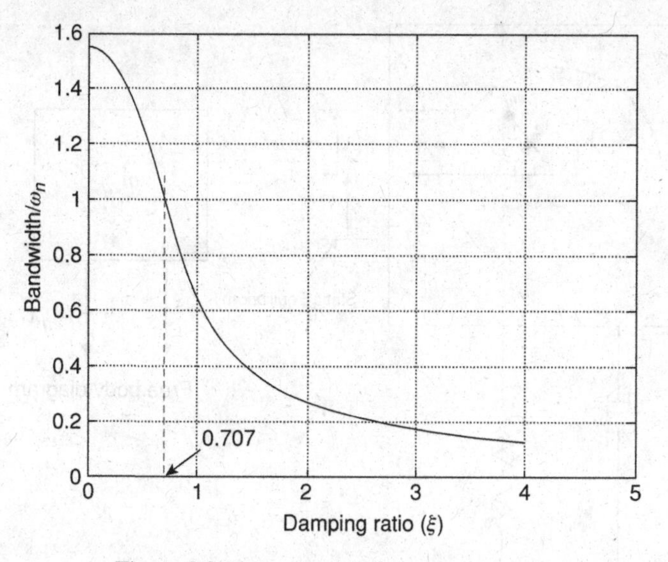

Figure 2.3.14 Bandwidth versus damping ratio

Therefore,

$$\frac{\text{bandwidth}}{\omega_n} = \sqrt{1 - 2\xi^2 + \sqrt{4\xi^4 - 4\xi^2 + 2}} \tag{2.3.85}$$

The bandwidth is plotted as a function of the damping ratio in Figure 2.3.14.

As the damping is increased, the bandwidth decreases. It is interesting to note that the bandwidth equals the undamped natural frequency ω_n when the damping ratio ξ equals $1/\sqrt{2}$ $(= 0.707)$.

2.4 ROTATING UNBALANCE

In a rotating machine such as a motor, a generator, a turbine, and so on, there is always an unbalance because the center of rotation never coincides with the center of the rotor mass. Even though this eccentricity is small, this results in a rotating centrifugal force on the rotor with a significant amplitude because the centrifugal force is proportional to the square of the angular speed. Further, a rotating

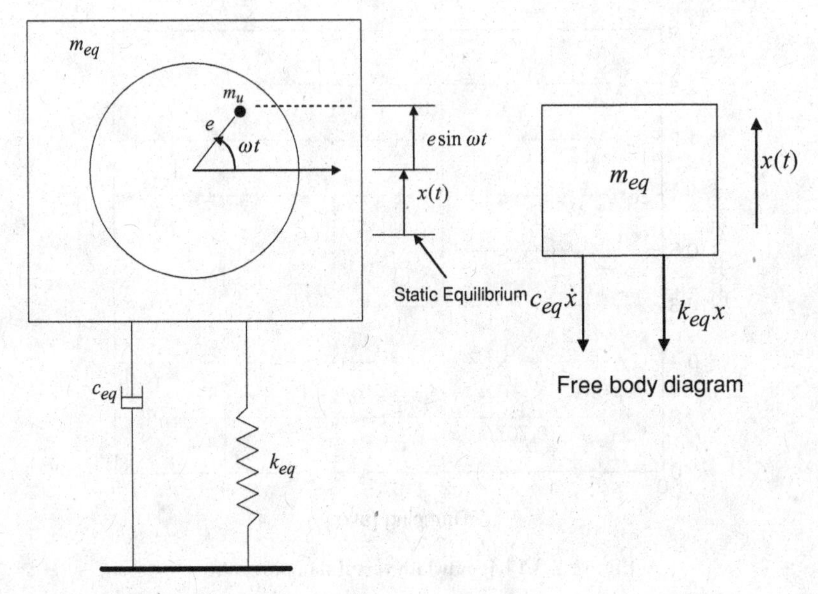

Figure 2.4.1 A spring–mass–damper system with rotating unbalance

centrifugal force results in a sinusoidal excitation on the structure, and hence can lead to a large magnitude of the structural vibration because of the resonance phenomenon.

Consider the system in which an unbalance mass m_u with an eccentricity e is rotating at a speed of ω rad/sec (Figure 2.4.1). Let the total mass be m_{eq}, which includes the unbalance mass m_u. From the free body diagram in Figure 2.4.1,

$$\text{net force in } x\text{-direction} = -k_{eq}x - c_{eq}\dot{x} \tag{2.4.1}$$

And the acceleration of the mass $(m_{eq} - m_u)$ is $\ddot{x}$. The displacement of the unbalance mass m_u is $x + e \sin \omega t$. As a result, the acceleration of the unbalance mass will be the second time derivative of $x + e \sin \omega t$. Applying Newton's second law of motion,

$$-k_{eq}x - c_{eq}\dot{x} = (m_{eq} - m_u)\ddot{x} + m_u \frac{d^2}{dt^2}(x + e \sin \omega t) \tag{2.4.2}$$

After some simple algebra,

$$m_{eq}\ddot{x} + c_{eq}\dot{x} + k_{eq}x = m_u e\omega^2 \sin\omega t \tag{2.4.3}$$

Comparing Equations 2.4.3 and 2.3.3,

$$f_0 = m_u e\omega^2 \tag{2.4.4}$$

Following the developments in Section 2.3, the steady state response $x_{ss}(t)$ is again given by

$$x_{ss}(t) = A\sin(\omega t - \phi) \tag{2.4.5}$$

where the amplitude A and the phase angle ϕ are given by Equations 2.3.25 and 2.3.26:

$$\frac{A}{f_0/k_{eq}} = \frac{1}{\sqrt{(1-r^2)^2 + (2\xi r)^2}} \tag{2.4.6}$$

and

$$\tan\phi = \frac{2\xi r}{1-r^2} \tag{2.4.7}$$

It should be noted that the complete solution is still given by Equations 2.3.29, 2.3.35, and 2.3.43 for underdamped, critically damped, and overdamped systems, respectively.

Using Equation 2.4.4,

$$\frac{f_0}{k_{eq}} = \frac{m_u e\omega^2}{k_{eq}} = \frac{em_u}{m_{eq}}\frac{m_{eq}\omega^2}{k_{eq}} = \frac{em_u}{m_{eq}}\frac{\omega^2}{\omega_n^2} = \frac{em_u}{m_{eq}}r^2 \tag{2.4.8}$$

Substituting Equation 2.4.8 into Equation 2.4.6, the steady state amplitude due to the rotating unbalance is given by

$$\frac{m_{eq}}{m_u}\frac{A}{e} = \frac{r^2}{\sqrt{(1-r^2)^2 + (2\xi r)^2}} \tag{2.4.9}$$

The steady state amplitude (A) due to the rotating unbalance is plotted in Figure 2.4.2 as a function of the frequency ratio for many values of the damping ratio. The plot for the phase angle is exactly the same as the one shown in Figure 2.3.4. Here, the amplitude (A) equals zero

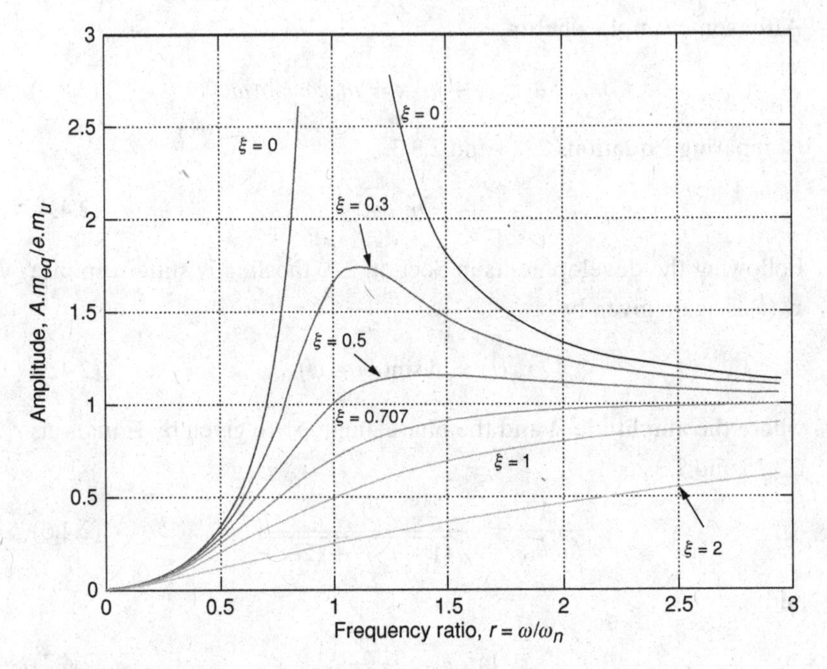

Figure 2.4.2 Steady state amplitude versus frequency ratio (rotating unbalance)

at $r = 0$ and equals $(m_u e/m_{eq})$ for all damping ratios as $r \to \infty$. Near resonance condition, the amplitude (A) can be large for small values of the damping ratio.

To find the maximum amplitude, Equation 2.4.9 is differentiated with respect to r,

$$\frac{m_{eq}}{m_u e}\frac{dA}{dr} = \frac{-r^2[2(1-r^2)(-2r)+(2\xi)^2 2r]}{2[(1-r^2)^2+(2\xi r)^2]^{1.5}} + \frac{2r}{[(1-r^2)^2+(2\xi r)^2]^{0.5}} \tag{2.4.10}$$

The condition for the maximum value of the amplitude A is

$$\frac{dA}{dr} = 0 \tag{2.4.11}$$

Therefore, from Equations 2.4.10 and 2.4.11,

$$-r^2\left[(1-r^2)(-2r)+(2\xi)^2 r\right] + 2r\left[(1-r^2)^2+(2\xi r)^2\right] = 0 \tag{2.4.12}$$

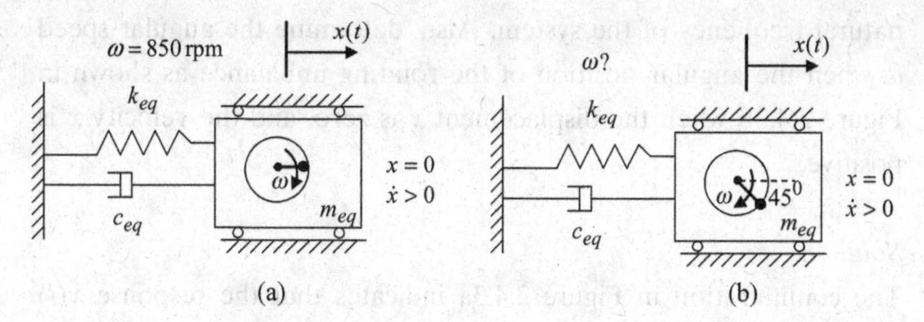

Figure 2.4.3 Rotating unbalance position under stroboscopic light

After some algebra, Equation 2.4.12 yields

$$r = \frac{1}{\sqrt{1-2\xi^2}} \text{ provided } \xi \leq \frac{1}{\sqrt{2}} = 0.707 \tag{2.4.13}$$

It should be noted that the maximum amplitude occurs for $r = \infty$ when $\xi \geq 0.707$.

For $\xi \leq 0.707$, substituting Equation 2.4.13 into Equation 2.4.9,

$$\frac{m_{eq}}{m_u}\frac{A_p}{e} = \frac{1}{2\xi\sqrt{1-\xi^2}} \tag{2.4.14}$$

where A_p is the peak amplitude. It is interesting to note that right-hand sides of Equations 2.3.58 and 2.4.14 are identical.

Using Equation 2.3.64, the amplitude f_{T0} of the force transmitted to the support in steady state will be given by

$$\frac{f_{T0}}{m_u e\omega^2} = \frac{\sqrt{1+(2\xi r)^2}}{\sqrt{(1-r^2)^2+(2\xi r)^2}} \tag{2.4.15}$$

Example 2.4.1: Identification of Damping Ratio and Natural Frequency

Consider an SDOF system with the rotating unbalance = 0.1 kg – meter and the equivalent mass $m_{eq} = 200$ kg (Figure 2.4.3a). At a speed of 850 rpm under stroboscopic light, the configuration of the eccentric mass is horizontal, the displacement x is zero, and the velocity $\dot{x}$ is positive. The steady state amplitude at the speed of 850 rpm is found to be 25 mm. Determine the damping ratio and the undamped natural frequency of the system. Also, determine the angular speed ω when the angular position of the rotating unbalance as shown in Figure 2.4.3b when the displacement x is zero, and the velocity $\dot{x}$ is positive.

Solution

The configuration in Figure 2.4.3a indicates that the response $x(t)$ lags behind the excitation force by 90°. Therefore, the frequency ratio $r = 1$ from Equation 2.4.7, that is,

$$\omega_n = \omega = 850 \text{ rpm} = 89.012 \text{ rad/sec}$$

Given:

$$m_u e = 0.1 \text{ kg-m} \quad \text{and} \quad A = 0.025 \text{ m}$$

From Equation 2.4.9,

$$\frac{m_{eq}}{m_u}\frac{A}{e} = \frac{1}{2\xi} \Rightarrow \xi = 0.01$$

The configuration in Figure 2.4.3b indicates that the response $x(t)$ lags behind the excitation force by 135°. Therefore,

$$\tan\phi = \frac{1}{-1}$$

From Equation 2.4.7,

$$\frac{2\xi r}{r^2 - 1} = 1 \Rightarrow r^2 - 0.02r - 1 = 0$$

Solving this quadratic equation: $r = 1.01$ and -0.99. Since the negative value of r is meaningless,

$$r = \frac{\omega}{\omega_n} = 1.01 \Rightarrow \omega = 1.01\omega_n = 858.5 \text{ rpm}$$

Example 2.4.2: A Rotor Shaft with Mass Unbalance

Consider the rotor–shaft system shown in Figure 1.3.1, but with a casing around the rotor as shown in Figure 2.4.4. The clearance δ between

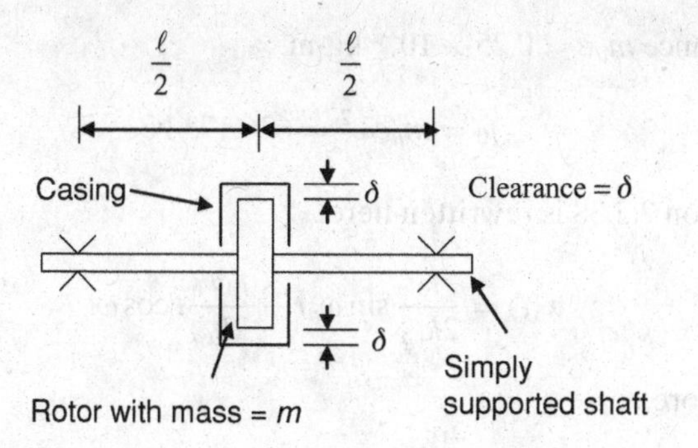

Figure 2.4.4 Rotor–shaft system with casing

the rotor and the casing is 10 mm. The length and the diameter of the circular **steel** shaft are 0.5 m and 3 cm, respectively. The mass and the unbalance of the turbine rotor are 12 kg and 0.25 kg-cm, respectively.

a. Determine the critical speed of the rotor.
b. If the rotor operates at the critical speed, find the time after which the rotor will hit the casing. Assume that the initial conditions are zeros.
c. Assume that the operating speed of the rotor is higher than its critical value. Then, the rotor must pass through its critical speed before it acquires the desired speed. Recommend a safe value of angular acceleration to cross the critical speed.

a. Area moment of inertia $I = \frac{\pi(0.03)^4}{64} = 3.9761 \times 10^{-8}\ \text{m}^4$
   For steel, $E = 2 \times 10^{11}\ \text{N/m}^2$

$$\ell = 0.5\ \text{m}$$

$$k_{eq} = \frac{48EI}{\ell^3} = 3.0536 \times 10^6\ \text{N/m}$$

$$\omega_n = \sqrt{\frac{k_{eq}}{m_{eq}}} = 504.45\ \text{rad/sec}$$

$$\text{critical speed } \omega = \omega_n = 504.45\ \text{rad/sec}$$

b. Unbalance $m_u e = 0.25 \times 10^{-2}$ kg-m

$$f_0 = m_u e \omega^2 = 636.173 \text{ N}$$

Equation 2.2.38 is rewritten here.

$$x(t) = \frac{f_0}{2k_{eq}} \sin \omega_n t - \frac{f_0 \omega_n}{2k_{eq}} t \cos \omega t$$

Therefore,

$$x(t) = 1.0417 \times 10^{-4} \sin \omega_n t - 0.0525t \cos \omega_n t$$

or

$$x(t) \approx -0.0525t \cos \omega_n t$$

Let the time after which the rotor hits the casing be t_h. Then,

$$0.0525t_h = 0.01 \quad \Rightarrow \quad t_h = 0.1905 \text{ sec}$$

c. Recommended acceleration $= \frac{\text{desired speed}-0}{(0.1905/2)}$ rad/sec^2.

The operating speed will reach the desired speed from rest in half the time that it takes for the rotor to hit the casing at the critical speed. Therefore, the rotor will not hit the casing.

2.5 BASE EXCITATION

There are many practical situations where the base is not fixed and the vibration is caused by the displacement of the base, for example, automobile vibration caused by uneven road profile, building vibration during earthquake, and so on.

Consider the SDOF spring–mass–damper system with the base having a displacement $y(t)$ (Figure 2.5.1). From the free body diagram,

$$\text{net force in } x\text{-direction} = -k_{eq}(x - y) - c_{eq}(\dot{x} - \dot{y}) \tag{2.5.1}$$

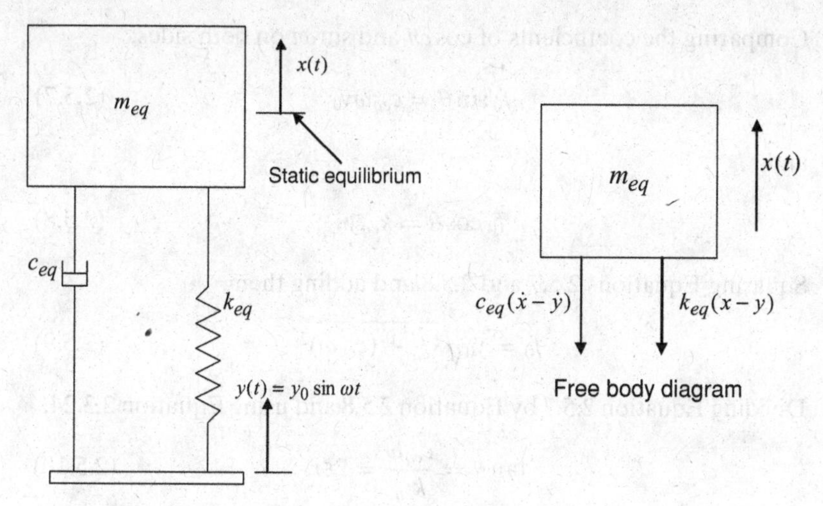

Figure 2.5.1 A spring–mass–damper system with base excitation

Applying Newton's second law of motion,

$$-k_{eq}(x - y) - c_{eq}(\dot{x} - \dot{y}) = m_{eq}\ddot{x} \tag{2.5.2}$$

or,

$$m_{eq}\ddot{x} + c_{eq}\dot{x} + k_{eq}x = c_{eq}\dot{y} + k_{eq}y \tag{2.5.3}$$

With $y(t) = y_0 \sin \omega t$, the right-hand side of Equation 2.5.3 can be written as

$$c_{eq}\dot{y} + k_{eq}y = c_{eq}\omega y_0 \cos \omega t + k_{eq} y_0 \sin \omega t \tag{2.5.4}$$

Let Equation 2.5.4 be represented as

$$c_{eq}\dot{y} + k_{eq}y = f_0 \sin(\omega t + \theta) \tag{2.5.5}$$

Using Equation 2.5.4,

$$\begin{aligned} c_{eq}\omega y_0 \cos \omega t + k_{eq} y_0 \sin \omega t &= f_0 \sin(\omega t + \theta) \\ &= f_0 \cos\theta \sin \omega t + f_0 \sin\theta \cos \omega t \end{aligned} \tag{2.5.6}$$

Comparing the coefficients of $\cos \omega t$ and $\sin \omega t$ on both sides,

$$f_0 \sin\theta = c_{eq}\omega y_0 \tag{2.5.7}$$

and

$$f_0 \cos\theta = k_{eq} y_0 \tag{2.5.8}$$

Squaring Equations 2.5.7 and 2.5.8 and adding them,

$$f_0 = y_0\sqrt{k_{eq}^2 + (c_{eq}\omega)^2} \tag{2.5.9}$$

Dividing Equation 2.5.7 by Equation 2.5.8 and using Equation 2.3.24,

$$\tan\theta = \frac{c_{eq}\omega}{k_{eq}} = 2\xi r \tag{2.5.10}$$

Dividing Equation 2.5.9 by k_{eq} and using Equation 2.3.24,

$$\frac{f_0}{k_{eq}} = y_0\sqrt{1 + \left(\frac{c_{eq}\omega}{k_{eq}}\right)^2} = y_0\sqrt{1 + (2\xi r)^2} \tag{2.5.11}$$

Using Equations 2.5.3 and 2.5.5, the differential equation of motion is written as

$$m_{eq}\ddot{x} + c_{eq}\dot{x} + k_{eq}x = f_0 \sin(\omega t + \theta) \tag{2.5.12}$$

where f_0 and θ are defined by Equations 2.5.11 and 2.5.10, respectively. Following the developments in Section 2.3, the steady state response $x_{ss}(t)$ is given by

$$x_{ss}(t) = A\sin(\omega t + \theta - \phi) \tag{2.5.13}$$

where the amplitude A and the phase angle ϕ are given by Equations 2.3.25 and 2.3.26.

$$\frac{A}{f_0/k_{eq}} = \frac{1}{\sqrt{(1-r^2)^2 + (2\xi r)^2}} \tag{2.5.14}$$

and

$$\tan\phi = \frac{2\xi r}{1 - r^2} \tag{2.5.15}$$

It should be noted that the complete solutions are given by Equations 2.3.29, 2.3.35, and 2.3.43 for underdamped, critically damped, and overdamped systems, respectively.

Substituting Equation 2.5.11 into Equation 2.5.14,

$$\frac{A}{y_0} = \frac{\sqrt{1+(2\xi r)^2}}{\sqrt{(1-r^2)^2+(2\xi r)^2}} \tag{2.5.16}$$

The ratio of the steady state amplitude A of the mass and the input base amplitude y_0 is known as the **displacement transmissibility**. It is important to note that the expressions for displacement and force transmissibilities are identical. Compare Equations 2.5.16 and 2.3.64. Therefore, the plot of transmissibility (Figure 2.3.10) holds for base excitation also. And a support system, that is, the spring stiffness and the damping constant, should be designed such that $r > \sqrt{2}$ and the damping constant c is as small as possible. Equation 2.3.70 is valid here also.

Equation 2.5.13 can be written as

$$x_{ss}(t) = A\sin(\omega t - \psi) \tag{2.5.17}$$

where

$$\psi = \phi - \theta \tag{2.5.18}$$

Since the input base displacement is $y(t) = y_0 \sin \omega t$, the phase ψ is the angle by which the steady state displacement $x_{ss}(t)$ lags behind the base displacement. From Equations 2.5.15 and 2.5.10,

$$\psi = \phi - \theta = \tan^{-1}\frac{2\xi r}{1-r^2} - \tan^{-1} 2\xi r = \tan^{-1}\frac{2\xi r^3}{1+r^2(4\xi^2-1)} \tag{2.5.19}$$

Again, note that the expression of the phase lag ψ for base excitation is the same as that for the force transmitted to the support when the mass is directly excited by the sinusoidal force. Compare Equations 2.3.65 and 2.5.19. The plot of the phase lag ψ (Figure 2.3.11) holds for base excitation also.

Example 2.5.1: Microgravity Isolation Systems
In a spacecraft, a number of scientific experiments are conducted to utilize the microgravity environment; that is, the acceleration due to gravity in space is $10^{-6}\,g$ where g is the acceleration due to gravity on earth. However, because of various disturbances such as crew motion, thruster firing, and so on, the acceleration due to gravity in space can be as high as $10^{-3}\,g$.

Design a spring–damper suspension system for an experiment module, which has a mass of 1.5 kg. The disturbance frequencies lie between 0.1 and 0.5 Hz, and assume that the damping ratio is $\xi = 0.2$.

Solution
In the context of microgravity isolation system, m_{eq} in Figure 2.5.1 is the mass of the experiment module. The base acceleration, $\ddot{y}(t)$ is caused by various disturbances on the spacecraft. The displacement and acceleration transmissibilities are identical because

$$\frac{\omega^2 A}{\omega^2 y_0} = \frac{A}{y_0}$$

The desired transmissibility is 10^{-3} because $\omega^2 A$ and $\omega^2 y_0$ are $10^{-6}\,g$ and $10^{-3}\,g$, respectively. Using Equation 2.5.16,

$$\frac{A}{y_0} = \frac{\sqrt{1 + (2\xi r)^2}}{\sqrt{(1 - r^2)^2 + (2\xi r)^2}} = 10^{-3}$$

This equation leads to the following quadratic equation in r^2 :

$$(r^2)^2 - 160001.84r^2 - 999999 = 0$$

There are two roots: $r^2 = 1.6 \times 10^5$ and -0.62. Since a negative value of r^2 is meaningless, $r^2 = 1.6 \times 10^5$ and $r = 400$.

When $r > \sqrt{2}$, the transmissibility decreases as r increases. Therefore,

$$\frac{\omega}{\omega_n} \geq 400 \quad \Rightarrow \quad \omega_n \leq \frac{\omega}{400}$$

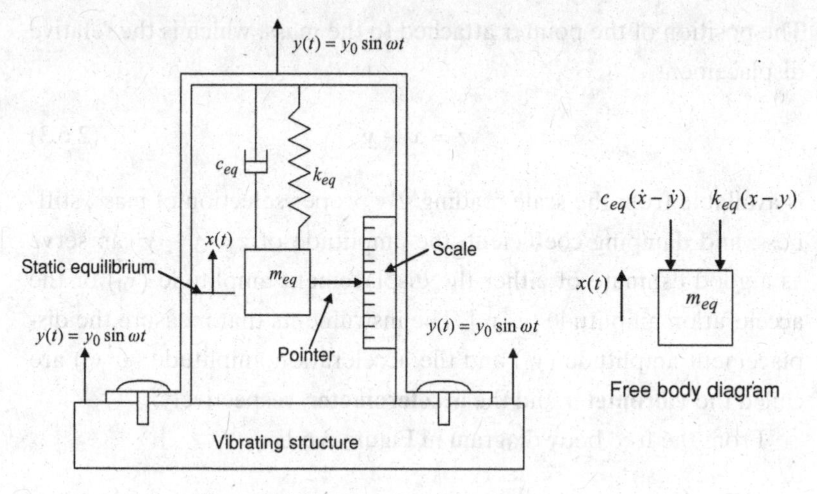

Figure 2.6.1 A vibration measuring instrument

Here, 0.1 Hz $\leq \omega \leq$ 0.5 Hz. Therefore,

$$\omega_n = \frac{0.1 \times 2\pi}{400} = 0.00157 \text{ rad/sec}$$

Therefore,

$$k_{eq} = m_{eq}\omega_n^2 = 3.6973 \times 10^{-6} \text{ N/m}$$

$$c_{eq} = 2\xi m_{eq}\omega_n = 9.42 \times 10^{-4} \text{ N-sec/m}$$

2.6 VIBRATION MEASURING INSTRUMENTS

Here, basic theories for designing instruments that measure amplitudes of vibratory displacements and acceleration are presented. These instruments are composed of a spring–mass–damper system as shown in Figure 2.6.1, and are rigidly attached to the vibratory structure with the displacement

$$y(t) = y_0 \sin \omega t \tag{2.6.1}$$

and therefore the acceleration

$$\ddot{y} = -\omega^2 y_0 \sin \omega t \tag{2.6.2}$$

The position of the pointer attached to the mass, which is the relative displacement

$$z = x - y \tag{2.6.3}$$

is available from the scale reading. By proper selection of mass, stiffness, and damping coefficient, the amplitude of $z = x - y$ can serve as a good estimate of either the displacement amplitude (y_0) or the acceleration amplitude ($\omega^2 y_0$). The instruments that measure the displacement amplitude (y_0) and the acceleration amplitude ($\omega^2 y_0$) are called the **vibrometer** and the **accelerometer**, respectively.

From the free body diagram in Figure 2.6.1,

$$\text{net force in } x\text{-direction} = -k_{eq}(x - y) - c_{eq}(\dot{x} - \dot{y}) \tag{2.6.4}$$

Applying Newton's second law of motion,

$$-k_{eq}(x - y) - c_{eq}(\dot{x} - \dot{y}) = m_{eq}\ddot{x} \tag{2.6.5}$$

From Equations 2.6.3 and 2.6.2,

$$\ddot{x} = \ddot{z} + \ddot{y} = \ddot{z} - \omega^2 y_0 \sin \omega t \tag{2.6.6}$$

Substituting Equations 2.6.3 and 2.6.6 into Equation 2.6.5,

$$m_{eq}\ddot{z} + c_{eq}\dot{z} + k_{eq}z(t) = f_0 \sin \omega t \tag{2.6.7}$$

where

$$f_0 = m_{eq}\omega^2 y_0 \tag{2.6.8}$$

Therefore,

$$\frac{f_0}{k_{eq}} = \frac{m_{eq}\omega^2 y_0}{k_{eq}} = \frac{\omega^2 y_0}{\omega_n^2} \tag{2.6.9}$$

The steady state

$$z_{ss}(t) = z_0 \sin(\omega t - \phi) \tag{2.6.10}$$

where the amplitude z_0 is obtained by using Equations 2.3.52 and 2.6.9:

$$z_0 = \frac{f_0/k_{eq}}{\sqrt{(1-r^2)^2 + (2\xi r)^2}} = \frac{\omega^2 y_0/\omega_n^2}{\sqrt{(1-r^2)^2 + (2\xi r)^2}} \tag{2.6.11}$$

where r is the frequency ratio ω/ω_n. The phase angle ϕ is given by Equation 2.3.53:

$$\phi = \tan^{-1} \frac{2\xi r}{1-r^2} \tag{2.6.12}$$

2.6.1 Vibrometer

From Equation 2.6.11,

$$\frac{z_0}{y_0} = \frac{r^2}{\sqrt{(1-r^2)^2 + (2\xi r)^2}} \tag{2.6.13}$$

The plot of the ratio in Equation 2.6.13 is shown in Figure 2.6.2. It should be noted that the right-hand side of Equation 2.6.13 is identical to the relationship for the steady state amplitude of the rotating unbalance problem (Equation 2.4.9).

For large r,

$$\frac{z_0}{y_0} \approx 1 \quad \text{or} \quad z_0 \approx y_0 \tag{2.6.14}$$

It should be recalled that z_0 is directly available from the scale reading. Equation 2.6.14 establishes the fact that z_0 can be a good estimate of the amplitude of vibration y_0 when the frequency ratio $r = \omega/\omega_n$ is large. A large value of $r = \omega/\omega_n$ implies a small value of the natural frequency ω_n, which is achieved by having a small stiffness and/or a large mass.

Example 2.6.1: Design of a Vibrometer

A vibrometer is to be designed such that the error in the estimated vibration amplitude is less than 4%. Determine the undamped natural frequency when the frequency of vibration lies between 20 and 50 Hz. Assume that the damping ratio is 0.3.

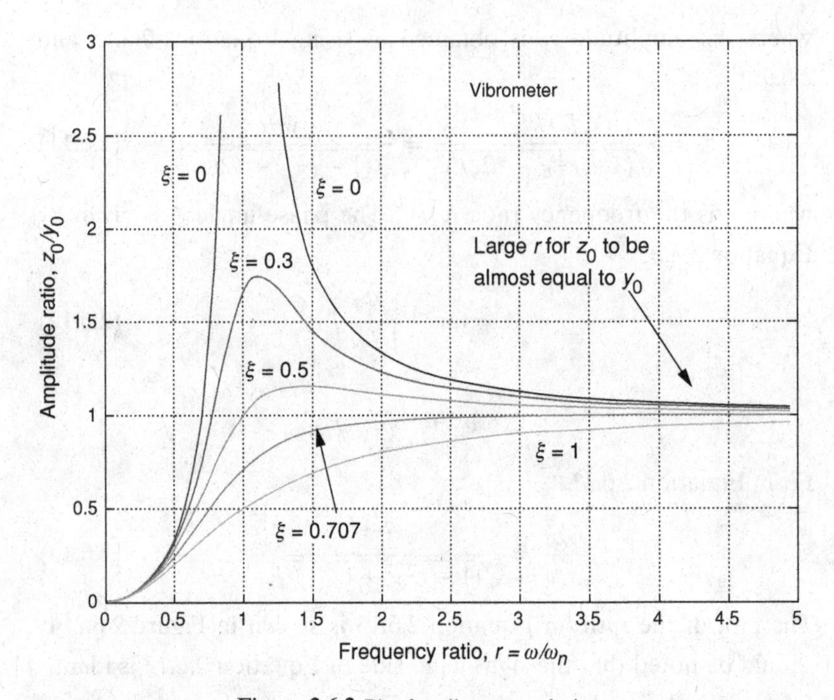

Figure 2.6.2 Plot for vibrometer design

Solution

It is required that

$$0.96 \leq \frac{z_0}{y_0} \leq 1.04$$

The plot of z_0/y_0 is shown in Figure 2.6.2 for $\xi = 0.3$. The line $z_0/y_0 = 1.04$ intersects the plot at two points, whereas $z_0/y_0 = 0.96$ intersects the plot at only one point. The intersection point corresponding to $z_0/y_0 = 0.96$ is not important for vibrometer design. The error is less than 4% when the frequency ratio is greater than the higher of the two values of the frequency ratio where $z_0/y_0 = 1.04$ intersects the plot. To determine this frequency ratio,

$$\frac{z_0}{y_0} = \frac{r^2}{\sqrt{(1-r^2)^2 + (2\xi r)^2}} = 1.04$$

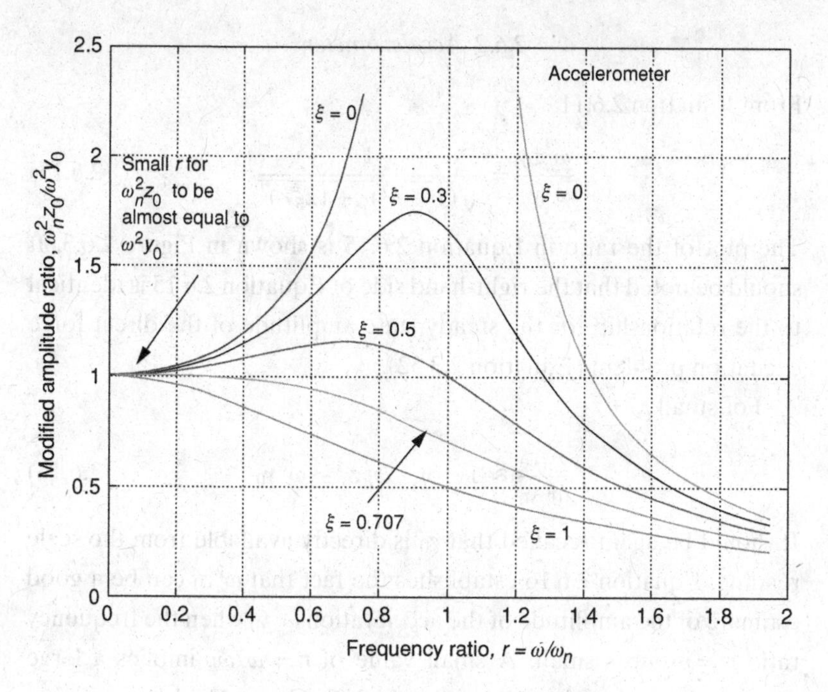

Figure 2.6.3 Plot for accelerometer design

With $\xi = 0.3$,

$$0.0816(r^2)^2 - 1.7738r^2 + 1.0816 = 0$$

This is a quadratic equation in r^2. Two roots are: $r^2 = 21.1098$ and 0.6279, or equivalently, $r = 4.5945$ and 0.7924.

Therefore,

$$r = \frac{\omega}{\omega_n} \geq 4.5945 \quad \Rightarrow \quad \omega_n \leq \frac{\omega}{4.5945}$$

Here 20 Hz $\leq \omega \leq$ 50 Hz. The undamped natural frequency ω_n which will satisfy the above inequality $\omega_n \leq \omega/4.5945$ for all signal frequencies ω is given as

$$\omega_n = \frac{20 \times 2\pi}{4.5945} \text{ rad/sec} = 27.3509 \text{ rad/sec}$$

2.6.2 Accelerometer

From Equation 2.6.11,

$$\frac{\omega_n^2 z_0}{\omega^2 y_0} = \frac{1}{\sqrt{(1-r^2)^2 + (2\xi r)^2}} \tag{2.6.15}$$

The plot of the ratio in Equation 2.6.15 is shown in Figure 2.6.3. It should be noted that the right-hand side of Equation 2.6.15 is identical to the relationship for the steady state amplitude of the direct force excitation problem (Equation 2.3.52).

For small r,

$$\frac{\omega_n^2 z_0}{\omega^2 y_0} \approx 1 \quad \text{or} \quad \omega_n^2 z_0 \approx \omega^2 y_0 \tag{2.6.16}$$

It should be again recalled that z_0 is directly available from the scale reading. Equation 2.6.16 establishes the fact that $\omega_n^2 z_0$ can be a good estimate of the amplitude of the acceleration $\omega^2 y_0$ when the frequency ratio $r = \omega/\omega_n$ is small. A small value of $r = \omega/\omega_n$ implies a large value of the natural frequency ω_n, which is achieved by having a large stiffness and/or a small mass.

Example 2.6.2: Design of an Accelerometer
An accelerometer is to be designed such that the error in the estimated acceleration amplitude is less than 4%. Determine the undamped natural frequency when the frequency of vibration lies between 20 and 50 Hz. Assume that the damping ratio is 0.3.

Solution
It is required that

$$0.96 \le \frac{\omega_n^2 z_0}{\omega^2 y_0} \le 1.04$$

The plot of $\omega_n^2 z_0/\omega^2 y_0$ is shown in Figure 2.6.4 for $\xi = 0.3$. The line $\omega_n^2 z_0/\omega^2 y_0 = 1.04$ intersects the plot at two points, whereas the line $\omega_n^2 z_0/\omega^2 y_0 = 0.96$ intersects the plot at only one point. The intersection point corresponding to $\omega_n^2 z_0/\omega^2 y_0 = 0.96$ is not important for the accelerometer design. The error is less than 4% when the frequency

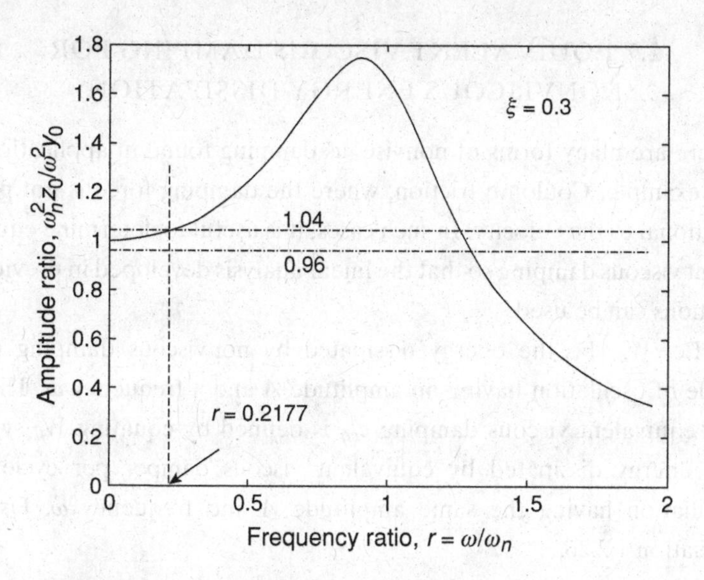

Figure 2.6.4 Accelerometer design with damping ratio $\xi = 0.3$

ratio is smaller than the lower of the two values of frequency ratio where $\omega_n^2 z_0/\omega^2 y_0 = 1.04$ intersects the plot. To determine this frequency ratio,

$$\frac{\omega_n^2 z_0}{\omega^2 y_0} = \frac{1}{\sqrt{(1-r^2)^2 + (2\xi r)^2}} = 1.04$$

With $\xi = 0.3$,

$$1.0816(r^2)^2 - 1.7738r^2 + 0.0816 = 0$$

This is a quadratic equation in r^2. Two roots are: $r^2 = 1.5926$ and 0.0474 or equivalently, $r = 1.2620$ and 0.2177.

Therefore,

$$r = \frac{\omega}{\omega_n} \le 0.2177 \quad \Rightarrow \quad \omega_n \ge \frac{\omega}{0.2177}$$

Here 20 Hz $\le \omega \le$ 50 Hz. The undamped natural frequency ω_n, which will satisfy the above inequality for all signal frequencies ω, is given as

$$\omega_n = \frac{50 \times 2\pi}{0.2177} \text{ rad/sec} = 1443.1 \text{ rad/sec}$$

2.7 EQUIVALENT VISCOUS DAMPING FOR NONVISCOUS ENERGY DISSIPATION

There are many forms of nonviscous damping found in applications; for example, Coulomb friction, where the damping force is not proportional to the velocity. In such cases, it is useful to determine equivalent viscous damping so that the linear analysis developed in previous sections can be used.

Let W_{nv} be the energy dissipated by nonviscous damping per cycle of oscillation having an amplitude A and a frequency ω. Then, the equivalent viscous damping c_{eq} is defined by equating W_{nv} with the energy dissipated by equivalent viscous damper per cycle of oscillation having the same amplitude A and frequency ω. Using Equation 1.2.26,

$$\pi c_{eq} \omega A^2 = W_{nv} \tag{2.7.1}$$

or

$$c_{eq} = \frac{W_{nv}}{\pi \omega A^2} \tag{2.7.2}$$

If the damping force is $f_d(t)$, then the energy dissipated per cycle of oscillation is calculated as follows:

$$W_{nv} = \oint f_d(t) dx \tag{2.7.3}$$

It should be noted that the integral in Equation 2.7.3 is evaluated for one complete cycle of oscillation. A typical plot of $f_d(t)$ versus displacement $x(t)$ is shown in Figure 2.7.1. Hence, the integral on the right-hand side of Equation 2.7.3 is the shaded area in Figure 2.7.1. In other words, the shaded area is the energy dissipated (W_{nv}) per cycle of oscillation.

Example 2.7.1 Forced Response of a Frictionally Damped Spring–Mass System

Consider a spring–mass system in Figure 2.7.2 where one side of the mass is pushed against the wall by the normal load N. Assuming that

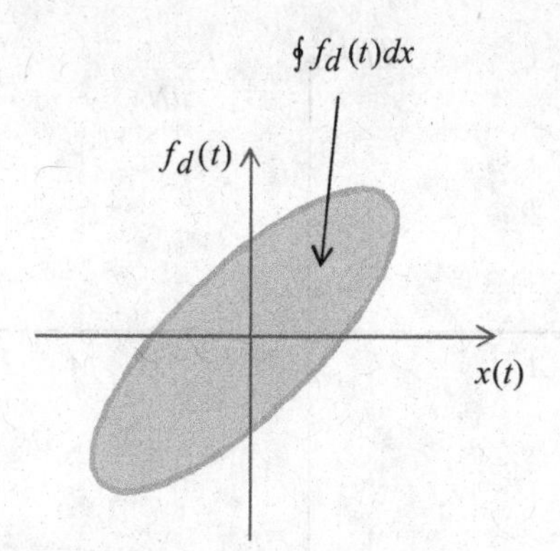

Figure 2.7.1 Damper force $f_d(t)$ versus oscillatory displacement $x(t)$

the coefficient of friction is μ, the friction force $f_d(t)$ will be μN in the direction opposite to the velocity $\dot{x}(t)$. The external force on the mass is $f_0 \sin \omega t$. It is assumed that the steady state response is sinusoidal, that is,

$$x(t) = A \sin(\omega t - \phi) \tag{2.7.4}$$

where amplitude A and phase ϕ are to be determined.

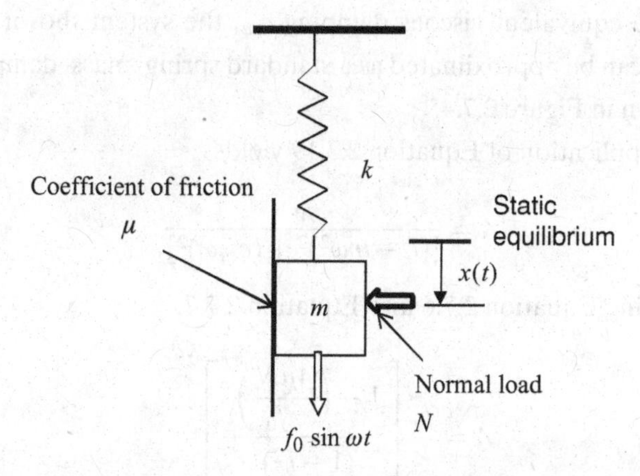

Figure 2.7.2 Frictionally damped spring–mass system under sinusoidal excitation

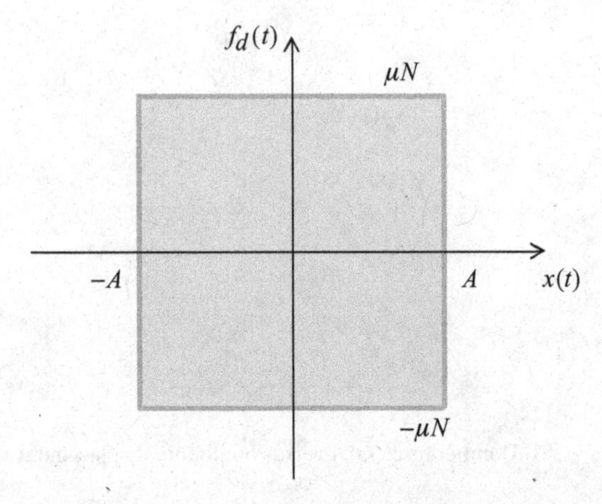

Figure 2.7.3 Coulomb friction force $f_d(t)$ versus steady state displacement $x(t)$

Figure 2.7.3 shows the friction force versus the steady state displacement plot. Therefore, the energy dissipated per cycle of oscillation is

$$W_{nv} = \oint f_d(t)dx = 4\mu NA \tag{2.7.5}$$

From Equation 2.7.2, the equivalent viscous damping is given by

$$c_{eq} = \frac{4\mu N}{\pi \omega A} \tag{2.7.6}$$

Using the equivalent viscous damping c_{eq}, the system shown in Figure 2.7.2 can be approximated as a standard spring–mass–damper system shown in Figure 2.7.4.

The application of Equation 2.3.19 yields

$$A = \frac{f_0}{[(k - m\omega^2)^2 + (c_{eq}\omega)^2]} \tag{2.7.7}$$

Substituting Equation 2.7.6 into Equation 2.7.7,

$$A = \frac{f_0}{k}\left[\frac{1 - \left(\frac{4\mu N}{\pi f_0}\right)^2}{(1 - r^2)^2}\right]^{0.5} \tag{2.7.8}$$

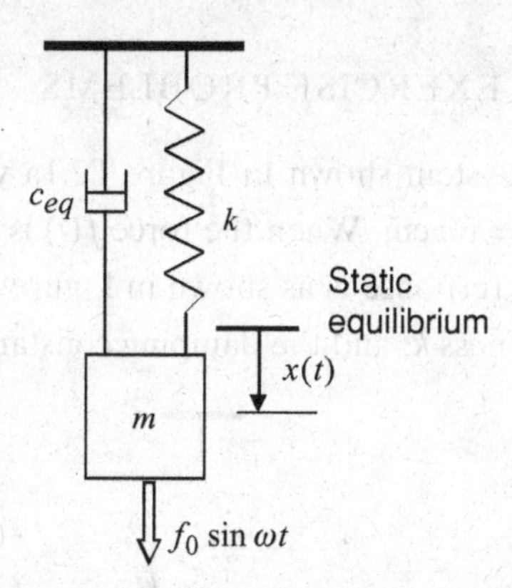

Figure 2.7.4 Spring–mass system with the equivalent viscous damper

where r is the frequency ratio ω/ω_n. For the amplitude A to be a real number, it is required that

$$f_0 > \frac{4\mu N}{\pi} \tag{2.7.9}$$

The inequality in Equation 2.7.9 is the condition for the validity of the equivalent viscous damping approach. From Equation 2.7.8, the following results are derived:

a. At resonance condition ($r = 1$ or $\omega = \omega_n$), the steady state amplitude of vibration is unbounded. In other words, Coulomb friction damping is unable to bring any change to the response of a spring–mass system. This is a reflection of the fact that the energy dissipated by Coulomb friction per cycle is proportional to the amplitude. *Therefore, the Coulomb friction is a weaker form of damping in comparison with the viscous damping for which the energy dissipated per cycle is proportional to the square of the amplitude.*
b. At nonresonance condition ($r \neq 1$ or $\omega \neq \omega_n$), the steady state amplitude of vibration does get reduced due to Coulomb friction, when compared with the amplitude of an undamped system.

EXERCISE PROBLEMS

P2.1 Consider the system shown in Figure P2.1a where $a = 25$ cm, $\ell_1 = 50$ cm, and $\ell_2 = 30$ cm. When the force $f(t)$ is a step function of magnitude 1 N, the response is as shown in Figure P2.1b. Determine the mass m, the stiffness k, and the damping constant c.

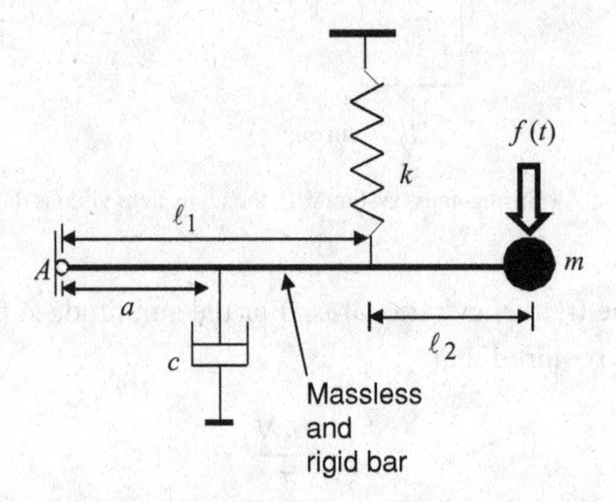

Figure P2.1a Spring–mass–damper system subjected to step forcing function

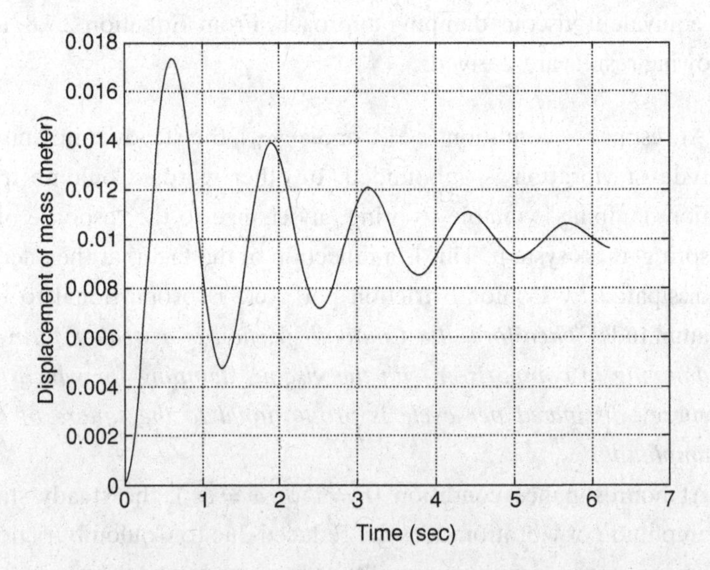

Figure P2.1b Unit step response for system in Figure P2.1a

P2.2 Consider a spring–mass-damper system (Figure P2.2) with $m_{eq} = 100\,\text{kg}$, $k_{eq} = 10{,}000\,\text{N/m}$, and $c_{eq} = 20\,\text{N-sec/m}$.

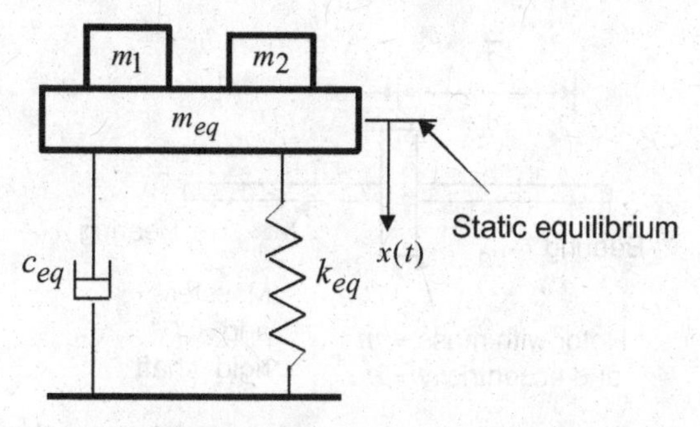

Figure P2.2 Spring–mass-damper system

It is required to place a 10 kg mass on the main mass m_{eq} such that the new static equilibrium is reached without any vibration. Develop a strategy to achieve this goal. You are allowed to use two separate masses m_1 and m_2 such that $m_1 + m_2 = 10\,\text{kg}$.

P2.3 Consider the system shown in Figure P2.1a where $a = 25$ cm, $\ell_1 = 50$ cm, and $\ell_2 = 30$ cm. Here, $k = 1{,}100\,\text{N/m}$ and $0.5\,\text{kg} < m < 2\,\text{kg}$. The force $f(t)$ is a step function.

Find a value of damping constant c such that the steady state is reached without any overshoot.

P2.4 Consider the system shown in Figure P1.12.

a. First the damper is detached and the mass is excited by a force $f(t) = 20 \sin \omega t\text{N}$. Find and plot responses when $\omega = 0.8\omega_n$, ω_n, and $1.5\omega_n$. Assume that the initial conditions are zero. Compare your results from analysis to those from MATLAB ODE23 or ODE45.
b. Reattach the damper, and the mass is again excited by a force $f(t) = 20 \sin \omega t\text{N}$. Find and plot responses when $\omega = 0.8\omega_n$, ω_n, and $1.5\omega_n$. Assume that the initial conditions are zero. Compare your results from the analysis to those from MATLAB ODE23 or ODE45.

Also, show the phase lags of the responses in your plots.

P2.5 Consider a rotor on a massless and rigid shaft, which is supported by ball bearings at the ends (Figure P2.5).

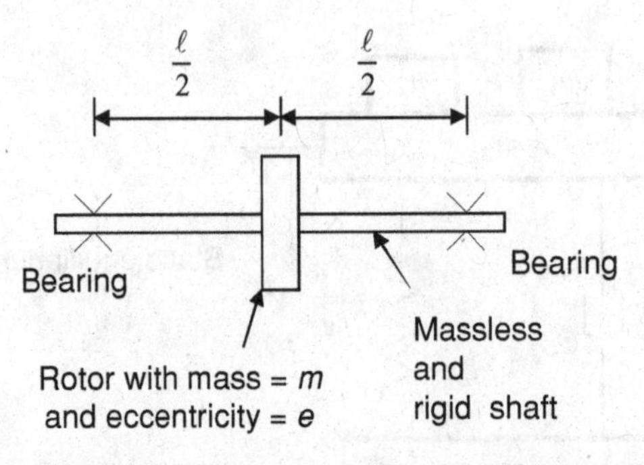

Figure P2.5 Rotor on massless and rigid shaft

The mass of the rotor is 10 kg and the eccentricity is 0.5 cm. If the operating speed of the rotor is 4,200 rpm, what should be the stiffnesses of the bearings such that the amplitude of rotor vibration does not exceed 0.05 cm?

P2.6 Consider a rotor on a massless and flexible steel shaft, which is simply supported at the ends (Figure P2.6).

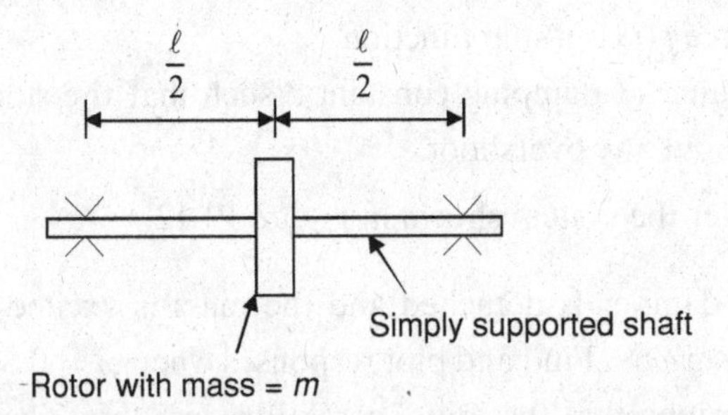

Figure P2.6 Rotor on massless and flexible shaft

The mass of the rotor is 10 kg and the eccentricity is 0.5 cm. The length and the diameter of the shaft are 50 cm and 5 cm, respectively.

a. Compute the critical speed of the rotor.
b. If the rotor operates at critical speed, how much time will it take for the maximum bending stress in the shaft to be about 70% of its yield stress? Assume that the rotor starts from rest.

c. Assuming that the rotor starts from rest, suggest an angular acceleration of the rotor to reach a speed that equals two times the critical speed. It is desired that the maximum bending stress in the shaft is below 50% of its yield stress. Verify your result using numerical integration of differential equation.

d. For part c, plot the force transmitted to each support as a function of time.

P2.7 An instrument with mass = 13 kg is to be isolated from aircraft engine vibrations ranging from 18,00 to 2,300 cpm. What should be the stiffness of an isolator for at least 65% isolation? Assume that the damping ratio is 0.045.

Assuming that the initial conditions are zero, demonstrate the performance of your isolator for both the extreme frequencies (1,800 and 2,300 cpm) using the MATLAB routine ODE23 or ODE45. For each frequency, plot displacements of the radio and the support in a single figure. Then, also verify the analytical expression of the phase of the steady state response of the instrument.

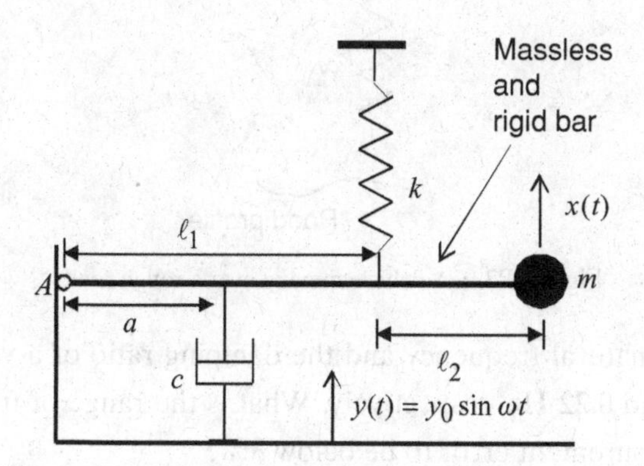

Figure P2.8 A spring–mass–damper system with sinusoidal base displacement

P2.8 Determine the amplitude and the phase of the steady state response of the mass m in Figure P2.8.

P2.9 A vehicle with mass $m_{eq} = 1{,}050$ kg and suspension stiffness $k_{eq} = 435{,}000$ N/m is traveling with a velocity V on a sinusoidal road surface with amplitude = 0.011 m and a wavelength of 5.3 m (Figure P2.9).

a. Determine the critical speed of the vehicle.
b. Select an appropriate amount of damping for the suspension system.
c. With the damping selected in part b, should the vehicle be operated above or below the critical speed so that the amplitude of the vehicle is small? Justify your answer.
d. With the damping selected in part b, should the vehicle be operated above or below the critical speed so that the amplitude of the vehicle is small relative to the road profile? Justify your answer.

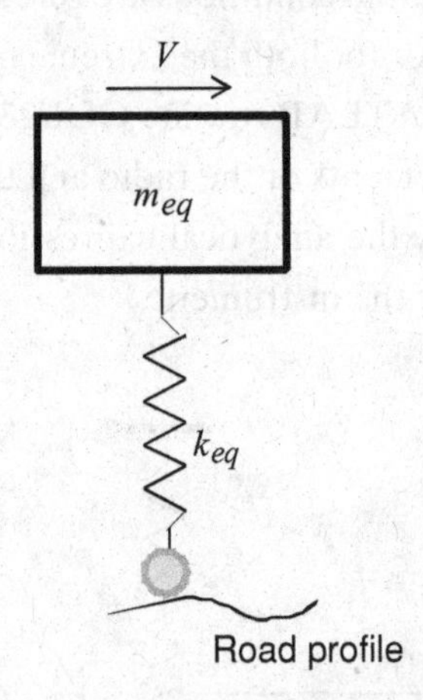

Figure P2.9 A vehicle moving over a rough road

P2.10 The natural frequency and the damping ratio of a **vibrometer** are 6 Hz and 0.22 Hz, respectively. What is the range of frequencies for the measurement error to be below 3%?

Corroborate the validity of your design using MATLAB ODE23 or ODE45 for a signal frequency. Show the signal that your instrument will produce after it has been attached to the vibrating structure.

P2.11 An **accelerometer** with mass = 0.01 kg and a damping ratio = 0.707 is to be designed. What should be the undamped natural frequency of the system so that the measurement error never exceeds 2%? The vibration signal, which is to be measured, can have a frequency as high as 200 Hz.

Corroborate the validity of your design using MATLAB ODE23 or ODE45 for the signal frequency = 80 Hz. Show the signal that your instrument will produce after it has been attached to the vibrating structure.

P2.12 The force-deflection curve for a structure is experimentally obtained (Figure P2.12). What is the equivalent viscous damping if the frequency of oscillation is 100 Hz?

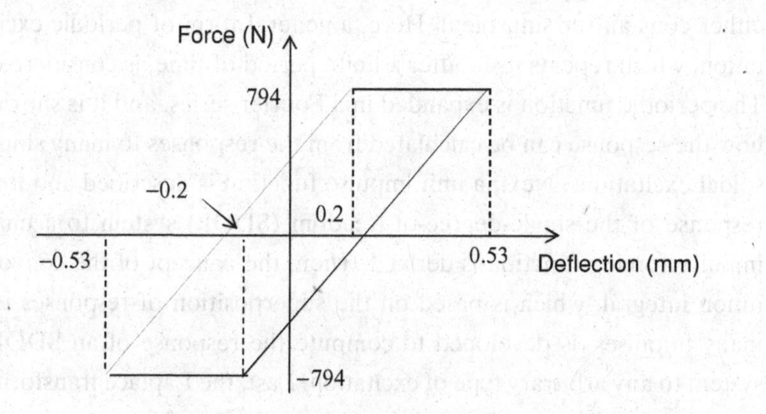

Figure P2.12 Force-deflection curve

3

RESPONSES OF AN SDOF SPRING–MASS–DAMPER SYSTEM TO PERIODIC AND ARBITRARY FORCES

In Chapter 2, the response has been calculated when the excitation is either constant or sinusoidal. Here, a general form of periodic excitation, which repeats itself after a finite period of time, is considered. The periodic function is expanded in a Fourier series, and it is shown how the response can be calculated from the responses to many sinusoidal excitations. Next, a unit impulse function is described and the response of the single-degree-of-freedom (SDOF) system to a unit impulse forcing function is derived. Then, the concept of the convolution integral, which is based on the superposition of responses to many impulses, is developed to compute the response of an SDOF system to any arbitrary type of excitation. Last, the Laplace transform technique is presented. The concepts of transfer function, poles, zeros, and frequency response function are also introduced. The connection between the steady-state response to sinusoidal excitation and the frequency response function is shown.

3.1 RESPONSE OF AN SDOF SYSTEM TO A PERIODIC FORCE

The procedure of a Fourier series expansion of a periodic function is described first. The concepts of odd and even functions are introduced next to facilitate the computation of the Fourier coefficients. It is also shown how can a Fourier series expansion be interpreted and used

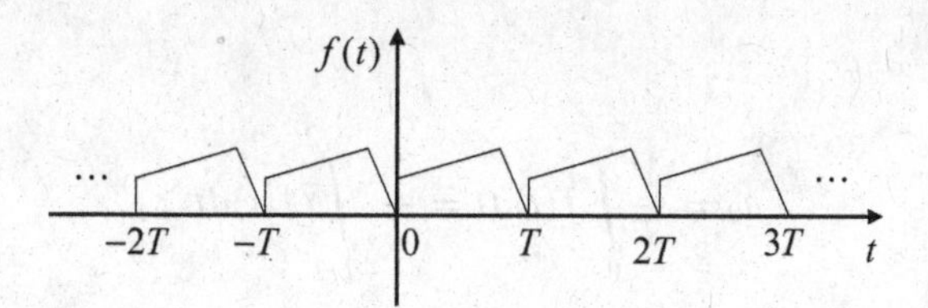

Figure 3.1.1 A periodic function

for a function with a finite duration. Last, the particular integral of an SDOF system subjected to a periodic excitation is obtained by computing the response due to each term in the Fourier series expansion and then using the principle of superposition.

3.1.1 Periodic Function and its Fourier Series Expansion

Consider a periodic function $f(t)$ with the time period T, that is, the function repeats itself after time T. Therefore,

$$f(t+nT) = f(t); \quad n = 1, 2, 3, \ldots, \tag{3.1.1}$$

and

$$f(t-nT) = f(t); \quad n = 1, 2, 3, \ldots, \tag{3.1.2}$$

It should be noted that a periodic function is defined for $-\infty \leq t \leq \infty$ (Figure 3.1.1). Sine and cosine functions are the simplest examples of periodic functions.

The fundamental frequency ω of a periodic function with the time period T is defined as follows:

$$\omega = \frac{2\pi}{T} \tag{3.1.3}$$

The Fourier series expansion of a periodic function $f(t)$ is defined as follows:

$$f(t) = a_0 + \sum_{n=1}^{\infty} a_n \cos(n\omega t) + \sum_{n=1}^{\infty} b_n \sin(n\omega t) \tag{3.1.4}$$

where

$$a_0 = \frac{1}{T}\int_0^T f(t)dt = \frac{1}{T}\int_{-T/2}^{T/2} f(t)dt \tag{3.1.5}$$

$$a_n = \frac{2}{T}\int_0^T f(t)\cos(n\omega t)dt = \frac{2}{T}\int_{-T/2}^{T/2} f(t)\cos(n\omega t)dt \tag{3.1.6}$$

and

$$b_n = \frac{2}{T}\int_0^T f(t)\sin(n\omega t)dt = \frac{2}{T}\int_{-T/2}^{T/2} f(t)\sin(n\omega t)dt \tag{3.1.7}$$

Derivations of Equations 3.1.5–3.1.7 are based on the following facts:

a. Integrals of cosine and sine functions over the time period T is zero, that is,

$$\int_0^T \cos(n\omega t)dt = \int_{-T/2}^{T/2} \cos(n\omega t)dt = 0; \quad n \neq 0 \tag{3.1.8}$$

$$\int_0^T \sin(\ell\omega t)dt = \int_{-T/2}^{T/2} \sin(\ell\omega t)dt = 0 \tag{3.1.9}$$

b. Orthogonality of $\cos n\omega t$ and $\sin \ell\omega t$ in the following sense:

$$\int_0^T \cos(n\omega t)\sin(\ell\omega t)dt = \int_{-T/2}^{T/2} \cos(n\omega t)\sin(\ell\omega t)dt = 0 \tag{3.1.10}$$

c. Orthogonality of $\sin n\omega t$ and $\sin \ell\omega t$ in the following sense:

$$\int_0^T \sin(n\omega t)\sin(\ell\omega t)dt = \int_{-T/2}^{T/2} \sin(n\omega t)\sin(\ell\omega t)dt = 0; \quad n \neq \ell \tag{3.1.11}$$

d. Orthogonality of $\cos n\omega t$ and $\cos \ell\omega t$ in the following sense:

$$\int_0^T \cos(n\omega t)\cos(\ell\omega t)dt = \int_{-T/2}^{T/2} \cos(n\omega t)\cos(\ell\omega t)dt = 0; \quad n \neq \ell \tag{3.1.12}$$

The expression for a_0 is derived by integrating both sides of Equation 3.1.4 over a full time period as follows:

$$\int_0^T f(t)dt = \int_0^T a_0 dt + \sum_{n=1}^{\infty} a_n \int_0^T \cos(n\omega t)dt + \sum_{n=1}^{\infty} b_n \int_0^T \sin(n\omega t)dt \tag{3.1.13}$$

Using Equations 3.1.8 and 3.1.9,

$$\int_0^T f(t)dt = a_0 T + 0 + 0 \tag{3.1.14}$$

It is easily seen that Equation 3.1.13 yields Equation 3.1.5.

The expression for a_ℓ is derived by multiplying both sides of Equation 3.1.4 by $\cos(\ell\omega t)$ and then integrating over a full time period as follows:

$$\int_0^T f(t)\cos(\ell\omega t)dt = \int_0^T a_0\cos(\ell\omega t)dt + \sum_{n=1}^{\infty} a_n \int_0^T \cos(n\omega t)\cos(\ell\omega t)dt + \sum_{n=1}^{\infty} b_n \int_0^T \sin(n\omega t)\cos(\ell\omega t)dt \tag{3.1.15}$$

Using Equations 3.1.8–3.1.12,

$$\int_0^T f(t)\cos(\ell\omega t)dt = 0 + a_\ell \int_0^T \cos^2(\ell\omega t)dt + 0 = \frac{a_\ell}{2}\int_0^T 1 + \cos(2\ell\omega t)dt = \frac{a_\ell}{2}T \tag{3.1.16}$$

It is easily seen that Equation 3.1.16 yields Equation 3.1.6. It should also be noted that the equation (3.1.4) is multiplied by a_ℓ where ℓ can be any value from 1 to ∞. The subscript ℓ is chosen to be different from n, which is a counter for the summation Σ.

The expression for b_ℓ is derived by multiplying both sides of Equation 3.1.4 by $\sin(\ell\omega t)$ and then integrating over a full time period as follows:

$$\int_0^T f(t)\sin(\ell\omega t)dt = \int_0^T a_0 \sin(\ell\omega t)dt + \sum_{n=1}^{\infty} a_n \int_0^T \cos(n\omega t)\sin(\ell\omega t)dt$$
$$+ \sum_{n=1}^{\infty} b_n \int_0^T \sin(n\omega t)\sin(\ell\omega t)dt \tag{3.1.17}$$

Using Equations 3.1.8–3.1.12,

$$\int_0^T f(t)\sin(\ell\omega t)dt = 0 + 0 + b_\ell \int_0^T \sin^2(\ell\omega t)dt$$
$$= \frac{b_\ell}{2}\int_0^T 1 - \cos(2\ell\omega t)dt = \frac{b_\ell}{2}T \tag{3.1.18}$$

It is easily seen that Equation 3.1.18 yields Equation 3.1.7.

3.1.2 Even and Odd Periodic Functions

For an even function $g_e(t)$,

$$g_e(t) = g_e(-t) \tag{3.1.19}$$

Because of the property in Equation 3.1.19,

$$\int_{-T/2}^{T/2} g_e(t)dt = 2\int_0^{T/2} g_e(t)dt \tag{3.1.20}$$

Cosine functions are even functions because $\cos(n\omega t) = \cos(-n\omega t)$.

For an odd function $g_o(t)$,

$$g_o(t) = -g_o(-t) \tag{3.1.21}$$

$$\text{Even} \times \text{Even} \equiv \text{Even}$$
$$\text{Even} \times \text{Odd} \equiv \text{Odd}$$
$$\text{Odd} \times \text{Even} \equiv \text{Odd}$$
$$\text{Odd} \times \text{Odd} \equiv \text{Even}$$

Figure 3.1.2 Multiplication of odd and even functions

Because of the property in Equation 3.1.21,

$$\int_{-T/2}^{T/2} g_o(t)dt = 0 \tag{3.1.22}$$

Sine functions are odd because $\sin(n\omega t) = -\sin(-n\omega t)$. When odd and even functions are multiplied among each other, the result can be either odd or even functions (Figure 3.1.2).

Fourier Coefficients for Even Periodic Functions

For an even periodic function, the computational effort needed to obtain the Fourier coefficients can be significantly reduced. First, using Equation 3.1.20,

$$a_0 = \frac{1}{T}\int_0^T f(t)dt = \frac{1}{T}\int_{-T/2}^{T/2} f(t)dt = \frac{2}{T}\int_0^{T/2} f(t)dt \tag{3.1.23}$$

Since cosine is an even function, $f(t)\cos(n\omega t)$ will be an even function according to the information in Figure 3.1.2. Therefore, using Equation 3.1.20 again,

$$a_n = \frac{2}{T}\int_0^T f(t)\cos(n\omega t)dt = \frac{2}{T}\int_{-T/2}^{T/2} f(t)\cos(n\omega t)dt$$
$$= \frac{4}{T}\int_0^{T/2} f(t)\cos(n\omega t)dt \tag{3.1.24}$$

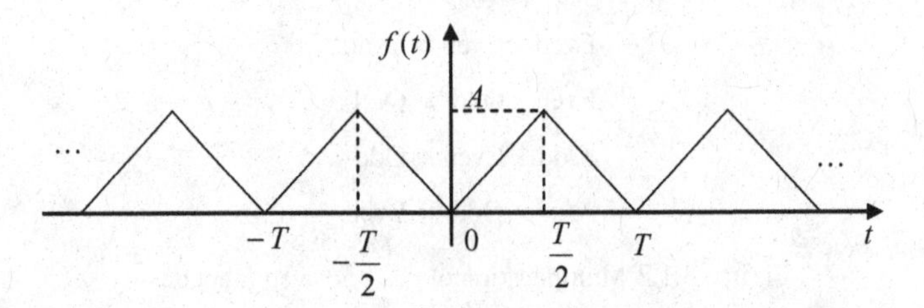

Figure 3.1.3 An even periodic function

Since sine is an odd function, $f(t)\sin(n\omega t)$ will be an odd function according to the information in Figure 3.1.2. Therefore, using Equation 3.1.22,

$$b_n = \frac{2}{T}\int_0^T f(t)\sin(n\omega t)dt = \frac{2}{T}\int_{-T/2}^{T/2} f(t)\sin(n\omega t)dt = 0 \qquad (3.1.25)$$

Example 3.1.2: Fourier Series Expansion of Triangular Waveform
Consider the even periodic function shown in Figure 3.1.3. Therefore,

$$b_n = 0; \quad n = 1, 2, 3, \ldots, \qquad (3.1.26)$$

To evaluate the integrals for a_0 and a_n in Equations 3.1.23 and 3.1.24, the function $f(t)$ is defined between 0 and $T/2$ as follows:

$$f(t) = \frac{A}{T/2}t = \frac{2A}{T}t; \quad 0 \le t \le \frac{T}{2} \qquad (3.1.27)$$

Using Equation 3.1.23,

$$a_0 = \frac{2}{T}\int_0^{T/2} \frac{2A}{T}tdt = \frac{4A}{T^2}\left(\frac{T}{2}\right)^2 \frac{1}{2} = \frac{A}{2} \qquad (3.1.28)$$

Using Equation 3.1.24,

$$a_n = \frac{4}{T}\int_0^{T/2} \frac{2A}{T}t\cos(n\omega t)dt = \frac{8A}{T^2}\left[\frac{t}{n\omega}\sin(n\omega t) + \frac{1}{(n\omega)^2}\cos(n\omega t)\right]\Bigg|_0^{T/2} \qquad (3.1.29)$$

or

$$a_n = 8A\left[\frac{1}{2n\omega T}\sin\left(\frac{n\omega T}{2}\right) + \frac{1}{(n\omega T)^2}\cos\left(\frac{n\omega T}{2}\right) - \frac{1}{(n\omega T)^2}\right]\Bigg|_0^{T/2} \tag{3.1.30}$$

Because $\omega T = 2\pi$ (Equation 3.1.3)

$$a_n = 8A\left[\frac{1}{4n\pi}\sin(n\pi) + \frac{1}{4\pi^2 n^2}\cos(n\pi) - \frac{1}{4\pi^2 n^2}\right] \tag{3.1.31}$$

For odd and even n, $\cos n\pi = -1$ and $+1$, respectively. And $\sin n\pi = 0$ for all n. Therefore,

$$a_n = -\frac{4A}{n^2\pi^2}; \quad n = 1, 3, 5, \ldots, \tag{3.1.32}$$

and

$$a_n = 0; \quad n = 2, 4, 6, \ldots, \tag{3.1.33}$$

Therefore, the Fourier series expansion is

$$f(t) = \frac{A}{2} - \frac{4A}{\pi^2}\cos\omega t - \frac{4A}{9\pi^2}\cos 3\omega t - \frac{4A}{25\pi^2}\cos 5\omega t - \cdots \tag{3.1.34}$$

Fourier Coefficients for Odd Periodic Functions

For an odd periodic function, the computational effort needed to obtain the Fourier coefficients can also be significantly reduced. First, using Equation 3.1.22,

$$a_0 = \frac{1}{T}\int_0^T f(t)dt = \frac{1}{T}\int_{-T/2}^{T/2} f(t)dt = 0 \tag{3.1.35}$$

Since cosine is an even function, $f(t)\cos(n\omega t)$ will be an odd function according to the information in Figure 3.1.2. Therefore, using Equation 3.1.22 again,

$$a_n = \frac{2}{T}\int_0^T f(t)\cos(n\omega t)dt = \frac{2}{T}\int_{-T/2}^{T/2} f(t)\cos(n\omega t)dt = 0 \tag{3.1.36}$$

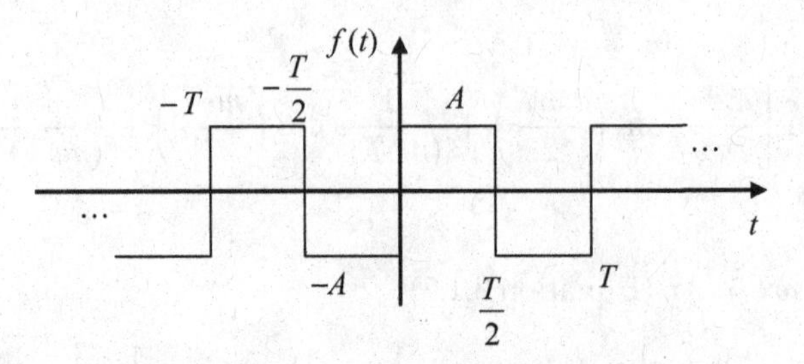

Figure 3.1.4 Square waveform: An odd periodic function

Since sine is an odd function, $f(t)\sin(n\omega t)$ will be an even function according to the information in Figure 3.1.2. Therefore, using Equation 3.1.20,

$$b_n = \frac{2}{T}\int_0^T f(t)\sin(n\omega t)dt = \frac{2}{T}\int_{-T/2}^{T/2} f(t)\sin(n\omega t)dt$$

$$= \frac{4}{T}\int_0^{T/2} f(t)\sin(n\omega t)dt \tag{3.1.37}$$

Example 3.1.3: Fourier Series Expansion of a Square Waveform

Consider the square waveform (Figure 3.1.4), which is an odd periodic function. Therefore,

$$a_0 = 0 \tag{3.1.38}$$

and

$$a_n = 0; \quad n = 1, 2, 3, \ldots, \tag{3.1.39}$$

To evaluate the integrals for b_n in Equation 3.1.37, the function $f(t)$ is defined between 0 and $T/2$ as follows:

$$f(t) = A; \quad 0 \leq t \leq \frac{T}{2} \tag{3.1.40}$$

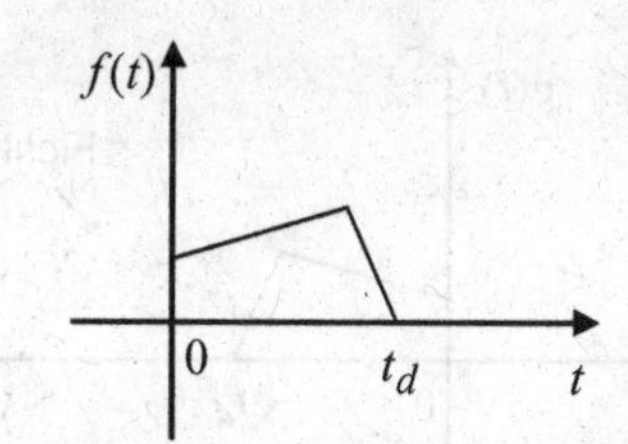

Figure 3.1.5 A finite duration function

Using Equation 3.1.37,

$$b_n = \frac{4}{T}\int_0^{T/2} A\sin(n\omega t)dt = \frac{4A}{T}\left[-\frac{\cos(n\omega t)}{n\omega}\right]\Bigg|_0^{T/2}$$

$$= \frac{4A}{n\omega T}\left[-\cos\left(\frac{n\omega T}{2}\right)+1\right] \tag{3.1.41}$$

Because $\omega T = 2\pi$ (Equation 3.1.3),

$$b_n = \frac{2A}{n\pi}(1 - \cos n\pi) \tag{3.1.42}$$

For odd and even n, $\cos n\pi = -1$ and $+1$, respectively. Therefore,

$$b_n = \frac{4A}{n\pi}; \quad n = 1, 3, 5, \ldots, \tag{3.1.43}$$

and

$$b_n = 0; \quad n = 2, 4, 6, \ldots, \tag{3.1.44}$$

Therefore, the Fourier series expansion is

$$f(t) = \frac{4A}{\pi}\sin\omega t + \frac{4A}{3\pi}\sin 3\omega t + \frac{4A}{5\pi}\sin 5\omega t + \cdots \tag{3.1.45}$$

3.1.3 Fourier Series Expansion of a Function with a Finite Duration

Consider a function $f(t)$ of finite duration t_d (Figure 3.1.5). Then, it can also be expanded in a Fourier series by treating this function as one period of a fictitious periodic function $g(t)$ (Figure 3.1.6).

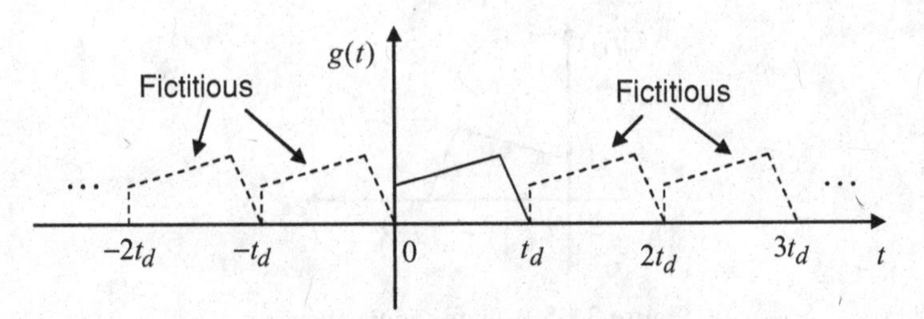

Figure 3.1.6 Fictitious periodic function corresponding to a finite duration function

The Fourier series expansion of this fictitious periodic function $g(t)$ can be defined as follows:

$$g(t) = a_0 + \sum_{n=1}^{\infty} a_n \cos(n\omega t) + \sum_{n=1}^{\infty} b_n \sin(n\omega t) \tag{3.1.46}$$

where the fundamental frequency ω is given by

$$\omega = \frac{2\pi}{t_d} \tag{3.1.47}$$

However, the Fourier series (Equation 3.1.46) should only be used for $0 \leq t \leq t_d$, as the actual function $f(t)$ is zero for $t < 0$ and $t > t_d$. Therefore,

$$f(t) = a_0 + \sum_{n=1}^{\infty} a_n \cos(n\omega t) + \sum_{n=1}^{\infty} b_n \sin(n\omega t); \quad 0 \leq t \leq t_d \tag{3.1.48}$$

Example 3.1.3: Fourier Series Expansion of a Triangular Pulse
Consider the triangular pulse $f(t)$ as shown in Figure 3.1.7. The corresponding fictitious periodic function $g(t)$ is shown in Figure 3.1.8, which is identical to the periodic function shown in Figure 3.1.3. Therefore, from Equation 3.1.34,

$$f(t) = \frac{A}{2} - \frac{4A}{\pi^2}\cos\omega t - \frac{4A}{9\pi^2}\cos 3\omega t - \frac{4A}{25\pi^2}\cos 5\omega t - \cdots; \; 0 \leq t \leq t_d \tag{3.1.49}$$

where

$$\omega = \frac{2\pi}{t_d} \tag{3.1.50}$$

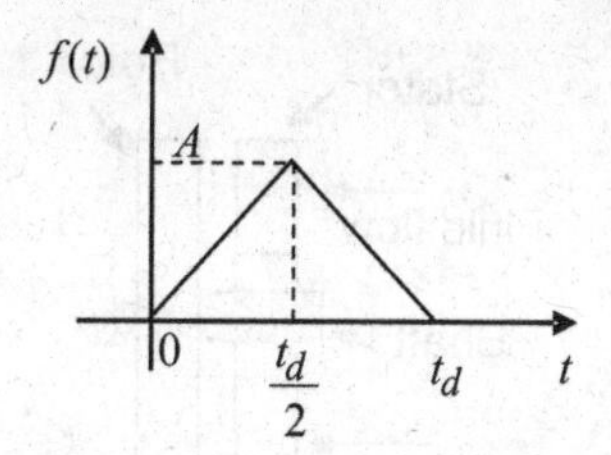

Figure 3.1.7 A triangular pulse

Example 3.1.4: Turbomachinery Blade Excitation
Consider the rotor and the stator of a turbomachine (Figure 3.1.9). The schematic drawings of the stator and the rotor are shown in Figure 3.1.10. The stator consists of four equi-spaced nozzles, with each nozzle having a 45 degrees sector. The rotor is a bladed disk with a rotational speed of Ω cps (Hz). When the blade comes in front of the nozzle, it experiences a constant force p due to fluid flow. The temporal variation of the force experienced by each blade is periodic and is shown in Figure 3.1.11.

The force pattern is repeated four times during one full rotation. As a result, the time period T of the forcing function is

$$T = \frac{1}{4\Omega}\ \text{sec}$$

And the fundamental frequency

$$\omega = \frac{2\pi}{T} = 8\pi\Omega\ \text{rad/sec}$$

The periodic force shown in Figure 3.1.11 is neither odd nor even. To evaluate the integrals for a_0, a_n, and b_n in Equation 3.1.4, the function

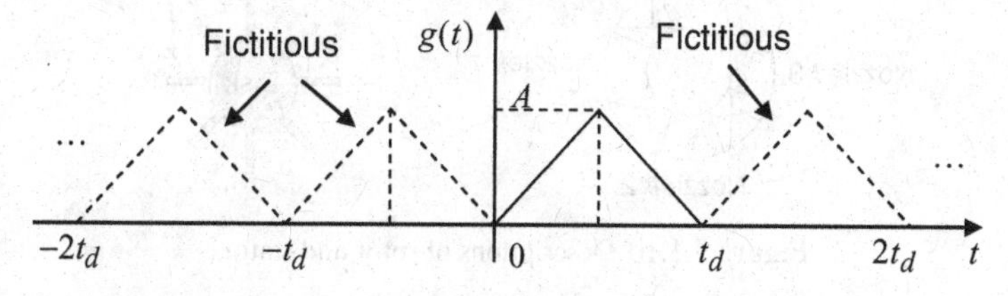

Figure 3.1.8 Fictitious periodic function for a triangular pulse

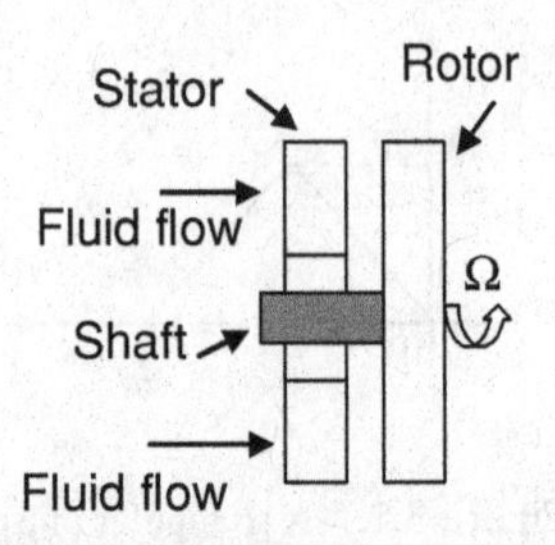

Figure 3.1.9 Fluid flow through a rotor/stator

$f(t)$ is defined as follows:

$$f(t) = \begin{cases} p & \text{for} \quad 0 \leq t < 0.5T \\ 0 & \text{for} \quad 0.5T \leq t < T \end{cases} \tag{3.1.51}$$

Using Equation 3.1.5,

$$a_0 = \frac{1}{T}\left[\int_0^{T/2} p dt + 0\right] = \frac{p}{2} \tag{3.1.52}$$

Using Equation 3.1.6,

$$a_n = \frac{2}{T}\left[\int_0^{T/2} p\cos(n\omega t)dt + 0\right] = \frac{2p}{T}\left[\frac{\sin(n\omega t)}{n\omega}\right]\Bigg|_0^{T/2} = \frac{2p}{n\omega T}\sin\left(\frac{n\omega T}{2}\right) \tag{3.1.53}$$

Because $\omega T = 2\pi$,

$$a_n = 0 \tag{3.1.54}$$

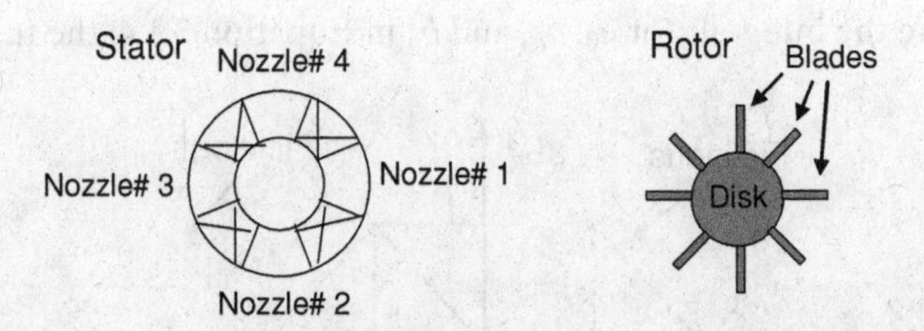

Figure 3.1.10 Descriptions of rotor and stator

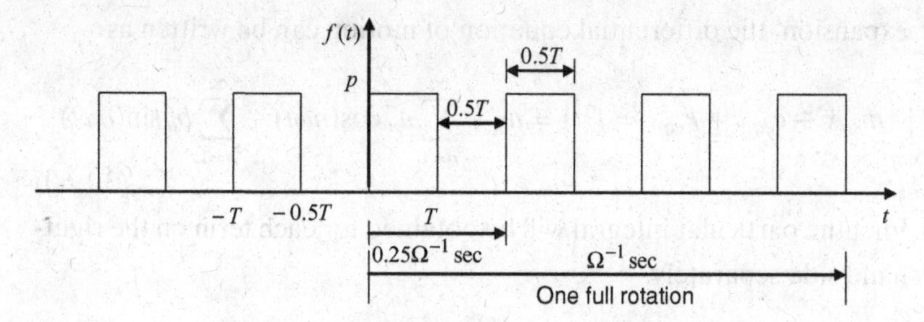

Figure 3.1.11 Force experienced by each blade

Next, using Equation 3.1.7,

$$b_n = \frac{2}{T}\left[\int_0^{T/2} p\sin(n\omega t)dt + 0\right] = \frac{2p}{T}\left[-\frac{\cos(n\omega t)}{n\omega}\right]_0^{T/2}$$

$$= \frac{2p}{n\omega T}\left[-\cos\left(\frac{n\omega T}{2}\right) + 1\right] \tag{3.1.55}$$

Because $\omega T = 2\pi$,

$$b_n = \frac{p}{n\pi}(1 - \cos n\pi) \tag{3.1.56}$$

For odd and even n, $\cos n\pi = -1$ and $+1$, respectively. Therefore,

$$b_n = \frac{2p}{n\pi}; \quad n = 1, 3, 5, \ldots, \tag{3.1.57}$$

and

$$b_n = 0; \quad n = 2, 4, 6, \ldots, \tag{3.1.58}$$

Therefore, the Fourier series expansion is

$$f(t) = \frac{p}{2} + \frac{2p}{\pi}\sin\omega t + \frac{2p}{3\pi}\sin 3\omega t + \frac{2p}{5\pi}\sin 5\omega t + \cdots \tag{3.1.59}$$

3.1.4 Particular Integral (Steady-State Response with Damping) Under Periodic Excitation

Consider the spring–mass–damper system subjected to a periodic force $f(t)$ with the fundamental frequency ω. Using the Fourier series expansion, the differential equation of motion can be written as

$$m_{eq}\ddot{x} + c_{eq}\dot{x} + k_{eq}x = f(t) = a_0 + \sum_{n=1}^{\infty} a_n \cos(n\omega t) + \sum_{n=1}^{\infty} b_n \sin(n\omega t) \tag{3.1.60}$$

First the particular integral will be obtained for each term on the right-hand side separately.

a. Constant term a_0
Let the particular integral due to the constant term be x_0. Then,

$$m_{eq}\ddot{x}_0 + c_{eq}\dot{x}_0 + k_{eq}x_0 = a_0 \tag{3.1.61}$$

The particular integral in this case will be a constant p, that is,

$$x_0 = p \tag{3.1.62}$$

Substituting Equation 3.1.62 into Equation 3.1.61,

$$k_{eq}p = a_0 \tag{3.1.63}$$

because $\dot{p} = 0$ and $\ddot{p} = 0$. In other words,

$$x_0 = p = \frac{a_0}{k_{eq}} \tag{3.1.64}$$

b. Term $a_n \cos(n\omega t)$
Let the particular integral be $x_{cn}(t)$. Then,

$$m_{eq}\ddot{x}_{cn} + c_{eq}\dot{x}_{cn} + k_{eq}x_{cn} = a_n \cos(n\omega t) \tag{3.1.65}$$

Using Equation 2.3.25,

$$x_{cn}(t) = \frac{a_n/k_{eq}}{\sqrt{(1-n^2r^2)^2 + (2\xi nr)^2}} \cos(n\omega t - \phi_n) \tag{3.1.66}$$

where

$$\phi_n = \tan^{-1} \frac{2\xi nr}{1-(nr)^2} \tag{3.1.67}$$

and

$$r = \frac{\omega}{\omega_n} \tag{3.1.68}$$

c. Term $b_n \sin(n\omega t)$

Let the particular integral be $x_{sn}(t)$. Then,

$$m_{eq}\ddot{x}_{sn} + c_{eq}\dot{x}_{sn} + k_{eq}x_{sn} = b_n \sin(n\omega t) \tag{3.1.69}$$

Using Equation 2.3.25,

$$x_{sn}(t) = \frac{b_n/k_{eq}}{\sqrt{(1-n^2r^2)^2 + (2\xi nr)^2}} \sin(n\omega t - \phi_n) \tag{3.1.70}$$

where ϕ_n and r are given by Equations 3.1.67 and 3.1.68.

Having obtained the steady-state response or the particular integral due to each term on the right-hand side of Equation 3.1.60, the complete particular integral $x_p(t)$ is obtained by using the principle of superposition:

$$x_p(t) = x_0 + \sum_{n=1}^{\infty} x_{cn}(t) + \sum_{n=1}^{\infty} x_{sn}(t) \tag{3.1.71}$$

or

$$x_p(t) = \frac{a_0}{k_{eq}} + \sum_{n=1}^{\infty} \frac{a_n/k_{eq}}{\sqrt{(1-n^2r^2)^2 + (2\xi nr)^2}} \cos(n\omega t - \phi_n) + \sum_{n=1}^{\infty} \frac{b_n/k_{eq}}{\sqrt{(1-n^2r^2)^2 + (2\xi nr)^2}} \sin(n\omega t - \phi_n) \tag{3.1.72}$$

where ϕ_n and r are given by Equations 3.1.67 and 3.1.68.

It should be noted that $x_p(t)$ will be the steady-state response when $c_{eq} > 0$.

Example 3.1.5: Steady-State Response of a Turbine Blade
Consider the turbomachinery problem of Example 3.1.3. The turbine rotates at 60 Hz. The natural frequency of the system is 160 Hz. Find the maximum steady-state amplitude in terms of p/k_{eq} assuming that the damping ratio ξ is 0.01.

Using Equations 3.1.59 and 3.1.72,

$$\frac{x_p(t)}{p/k_{eq}} = \frac{1}{2} + \sum_{n=1,3,\ldots}^{\infty} \frac{2/(n\pi)}{\sqrt{(1-n^2r^2)^2+(2\xi nr)^2}} \sin(n\omega t - \phi_n) \quad (3.1.73)$$

where

$$r = \frac{\omega}{\omega_n} = \frac{60}{160} = \frac{3}{8} \quad \text{and } \xi = 0.01$$

$n = 1$:

$$\frac{2/n}{\sqrt{(1-n^2r^2)^2+(2\xi nr)^2}} = 2.3272; \quad \phi_1 = 0.0087 \text{ rad}$$

$n = 3$:

$$\frac{2/n}{\sqrt{(1-n^2r^2)^2+(2\xi nr)^2}} = 2.5008; \quad \phi_3 = 3.0571 \text{ rad}$$

$n = 5$:

$$\frac{2/n}{\sqrt{(1-n^2r^2)^2+(2\xi nr)^2}} = 0.1590; \quad \phi_5 = 3.1267 \text{ rad}$$

$n = 7$:

$$\frac{2/n}{\sqrt{(1-n^2r^2)^2+(2\xi nr)^2}} = 0.0485; \quad \phi_7 = 3.1327 \text{ rad}$$

Neglecting $n = 7$ and higher terms,

$$\frac{x_p(t)}{p/k_{eq}} = 0.5 + \frac{2.3272}{\pi}\sin(\omega t - \phi_1) + \frac{2.5008}{\pi}\sin(3\omega t - \phi_3) + \frac{0.1590}{\pi}\sin(5\omega t - \phi_5)$$

The plot of $x_p(t)/(p/k_{eq})$ is shown in Figure 3.1.12. The maximum value of the steady-state response $x_p(t)/(p/k_{eq})$ is 2.3.

3.2 RESPONSE TO AN EXCITATION WITH ARBITRARY NATURE

The response to an arbitrary type of excitation is obtained via an impulsive force, which has a large magnitude and a small duration.

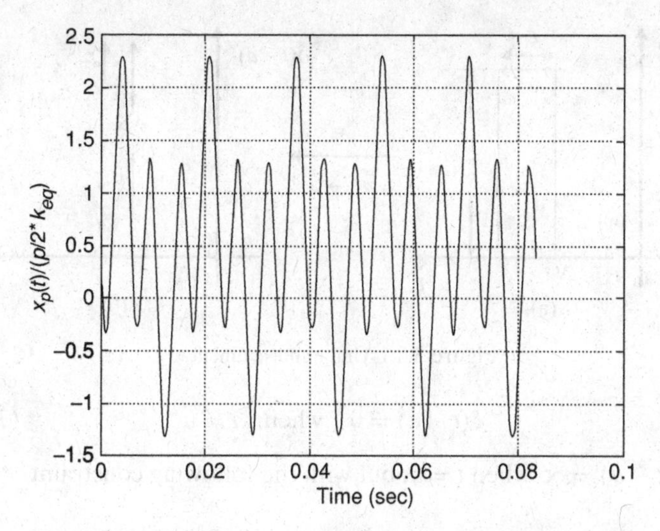

Figure 3.1.12 Steady-state response with $n = 5$ in Equation 3.1.73

The time integral $\hat{f}$ of an impulsive force $f(t)$ is finite, and is defined as the *impulse* of the force:

$$\hat{f} = \int f(t)dt \tag{3.2.1}$$

In other words, the impulse of a force $\hat{f}$ is the area under the force–time plot. For a unit impulse, $\hat{f} = 1$. Drawing an analogy with this impulsive force, a mathematical impulse unit function is defined as follows.

3.2.1 Unit Impulse Function $\delta(t - a)$

Consider a constant function of the magnitude $1/\varepsilon$ of duration ε (Figure 3.2.1a). Therefore,

$$\hat{f} = \int f(t)dt = \frac{1}{\varepsilon}\varepsilon = 1 \tag{3.2.2}$$

The function shown in Figure 3.2.1a has a unit impulse. Having $\varepsilon \to 0$, the unit impulse function $\delta(t - a)$ is obtained (Figure 3.2.1b) with the following properties:

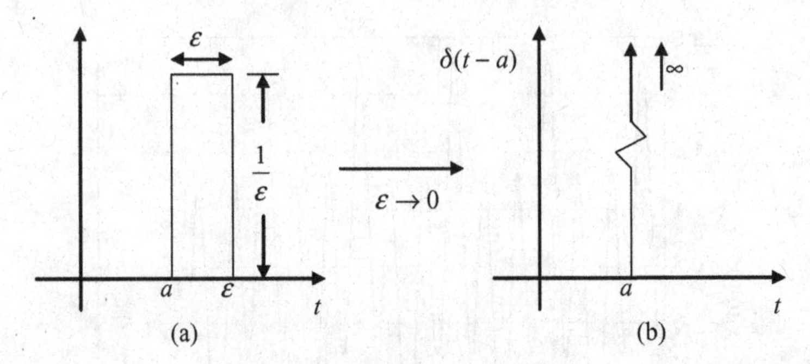

Figure 3.2.1 Unit impulse function

1.
$$\delta(t-a)=0 \quad \text{when} \quad t\neq a \tag{3.2.3}$$

2. $\delta(t-a)=\infty$ when $t=a$, but with the following constraint

$$\int_0^\infty \delta(t-a)dt=1 \tag{3.2.4}$$

3. For any function $g(t)$,

$$\int_0^\infty g(t)\delta(t-a)dt=g(a) \tag{3.2.5}$$

It should be noted that the unit impulse function is not necessarily a force function. It could be defined for any variable, such as displacement and velocity.

3.2.2 Unit Impulse Response of an SDOF System with Zero Initial Conditions

Consider a spring–mass–damper system subjected to the unit impulse force $\delta(t)$ (Figure 3.2.2). The differential equation of motion is

$$m_{eq}\ddot{x}+c_{eq}\dot{x}+k_{eq}x=\delta(t) \tag{3.2.6}$$

This unit impulse force is applied at $t=0$. Since the duration of this impulse function is zero, symbols 0^- and 0^+ are introduced to denote instants just before and after the application of the force, respectively.

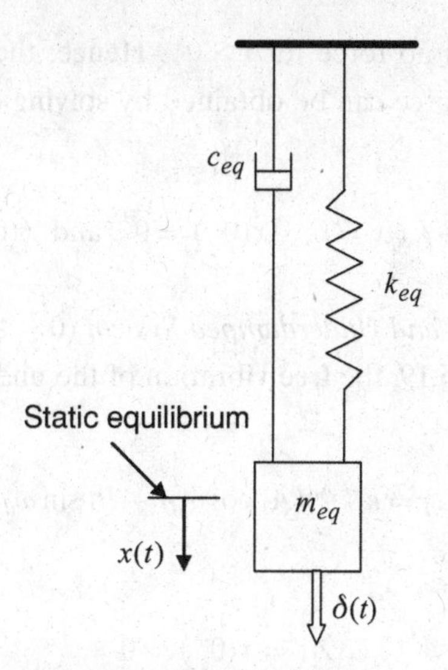

Figure 3.2.2 Spring–mass–damper subjected to unit impulse force

Zero initial conditions will then be represented as

$$x(0^-) = 0 \quad \text{and} \quad \dot{x}(0^-) = 0 \tag{3.2.7}$$

Due to the impulsive force,

$$\text{change in momentum} = m\,\dot{x}(0^+) - m\dot{x}(0^-) = m\dot{x}(0^+) \tag{3.2.8}$$

Since the change in the momentum equals the magnitude of the impulse,

$$m\dot{x}(0^+) = 1 \tag{3.2.9}$$

or,

$$\dot{x}(0^+) = \frac{1}{m} \tag{3.2.10}$$

In other words, the velocity has changed instantaneously from 0 to $1/m$. It can be shown that the displacement remains unchanged, that is,

$$x(0^+) = x(0^-) = 0 \tag{3.2.11}$$

Note that there is no force for $t \geq 0^+$. Hence, the response due to the unit impulse force can be obtained by solving the following free vibration problem:

$$m_{eq}\ddot{x} + c_{eq}\dot{x} + k_{eq}x = 0; \quad x(0^+) = 0 \quad \text{and} \quad \dot{x}(0^+) = \frac{1}{m} \qquad (3.2.12)$$

Case I: Undamped and Underdamped System $(0 \leq \xi < 1)$
From Equation 1.5.19, the free vibration of the underdamped system is given by

$$x(t) = e^{-\xi\omega_n t}(A_1 \cos\omega_d t + B_1 \sin\omega_d t) \qquad (3.2.13)$$

where

$$A_1 = x(0^+) = 0 \qquad (3.2.14)$$

$$B_1 = \frac{\dot{x}(0^+) + \xi\omega_n x(0^+)}{\omega_d} = \frac{1}{m\omega_d} \qquad (3.2.15)$$

Substituting Equations 3.2.14 and 3.2.15 into Equation 3.2.13,

$$x(t) = \frac{1}{m\omega_d} e^{-\xi\omega_n t} \sin\omega_d t \qquad (3.2.16)$$

The unit impulse response (the response due to the unit impulse force) of an underdamped system is given by Equation 3.2.16 and will be represented by a new symbol $g(t)$, that is,

$$g(t) = \frac{1}{m\omega_d} e^{-\xi\omega_n t} \sin\omega_d t \qquad (3.2.17)$$

For an undamped system $(\xi = 0)$,

$$g(t) = \frac{1}{m\omega_n} \sin\omega_n t \qquad (3.2.18)$$

Case II: Critically Damped $(\xi = 1 \text{ or } c_{eq} = c_c)$
From Equation 1.5.38, the free vibration of the critically damped system is given by

$$x(t) = x(0^+)e^{-\omega_n t} + [\dot{x}(0^+) + \omega_n x(0^+)]te^{-\omega_n t} = \frac{1}{m}te^{-\omega_n t} \qquad (3.2.19)$$

The unit impulse response (the response due to the unit impulse force) of a critically damped system is given by Equation 3.2.19 and will be represented by a new symbol $g(t)$, that is,

$$g(t) = \frac{1}{m} t e^{-\omega_n t} \tag{3.2.20}$$

Case III: Overdamped ($\xi > 1$ *or* $c_{eq} > c_c$)

From Equation 1.5.41, the free vibration of the overdamped system is given by

$$x(t) = A_1 e^{s_1 t} + B_1 e^{s_2 t} \tag{3.2.21}$$

where

$$s_1 = -\xi\omega_n + \omega_n\sqrt{\xi^2 - 1} < 0 \tag{3.2.22}$$

$$s_2 = -\xi\omega_n - \omega_n\sqrt{\xi^2 - 1} < 0 \tag{3.2.23}$$

Using Equations 1.5.45 and 1.5.46,

$$A_1 = \frac{s_2 x(0^+) - \dot{x}(0^+)}{s_2 - s_1} = -\frac{1}{m(s_2 - s_1)} = \frac{1}{2m\omega_n\sqrt{\xi^2 - 1}} \tag{3.2.24}$$

$$B_1 = \frac{-s_1 x(0^+) + \dot{x}(0^+)}{s_2 - s_1} = \frac{1}{m(s_2 - s_1)} = -\frac{1}{2m\omega_n\sqrt{\xi^2 - 1}} \tag{3.2.25}$$

The unit impulse response (the response due to the unit impulse force) of an overdamped system is given by Equation 3.2.21 and will be represented by a new symbol $g(t)$, that is,

$$g(t) = \frac{1}{2m\omega_n\sqrt{\xi^2 - 1}}\left(e^{s_1 t} - e^{s_2 t}\right) \tag{3.2.26}$$

Nondimensional unit impulse responses, $g(t)m\omega_n$, are plotted in Figure 3.2.3 for all cases of damping.

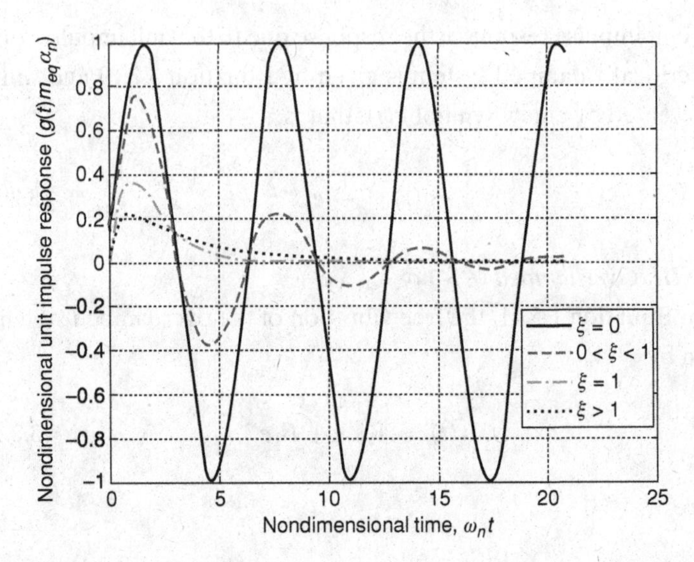

Figure 3.2.3 Unit impulse response for undamped, underdamped, critically damped, and overdamped systems

3.2.3 Convolution Integral: Response to an Arbitrary Excitation with Zero Initial Conditions

Consider a forcing function $f(t)$ of an arbitrary nature (Figure 3.2.4). Let us determine the response at $t = t_0$. First, the following points should be noted:

a. The entire forcing function before $t = t_0$ will have an influence on the response at $t = t_0$.

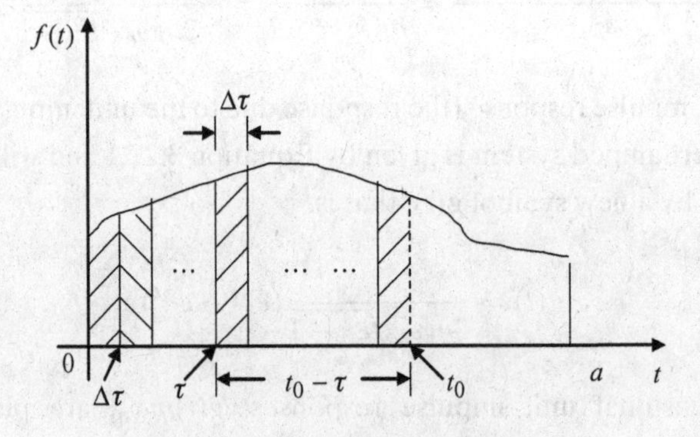

Figure 3.2.4 An arbitrary forcing function

b. The forcing function after $t = t_0$ will **not** have any influence on the response at $t = t_0$.

The forcing function before the instant $t = t_0$ is divided into many strips of small time width $\Delta\tau$. Each strip can be viewed as an impulse function as shown in Figure 3.2.4. Let us consider one such impulse function at $t = \tau$. The magnitude of the impulse is

$$\hat{f}_\tau = f(\tau)\Delta\tau \tag{3.2.27}$$

The contribution of this impulse to the response at $t = t_0$ is

$$x_\tau(t_0) = g(t_0 - \tau)\hat{f}_\tau = g(t_0 - \tau)f(\tau)\Delta\tau \tag{3.2.28}$$

where $g(.)$ is the unit impulse response. The response at $t = t_0$ is then found by summing up the contributions from all the impulses (strips of width $\Delta\tau$):

$$x(t_0) = \sum_{\tau=0}^{\tau=t_0-\Delta\tau} x_\tau(t_0) = \sum_{\tau=0}^{\tau=t_0-\Delta\tau} g(t_0 - \tau)f(\tau)\Delta\tau \tag{3.2.29}$$

When $\Delta\tau \to 0$, the summation in Equation 3.2.29 becomes the integral as follows:

$$x(t_0) = \int_{\tau=0}^{\tau=t_0} g(t_0 - \tau)f(\tau)d\tau \tag{3.2.30}$$

Since t_0 is arbitrarily chosen, the response at any time t is given by

$$x(t) = \int_{\tau=0}^{\tau=t} g(t - \tau)f(\tau)d\tau \tag{3.2.31}$$

This is the **convolution integral** and yields the response for zero initial conditions. This represents the complete solution, that is, it contains both the homogeneous part and the particular integral.

Example 3.2.1: Step Response of an Undamped System
Consider the undamped spring–mass system subjected to a step forcing function (Figure 3.2.5). Assume that the initial conditions are

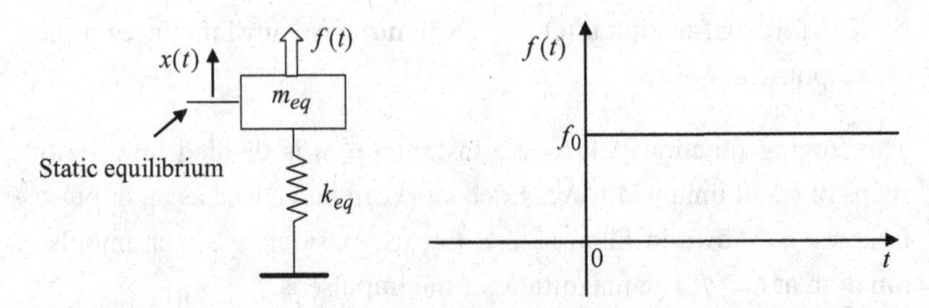

Figure 3.2.5 Undamped SDOF system subjected to step forcing function

zeros. Then,

$$f(t) = f_0 \quad \text{for } t > 0 \tag{3.2.32}$$

Using Equations 3.2.18, 3.2.31, and 3.2.32,

$$x(t) = \int_{\tau=0}^{\tau=t} \frac{1}{m\omega_n} \sin(\omega_n(t-\tau)) f_0 d\tau = \frac{f_0}{m\omega_n} \left[\frac{\cos(\omega_n(t-\tau))}{\omega_n} \right]_{\tau=0}^{\tau=t} \tag{3.2.33}$$

After some simple algebra,

$$x(t) = \frac{f_0}{m\omega_n^2} [1 - \cos(\omega_n t)] = \frac{f_0}{k} [1 - \cos(\omega_n t)]; \quad t \geq 0 \tag{3.2.34}$$

Note that Equation 3.2.34 is identical to Equation 2.1.21 with $\xi = 0$.

Example 3.2.2: Underdamped SDOF System Subjected to a Rectangular Pulse

Consider an underdamped spring–mass–damper system in which the mass is subjected to the force $f(t)$ as shown in Figure 3.2.6. Assuming that all initial conditions are zero, find the response using the convolution integral.

The differential equation of motion is

$$m_{eq}\ddot{x} + c_{eq}\dot{x} + k_{eq}x(t) = f(t); \quad x(0) = 0, \quad \dot{x}(0) = 0 \tag{3.2.35}$$

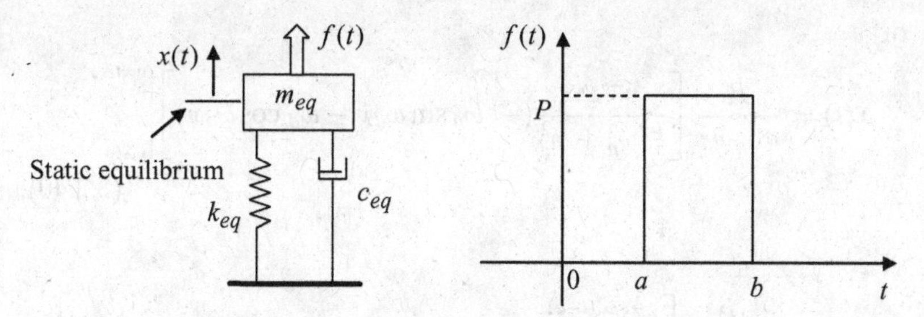

Figure 3.2.6 Underdamped SDOF system and rectangular pulse

Case I: $0 \le t \le a$

$$x(t) = \int_{\tau=0}^{\tau=t} g(t-\tau)f(\tau)d\tau = \int_{\tau=0}^{\tau=t} g(t-\tau)\,0d\tau = 0 \qquad (3.2.36)$$

Case II: $a \le t \le b$

$$\begin{aligned} x(t) &= \int_{\tau=0}^{\tau=t} g(t-\tau)f(\tau)d\tau = \int_{\tau=0}^{\tau=a} g(t-\tau)\,0d\tau + \int_{\tau=a}^{\tau=t} g(t-\tau)\,Pd\tau \\ &= \int_{\tau=a}^{\tau=t} g(t-\tau)\,Pd\tau \end{aligned} \qquad (3.2.37)$$

Therefore,

$$x(t) = \int_{\tau=a}^{\tau=t} g(t-\tau)\,Pd\tau = \frac{P}{m_{eq}\omega_d} \int_{\tau=a}^{\tau=t} e^{-\xi\omega_n(t-\tau)} \sin(\omega_d(t-\tau))d\tau \qquad (3.2.38)$$

Substituting $\nu = t - \tau$ into Equation 3.2.38,

$$x(t) = \frac{P}{m_{eq}\omega_d} \int_{\nu=0}^{\nu=t-a} e^{-\xi\omega_n\nu} \sin(\omega_d\nu)d\nu \qquad (3.2.39)$$

or

$$x(t) = \frac{P}{m_{eq}\omega_d}\left[\frac{e^{-\xi\omega_n\nu}}{\xi^2\omega_n^2+\omega_d^2}(-\xi\omega_n\sin\omega_d\nu - \omega_d\cos\omega_d\nu)\right]\Bigg|_{\nu=0}^{\nu=t-a} \tag{3.2.40}$$

or

$$x(t) = -\frac{P}{m_{eq}\omega_d}\left[\frac{e^{-\xi\omega_n(t-a)}}{\omega_n^2}(\xi\omega_n\sin\omega_d(t-a)+\omega_d\cos\omega_d(t-a))\right] + \frac{P}{m_{eq}\omega_n^2} \tag{3.2.41}$$

Simplifying Equation 3.2.41,

$$x(t) = \frac{P}{m_{eq}\omega_n^2}\left[1-\frac{\omega_n}{\omega_d}e^{-\xi\omega_n(t-a)}\sin(\omega_d(t-a)+\phi)\right] \tag{3.2.42}$$

where $\cos\phi = \xi$

Case III: $t \geq b$

$$\begin{aligned} x(t) &= \int_{\tau=0}^{\tau=t} g(t-\tau)f(\tau)d\tau \\ &= \int_{\tau=0}^{\tau=a} g(t-\tau)0d\tau + \int_{\tau=a}^{\tau=b} g(t-\tau)\,Pd\tau + \int_{\tau=b}^{\tau=t} g(t-\tau)\,0d\tau \end{aligned} \tag{3.2.43}$$

Therefore,

$$x(t) = \int_{\tau=a}^{\tau=b} g(t-\tau)\,Pd\tau = \frac{P}{m_{eq}\omega_d}\int_{\tau=a}^{\tau=b} e^{-\xi\omega_n(t-\tau)}\sin(\omega_d(t-\tau))d\tau \tag{3.2.44}$$

or

$$x(t) = \frac{P}{m_{eq}\omega_d}\left[\frac{e^{-\xi\omega_n\nu}}{\xi^2\omega_n^2+\omega_d^2}(-\xi\omega_n\sin\omega_d\nu - \omega_d\cos\omega_d\nu)\right]\Bigg|_{\nu=t-b}^{\nu=t-a} \tag{3.2.45}$$

Simplifying Equation 3.2.45,

$$x(t) = \frac{P}{m_{eq}\omega_d\omega_n}\Big[e^{-\xi\omega_n(t-b)}\sin(\omega_d(t-b)+\phi) - e^{-\xi\omega_n(t-a)}\sin(\omega_d(t-a)+\phi)\Big] \tag{3.2.46}$$

where $\cos\phi = \xi$.

3.2.4 Convolution Integral: Response to an Arbitrary Excitation with Nonzero Initial Conditions

Consider again the same equivalent SDOF system considered in Section 3.2.3, that is,

$$m_{eq}\ddot{x} + c_{eq}\dot{x} + k_{eq}x = f(t) \tag{3.2.47}$$

Assume that $x(0)$ and/or $\dot{x}(0)$ are not zero. In this case, the following two problems are separately solved:

1. Forcing function with zero initial conditions

$$m_{eq}\ddot{x}_1 + c_{eq}\dot{x}_1 + k_{eq}x_1 = f(t); \quad x_1(0) = 0 \quad \text{and } \dot{x}_1(0) = 0 \tag{3.2.48}$$

This problem has already been solved in Section 3.2.3. Therefore,

$$x_1(t) = \int_{\tau=0}^{\tau=t} g(t-\tau)f(\tau)d\tau \tag{3.2.49}$$

2. Nonzero initial conditions without any forcing function

$$m_{eq}\ddot{x}_2 + c_{eq}\dot{x}_2 + k_{eq}x_2(t) = 0; \quad x_2(0) = x(0) \quad \text{and } \dot{x}_2(0) = \dot{x}(0) \tag{3.2.50}$$

The problem in Equation 3.2.50 represents the free vibration of a damped or undamped SDOF system. As seen in Chapter 2, the solution to Equation 3.2.50 depends on the damping values as follows.

Case I: Undamped and Underdamped ($0 \leq \xi < 1$ *or* $0 \leq c_{eq} < c_c$)

$$x_2(t) = e^{-\xi\omega_n t}(A_1 \cos \omega_d t + B_1 \sin \omega_d t) \tag{3.2.51}$$

where

$$A_1 = x(0) \tag{3.2.52}$$

and

$$B_1 = \frac{\dot{x}(0) + \xi\omega_n x(0)}{\omega_d} \tag{3.2.53}$$

Case II: Critically Damped ($\xi = 1$ *or* $c_{eq} = c_c$)

$$x_2(t) = x(0)e^{-\omega_n t} + [\dot{x}(0) + \omega_n x(0)]te^{-\omega_n t} \tag{3.2.54}$$

Case III: Overdamped ($\xi > 1$ *or* $c_{eq} > c_c$)

$$x_2(t) = A_1 e^{s_1 t} + B_1 e^{s_2 t} \tag{3.2.55}$$

where

$$s_1 = -\xi\omega_n + \omega_n\sqrt{\xi^2 - 1} < 0 \tag{3.2.56}$$

$$s_2 = -\xi\omega_n - \omega_n\sqrt{\xi^2 - 1} < 0 \tag{3.2.57}$$

$$A_1 = \frac{s_2 x(0) - \dot{x}(0)}{s_2 - s_1} \tag{3.2.58}$$

and

$$B_1 = \frac{-s_1 x(0) + \dot{x}(0)}{s_2 - s_1} \tag{3.2.59}$$

Last, **the solution to** Equation 3.2.47 with an arbitrary forcing function and nonzero initial conditions is

$$x(t) = x_1(t) + x_2(t) \tag{3.2.60}$$

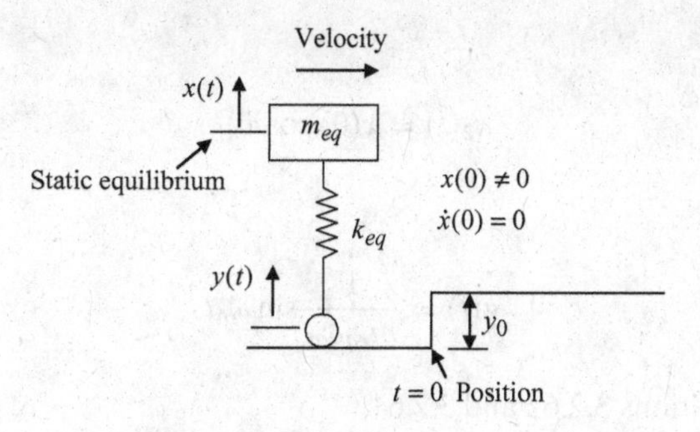

Figure 3.2.7 Vehicle moving over a step bump

where $x_1(t)$ is given by Equation 3.2.49 and $x_2(t)$ is given by Equations 3.2.51, 3.2.54, and 3.2.55 for $0 \leq \xi < 1$, $\xi = 1$, and $\xi > 1$, respectively. It should also be remembered that the expression of the unit impulse response $g(t)$, in the solution to Equation 3.2.49 for $x_1(t)$ is given by Equations 3.2.17, 3.2.20, and 3.2.26 for $0 \leq \xi < 1$, $\xi = 1$, and $\xi > 1$, respectively.

Example 3.2.3: A Vehicle Past a Step Bump
The differential equation of motion (Figure 3.2.7) is

$$m_{eq}\ddot{x} + k_{eq}x = f(t) \tag{3.2.61}$$

where

$$f(t) = k_{eq}y(t) \tag{3.2.62}$$

Using Equations 3.2.60, 3.2.49, and 3.2.51–3.2.53, the response is

$$x(t) = x_1(t) + x_2(t) \tag{3.2.63}$$

where

$$x_1(t) = \int_{\tau=0}^{\tau=t} g(t-\tau)f(\tau)d\tau \tag{3.2.64}$$

and

$$x_2(t) = x(0)\cos\omega_n t \tag{3.2.65}$$

Here,

$$g(t) = \frac{1}{m_{eq}\omega_n}\sin\omega_n t \tag{3.2.66}$$

From Equations 3.2.62 and 3.2.64,

$$\begin{aligned} x_1(t) &= \frac{1}{m_{eq}\omega_n}\int_{\tau=0}^{\tau=t}\sin(\omega_n(t-\tau))k_{eq}y(\tau)d\tau \\ &= \frac{1}{m_{eq}\omega_n}\int_{\tau=0}^{\tau=t}\sin(\omega_n(t-\tau))k_{eq}y_0 d\tau \end{aligned} \tag{3.2.67}$$

Evaluating the integral in Equation 3.2.67,

$$x_1(t) = y_0(1-\cos\omega_n t) \tag{3.2.68}$$

Last, from Equations 3.2.63, 3.2.65, and 3.2.68,

$$x(t) = y_0(1-\cos\omega_n t) + x(0)\cos\omega_n t \tag{3.2.69}$$

3.3 LAPLACE TRANSFORMATION

Consider a function $f(t)u_s(t)$ in time-domain where $u_s(t)$ is the unit step function (Figure 3.3.1). In other words,

$$f(t)u_s(t) = \begin{cases} f(t) & \text{for} \quad t \geq 0 \\ 0 & \text{for} \quad t < 0 \end{cases} \tag{3.3.1}$$

The Laplace transform of $f(t)u_s(t)$ is defined as

$$f(s) = L(f(t)u_s(t)) = \int_0^\infty f(t)e^{-st}dt \tag{3.3.2}$$

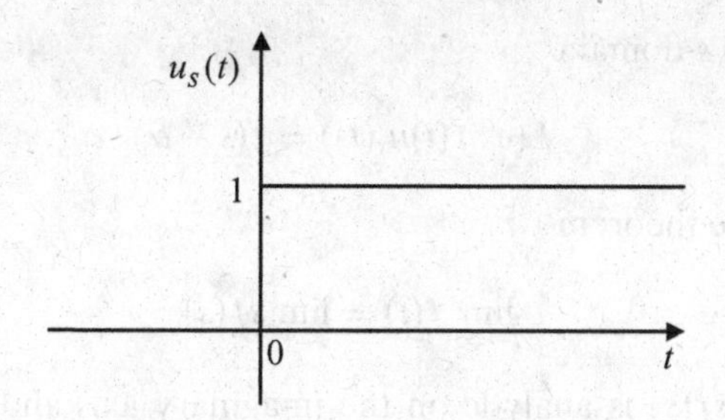

Figure 3.3.1 Unit step function

And $f(t)u_s(t)$ is defined as the inverse Laplace transformation of $f(s)$, that is,

$$L^{-1}(f(s)) = f(t)u_s(t) \tag{3.3.3}$$

3.3.1 Properties of Laplace Transformation

a. Linearity

$$L(\alpha f_1(t) + \beta f_2(t)) = \alpha L(f_1(t)) + \beta L(f_2(t)) \tag{3.3.4}$$

where α and β are any real or complex numbers.

b. Derivatives

$$L\left(\frac{df}{dt}\right) = sL(f(t)) - f(0) \tag{3.3.5}$$

$$L\left(\frac{d^2f}{dt^2}\right) = s^2L(f(t)) - sf(0) - \frac{df}{dt}(0) \tag{3.3.6}$$

c. Integrals

$$L\left(\int_0^t f(\tau)d\tau\right) = \frac{1}{s}L(f(t)) \tag{3.3.7}$$

d. Shifting in time-domain

$$L(f(t-a)u_s(t-a)) = e^{-as}L(f(t)) \tag{3.3.8}$$

e. Shifting in s-domain

$$L(e^{\alpha t} f(t) u_s(t)) = f(s - \alpha) \tag{3.3.9}$$

f. Final value theorem

$$\lim_{t \to \infty} f(t) = \lim_{s \to 0} s f(s) \tag{3.3.10}$$

provided $sf(s)$ is analytic on the imaginary axis and in the right half of the complex s-plane.

g. Convolution integral

$$L\left(\int_0^t f_1(t-\tau) f_2(\tau) d\tau \right) = L(f_1(t)).L(f_2(t)) = f_1(s) f_2(s) \tag{3.3.11}$$

Therefore, the convolution integral in time-domain is equivalent to a simple multiplication in s-domain.

3.3.2 Response of an SDOF System via Laplace Transformation

The differential equation of motion for the system shown in Figure 2.1.1 is

$$m_{eq} \ddot{x} + c_{eq} \dot{x} + k_{eq} x = f(t) \tag{3.3.12}$$

where the nature of forcing function $f(t)$ is arbitrary.

Taking Laplace transformation on both sides,

$$L(m_{eq} \ddot{x} + c_{eq} \dot{x} + k_{eq} x) = L(f(t)) \tag{3.3.13}$$

Using the linearity property in Equation 3.3.4,

$$m_{eq} L(\ddot{x}(t)) + c_{eq} L(\dot{x}(t)) + k_{eq} L(x(t)) = L(f(t))$$

Using properties in Equations 3.3.5 and 3.3.6,

$$m_{eq}[s^2 x(s) - s x(0) - \dot{x}(0)] + c_{eq}[s x(s) - x(0)] + k_{eq} x(s) = f(s) \tag{3.3.14}$$

or

$$(m_{eq} s^2 + c_{eq} s + k_{eq}) x(s) - m_{eq} \dot{x}(0) - (m_{eq} s + c_{eq}) x(0) = f(s) \tag{3.3.15}$$

Solving the algebraic Equation 3.3.15,

$$x(s) = \frac{f(s)}{m_{eq}s^2 + c_{eq}s + k_{eq}} + \frac{(m_{eq}s + c_{eq})x(0)}{m_{eq}s^2 + c_{eq}s + k_{eq}} + \frac{m_{eq}\dot{x}(0)}{m_{eq}s^2 + c_{eq}s + k_{eq}} \tag{3.3.16}$$

or,

$$x(s) = \frac{f(s)}{m_{eq}\left[s^2 + \frac{c_{eq}}{m_{eq}}s + \frac{k_{eq}}{m_{eq}}\right]} + \frac{\left(s + \frac{c_{eq}}{m_{eq}}\right)x(0)}{\left[s^2 + \frac{c_{eq}}{m_{eq}}s + \frac{k_{eq}}{m_{eq}}\right]} + \frac{\dot{x}(0)}{\left[s^2 + \frac{c_{eq}}{m_{eq}}s + \frac{k_{eq}}{m_{eq}}\right]} \tag{3.3.17}$$

Using Equations 1.4.46 and 1.5.12,

$$x(s) = \frac{f(s)}{m_{eq}(s^2 + 2\xi\omega_n s + \omega_n^2)} + \frac{(s + 2\xi\omega_n)x(0)}{(s^2 + 2\xi\omega_n s + \omega_n^2)} + \frac{\dot{x}(0)}{(s^2 + 2\xi\omega_n s + \omega_n^2)} \tag{3.3.18}$$

Last, the response $x(t)$ is obtained by taking the inverse Laplace transformation of $x(s)$.

Example 3.3.1: Rectangular Pulse as the Force

Consider an underdamped SDOF system subjected to the force shown in Figure 3.3.2. To determine the Laplace transform of the forcing function $f(s)$, via the table in Appendix C, the function $f(t)$ is represented as a difference between the two forcing functions $f_1(t)$ and $f_2(t)$ shown in Figure 3.3.3, that is,

$$f(t) = f_1(t) - f_2(t) \tag{3.3.19}$$

Using the property in Equation 3.3.4,

$$f(s) = f_1(s) - f_2(s) \tag{3.3.20}$$

From the table in Appendix C,

$$L(u_s(t)) = \frac{1}{s} \tag{3.3.21}$$

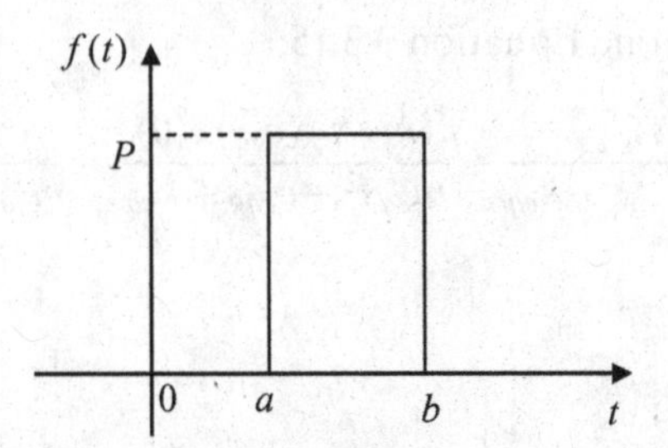

Figure 3.3.2 Rectangular pulse as a forcing function for SDOF System

Therefore, using the property in Equation 3.3.8,

$$L(f_1(t)) = L(Pu_s(t-a)) = \frac{P}{s}e^{-as} \tag{3.3.22}$$

and

$$L(f_2(t)) = L(Pu_s(t-b)) = \frac{P}{s}e^{-bs} \tag{3.3.23}$$

Therefore, from Equation 3.3.20,

$$f(s) = \frac{P}{s}(e^{-as} - e^{-bs}) \tag{3.3.24}$$

With zero initial conditions, Equation 3.3.18 yields

$$x(s) = \frac{Pe^{-as}}{m_{eq}s(s^2 + 2\xi\omega_n s + \omega_n^2)} - \frac{Pe^{-bs}}{m_{eq}s(s^2 + 2\xi\omega_n s + \omega_n^2)} \tag{3.3.25}$$

From the table in Appendix C,

$$L^{-1}\left[\frac{\omega_n^2}{s(s^2 + 2\xi\omega_n s + \omega_n^2)}\right] = \left[1 - \frac{\omega_n}{\omega_d}e^{-\xi\omega_n t}\sin(\omega_d t + \phi)\right]u_s(t) \tag{3.3.26}$$

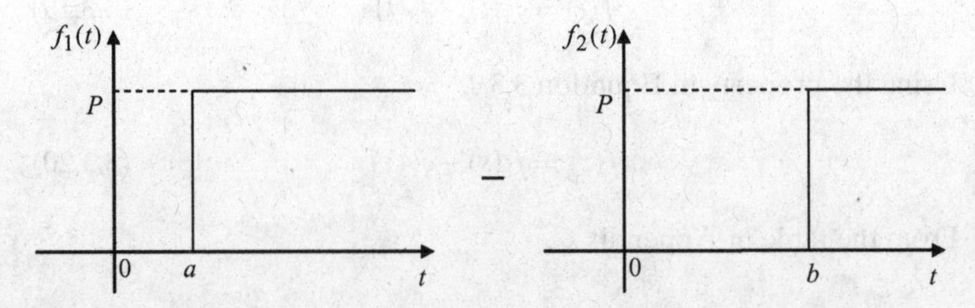

Figure 3.3.3 Rectangular pulse as a difference between two delayed step functions

where $\phi = \cos^{-1}\xi$. Using the property in Equation 3.3.8 and the result in Equation 3.3.26, the inverse Laplace transformation of Equation 3.3.25 yields

$$x(t) = \left[1 - \frac{\omega_n}{\omega_d} e^{-\xi\omega_n(t-a)} \sin(\omega_d(t-a)+\phi)\right] \frac{P}{m_{eq}\omega_n^2} u_s(t-a) - \left[1 - \frac{\omega_n}{\omega_d} e^{-\xi\omega_n(t-b)} \sin(\omega_d(t-b)+\phi)\right] \frac{P}{m_{eq}\omega_n^2} u_s(t-b) \quad (3.3.27)$$

3.3.3 *Transfer Function and Frequency Response Function*

A general second-order differential equation can be written as

$$\ddot{x} + a_1\dot{x} + a_0 x(t) = b_2\ddot{u} + b_1\dot{u} + b_0 u(t) \quad (3.3.28)$$

where $x(t)$ and $u(t)$ are defined as the output and the input of the system, respectively. Taking Laplace transform of Equation 3.3.28,

$$s^2x(s) - sx(0) - \dot{x}(0) + a_1(sx(s) - x(0)) + a_0x(s) = s^2u(s) - su(0) - \dot{u}(0) + b_1(su(s) - u(0)) + b_0u(s) \quad (3.3.29)$$

Setting all initial conditions ($x(0)$, $\dot{x}(0)$, $u(0)$, and $\dot{u}(0)$) to be zero, Equation 3.3.29 yields

$$s^2x(s) + a_1sx(s) + a_0x(s) = b_2s^2u(s) + b_1su(s) + b_0u(s) \quad (3.3.30)$$

The **transfer function** $G(s)$ is then defined as

$$G(s) = \frac{x(s)}{u(s)} = \frac{b_2s^2 + b_1s + b_0}{s^2 + a_1s + a_0} \quad (3.3.31)$$

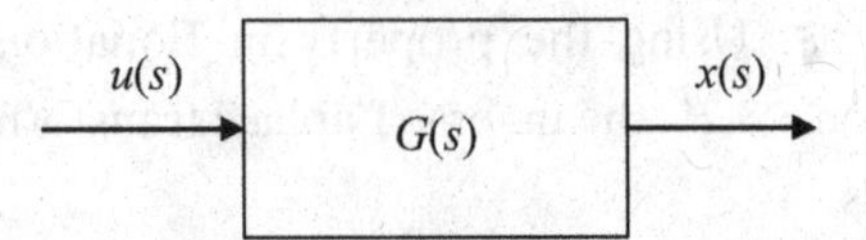

Figure 3.3.4 Block diagram

Equation 3.3.11 is represented by a block diagram, shown in Figure 3.3.4, where

$$x(s) = G(s)u(s) \tag{3.3.32}$$

Example 3.3.2: Direct Excitation Problem
For the direct excitation problem, the governing differential equation is given by Equation 3.3.12. Comparing Equations 3.3.12 and 3.3.28,

$$u(t) = f(t) \tag{3.3.33}$$

$$b_0 = \frac{1}{m_{eq}}, \quad b_1 = 0, \quad b_2 = 0 \tag{3.3.34}$$

$$a_0 = \frac{k_{eq}}{m_{eq}}, \quad a_1 = \frac{c_{eq}}{m_{eq}} \tag{3.3.35}$$

The transfer function is given by

$$G(s) = \frac{x(s)}{u(s)} = \frac{b_0}{s^2 + a_1 s + a_0} \tag{3.3.36}$$

Example 3.3.3: Base Excitation Problem
For the base excitation problem, governing differential equation is Equation 2.5.3. Comparing Equations 2.5.3 and 3.3.28,

$$u(t) = y(t) \tag{3.3.37}$$

$$b_2 = 0, \quad b_1 = \frac{c_{eq}}{m_{eq}}, \quad b_0 = \frac{k_{eq}}{m_{eq}} \tag{3.3.38}$$

$$a_0 = \frac{k_{eq}}{m_{eq}}, \quad a_1 = \frac{c_{eq}}{m_{eq}} \tag{3.3.39}$$

The transfer function is given by

$$G(s) = \frac{x(s)}{u(s)} = \frac{b_1 s + b_0}{s^2 + a_1 s + a_0} \tag{3.3.40}$$

Significance of Transfer Function
Let $u(t) = \delta(t)$, the unit impulse function. In this case, $u(s) = 1$ and Equation 3.3.32 yields $x(s) = G(s)$, where $x(s)$ is the Laplace transform of the response to the unit impulse input under zero initial conditions, known as the unit impulse response. In other words, the transfer function is the Laplace transform of the unit impulse response of the system. Taking the inverse Laplace transform of Equation 3.3.32,

$$x(t) = \int_0^t g(t-\tau)u(\tau)d\tau \tag{3.3.41}$$

Equation 3.3.41 is the same convolution integral that has been derived in Section 3.2.3.

Poles and Zeros of Transfer Function
A transfer function $G(s)$ can be written as a ratio of two polynomials $N(s)$ and $D(s)$:

$$G(s) = \frac{N(s)}{D(s)} \tag{3.3.42}$$

Poles are defined as the roots of the denominator polynomial equation $D(s) = 0$, whereas zeros are defined as the roots of the numerator polynomial equation $N(s) = 0$ In general, $D(s) = 0$ is also the characteristic equation. For example, poles are the roots of $s^2 + a_1 s + a_0 = 0$ for both direct force and base excitation transfer functions (Equations 3.3.36 and 3.3.40). They are the same as the characteristic roots defined in Chapter 1. There is no finite zero for the direct force excitation transfer function (Equation 3.3.36). However, there is a real zero at $-b_0/b_1$ for the base excitation transfer function (Equation 3.3.40).

The transfer function is called **stable** when all poles are located in the left half of the complex s-plane (Figure 1.6.1).

Frequency Response Function

The frequency response function $G(j\omega)$ is obtained by substituting $s = j\omega$ into the transfer function $G(s)$, where ω is the frequency. The frequency response function for the transfer function in Equation 3.3.40 is given by

$$G(j\omega) = \frac{b_1 j\omega + b_0}{(j\omega)^2 + a_1 j\omega + a_0} = \frac{b_0 + jb_1\omega}{(a_0 - \omega^2) + ja_1\omega} \tag{3.3.43}$$

For a specified frequency ω, $G(j\omega)$ is a complex number, for which the magnitude and the phase (or angle) can be expressed as

$$\left|G(j\omega)\right| = \frac{\sqrt{b_0^2 + b_1^2\omega^2}}{\sqrt{(a_0 - \omega^2)^2 + (a_1\omega)^2}} \tag{3.3.44}$$

$$\angle G(j\omega) = \tan^{-1}\frac{b_1\omega}{b_0} - \tan^{-1}\frac{a_1\omega}{(a_0 - \omega^2)} \tag{3.3.45}$$

When the magnitude in decibels, dB ($20\log_{10}\left|G(j\omega)\right|$) and the phase are plotted as a function of the frequency ω, the resulting diagrams are known as Bode plots. The MATLAB routine "**bode**" can be readily used to make Bode plots.

For a stable $G(s)$, the magnitude and the phase of the frequency response function have the following physical significance:

a. The magnitude $\left|G(j\omega)\right|$ is the ratio of the amplitudes of the steady-state output and the sinusoidal input with frequency ω.
b. The angle $\angle G(j\omega)$ is the phase difference between the steady-state sinusoidal output and the sinusoidal input with frequency ω.

Example 3.3.4: Frequency Response Function for Direct Excitation Problem

The frequency response function for the direct force transfer function (Equation 3.3.36) is as follows:

$$G(j\omega) = \frac{b_0}{(j\omega)^2 + ja_1\omega + a_0} \tag{3.3.46}$$

For a linear differential equation with characteristic roots in the left half of the complex plane, the steady-state response $x(t)$ is sinusoidal when the input $u(t)$ is sinusoidal. Let

$$u(t) = u_0 \sin \omega t \tag{3.3.47}$$

Then the steady-state response $x(t)$ will be expressed as

$$x(t) = A \sin(\omega t - \phi) \tag{3.3.48}$$

where ϕ is the phase lag between the steady-state response $x(t)$ and the input $u(t)$. The magnitude and the phase of the complex number $G(j\omega)$ are related to the steady-state amplitude and the phase as follows:

$$\left|G(j\omega)\right| = \frac{A}{u_0} \tag{3.3.49}$$

and

$$\angle G(j\omega) = -\phi \tag{3.3.50}$$

For the direct force excitation problem, substitution of the parameters in Equations 3.3.34 and 3.3.35 yields

$$G(j\omega) = \frac{1}{(k_{eq} - m_{eq}\omega^2) + jc_{eq}\omega} \tag{3.3.51}$$

From Equation 3.3.49,

$$\frac{A}{u_0} = \left|G(j\omega)\right| = \frac{1}{\sqrt{\left(k_{eq} - m_{eq}\omega^2\right)^2 + \left(c_{eq}\omega\right)^2}} \tag{3.3.52}$$

From Equation 3.3.50,

$$\phi = -\angle G(j\omega) = \tan^{-1} \frac{c_{eq}\omega}{k_{eq} - m_{eq}\omega^2} \tag{3.3.53}$$

Note that Equations 3.3.52 and 3.3.53 are same as Equations 2.3.19 and 2.3.20.

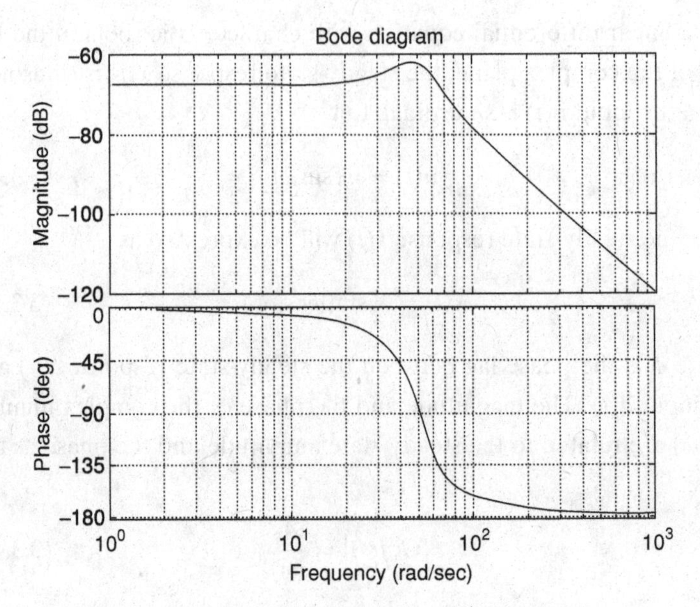

Figure 3.3.5 Bode plot for direct excitation problem

Example 3.3.5: Bode Plot via MATLAB
Consider the Example 3.3.2 for which $m_{eq} = 1$ kg, $k_{eq} = 2,500$ N/m, and $c_{eq} = 25$ N-sec/meter. Construct Bode plots for this system.

Using Equation 3.3.36, the transfer function of the system is

$$G(s) = \frac{1}{s^2 + 25s + 2,500} \tag{3.3.54}$$

The Bode plots are shown in Figure 3.3.5. The MATLAB program is given in Program 3.3.1.

```
MATLAB Program 3.3.1: Bode Plots
num=1;
den=[1 25 2500];
sysG=tf(num,den);
bode(sysG)
grid
```

EXERCISE PROBLEMS

P3.1 Find the Fourier series expansions of the periodic functions shown in Figures P3.1a–c.

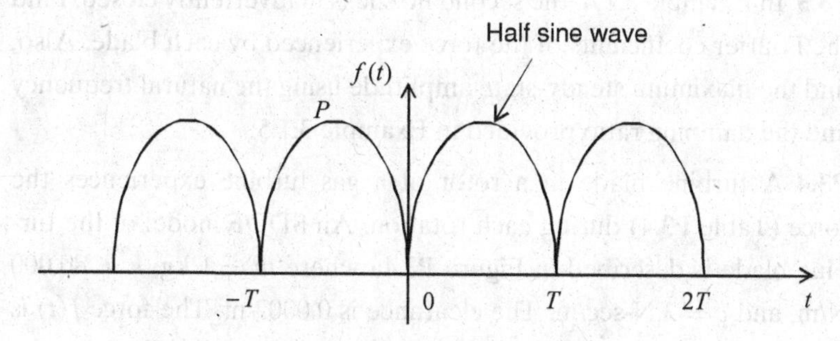

Figure P3.1a Periodic function with half sine waves

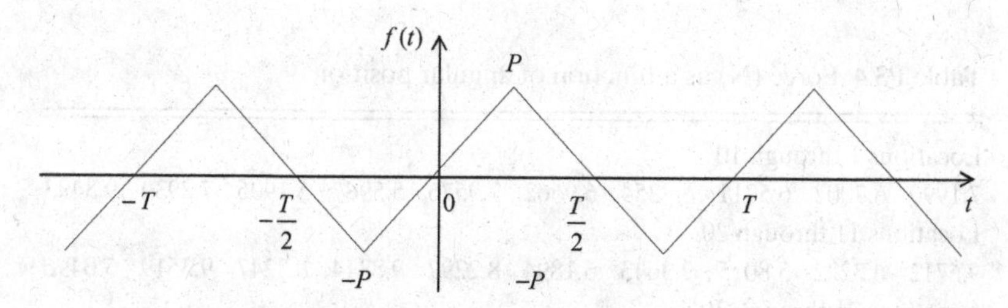

Figure P3.1b Periodic function with triangular waves

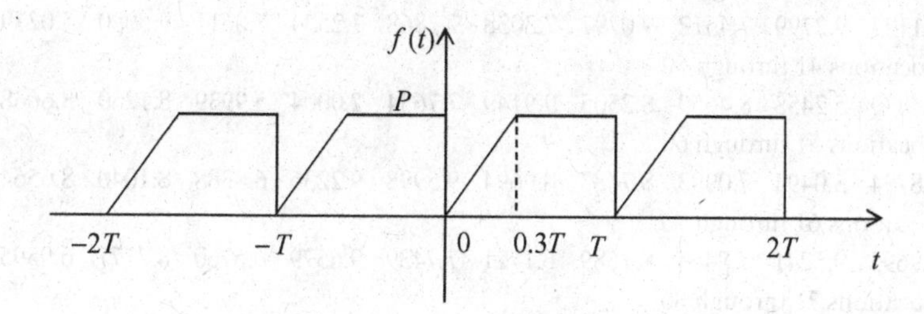

Figure P3.1c A periodic function with ramps

P3.2 Consider a spring–mass–damper system with mass = 1.2 kg and damping ratio = 0.05, which is subjected to the periodic force shown in Figures P3.1a–c. Let the natural frequency of the system be π/T.

a. Determine the steady-state response for each forcing function.

b. Verify your analytical results by numerical integration of the differential equations.

P3.3 In Example 3.1.4, the second nozzle is inadvertently closed. Find the Fourier coefficients of the force experienced by each blade. Also, find the maximum steady-state amplitude using the natural frequency and the damping ratio provided in Example 3.1.5.

P3.4 A turbine blade in a rotor of a gas turbine experiences the force (Table P3.4) during each rotation. An SDOF model of the turbine blade is described in Figure P3.4a where $m = 1$ kg, $k = 80{,}000$ N/m, and $c = 3$ N-sec/m. The clearance is 0.0002 m. The force $f(t)$ is described by data in Table P3.4, where the locations of the pressure measurements are shown in P3.4b.

Table P3.4 Force (N) as a function of angular position

Locations 1 through 10									
7.1990	6.7002	6.5711	6.8254	6.9662	7.9576	5.5987	5.1906	7.2930	9.3493
Locations 11 through 20									
9.6712	6.3222	5.8015	9.3643	6.1894	8.2292	9.8344	8.3247	9.3519	5.0496
Locations 21 through 30									
5.6850	9.0938	7.1508	9.4516	8.6745	8.4366	6.7306	5.8302	5.7781	5.9556
Locations 31 through 40									
7.1123	9.2799	7.4512	9.0797	7.3038	7.2868	7.2534	7.0611	9.5080	5.0279
Locations 41 through 50									
6.4870	5.2458	8.4659	8.2505	9.9149	7.7634	7.0004	5.9939	8.1260	8.6668
Locations 51 through 60									
6.8794	5.0494	7.0993	8.7683	8.9694	9.5998	9.2236	6.8388	8.1040	8.6564
Locations 61 through 70									
5.9695	9.5241	7.8460	8.1589	6.1721	7.7439	9.6579	6.6760	8.2777	6.9595
Locations 71 through 80									
8.1366	8.4954	6.9859	7.0681	8.2761	9.1879	6.8580	7.1263	7.9733	7.8287
Locations 81 through 90									
8.5827	7.5566	8.8820	7.4467	5.9295	8.5032	9.9135	9.0332	8.5178	7.4248
Locations 91 through 99									
5.5731	8.3243	6.8269	5.7002	7.8339	9.1150	8.3697	9.9972	9.8082	

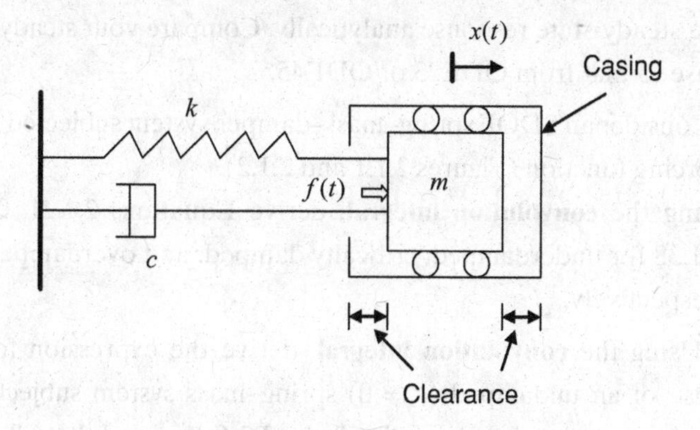

Figure P3.4a Model of a turbine blade

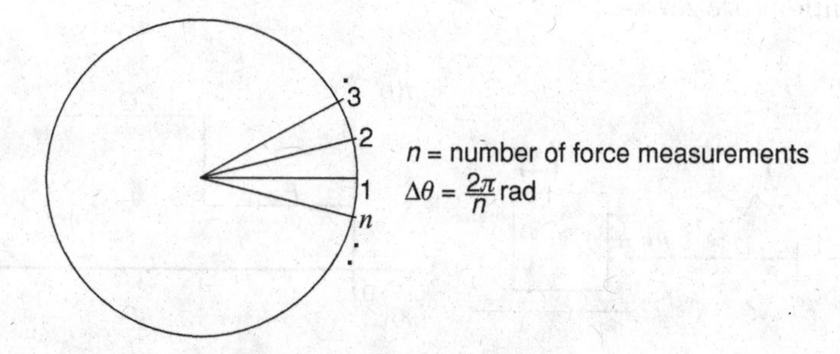

Figure P3.4b Locations of force measurements

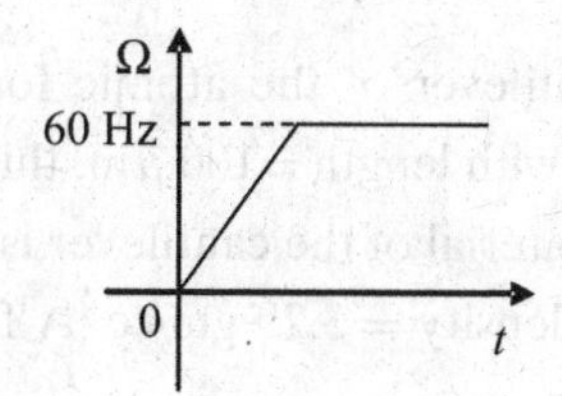

Figure P3.4c Angular velocity profile

The gas turbine is stationary. You are in charge of starting the engine and operating the gas turbine at a steady rotational speed equal to 60 Hz. Using the velocity profile (Figure P3.4c), determine the smallest angular acceleration of the turbine so that the blade never hits the casing.

At the steady speed of 60 Hz, compute the Fourier coefficients and the steady-state response analytically. Compare your steady-state response to that from ODE23 or ODE45.

P3.5 Consider an SDOF spring–mass–damper system subjected to the step forcing function (Figures 2.1.1 and 2.1.2).

Using the **convolution integral**, derive Equations 2.1.21, 2.1.28, and 2.1.38 for underdamped, critically damped, and overdamped system, respectively.

P.3.6 Using the **convolution integral**, derive the expression for the response of an undamped ($c = 0$) spring–mass system subjected to the forcing function $f(t)$ shown in Figure P3.6. Assume that all initial conditions are zero.

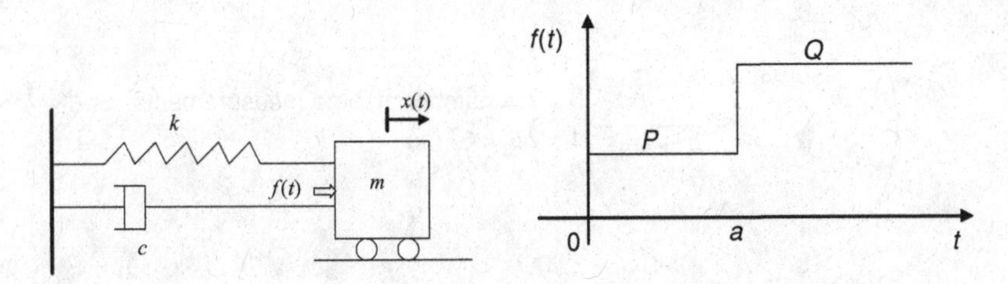

Figure P3.6 Spring–mass system subjected to staircase function with two steps

P3.7 Consider the cantilever of the atomic force microscope (Binning and Quate, 1986) with length $=100\ \mu$m, thickness $= 0.8\ \mu$m, and width $= 20\ \mu$m. The material of the cantilever is silicon nitride having $E = 310 \times 10^9$Pa and density $= 3.29$ gm/cc. A force $f(t)$ is applied at the tip of the cantilever.

The force $f(t)$ is as shown in Figure P3.7 for which $P = 1\ \mu$N, $Q = 0.5\ \mu$N, $a = 1$ sec, $b = 2$ sec, $d = 3$ sec, and $g = 5$ sec.

a. On the basis of an equivalent SDOF model, find the unit impulse response function for zero initial conditions. Assume the damping ratio $= 0.001$.

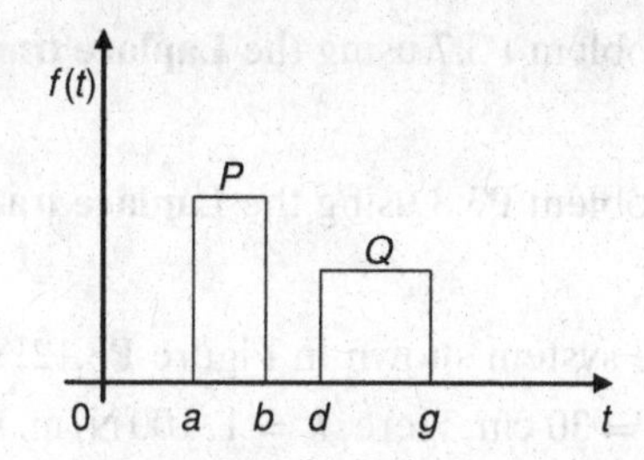

Figure P3.7 Forcing function for atomic force microscope problem

b. Using the **convolution integral**, determine $x(t)$ for $t > 0$. Assume that $x(0) = 0.5\ \mu\text{m}$, and $dx/dt = 0$.

Using MATLAB, plot $x(t)$ versus t.

P3.8 The base displacement $y(t)$ in Figure P3.8a is described in Figure P3.8b. Using the convolution integral, determine the response $x(t)$.

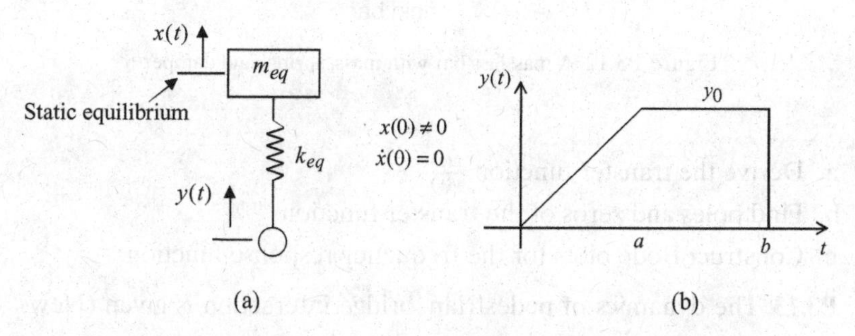

Figure P3.8 (a) Base excitation problem (b) Specified base displacement

P3.9 Consider an SDOF spring–mass–damper system subjected to the step forcing function (Figures 2.1.1 and 2.1.2).

Using the **Laplace transformation technique**, derive Equations 2.1.21, 2.1.28, and 2.1.38 for underdamped, critically damped, and overdamped systems, respectively.

P3.10 Solve the problem P3.7 using the **Laplace transformation technique**.

P3.11 Solve the problem P3.8 using the **Laplace transformation technique**.

P3.12 Consider the system shown in Figure P3.12, where $a = 25$ cm, $\ell_1 = 50$ cm, and $\ell_2 = 30$ cm. Here, $k = 1,100\,\text{N/m}$, $m = 2$ kg, and the damping ratio is 0.1.

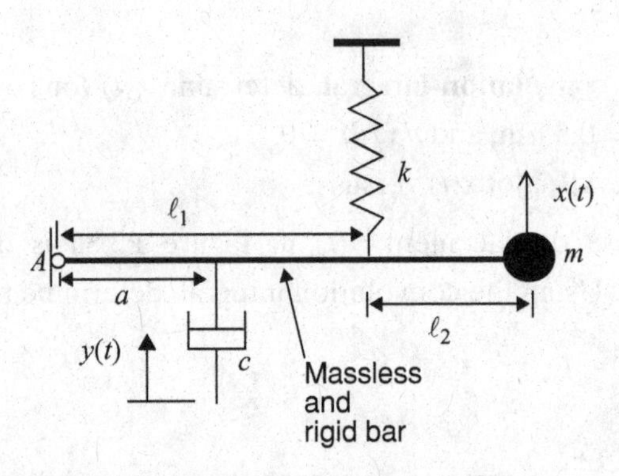

Figure P3.12 A massless bar with mass, spring, and damper

a. Derive the transfer function $\frac{x(s)}{y(s)}$.
b. Find poles and zeros of the transfer function.
c. Construct Bode plots for the frequency response function.

P3.13 The dynamics of pedestrian–bridge interaction is given (Newland, 2004) by

$$M\ddot{y} + Ky(t) + m\alpha\beta\ddot{y}(t - \Delta) + 2\xi\sqrt{KM}\dot{y} = -m\beta\ddot{x}(t)$$

where M : Mass of the bridge, K : Stiffness of the bridge, ξ : damping ratio, Y : displacement of the pavement, x : movement of the center of mass caused by pedestrian walking, m : mass of the pedestrian, and Δ : time lag. α and β are constants.

a. Derive the transfer function $G(s) = \frac{y(s)}{x(s)}$

b. Plot $|\alpha G(j\omega)|$ versus frequency ratio $\frac{\omega}{\omega_n}$ where ω : input frequency, $\omega_n = \sqrt{\frac{K}{M}}$ for $\omega\Delta = 0, -\frac{\pi}{2}, -\pi$ and $\frac{\pi}{2}$. Assume that $\frac{\alpha\beta m}{M} = 0.1$ and $\xi = 0.1$.

4

VIBRATION OF TWO-DEGREE-OF-FREEDOM-SYSTEMS

For a two-degree-of-freedom (2DOF) system, the number of independent second-order differential equations is two. With respect to a vector composed of the displacements associated with each degree of freedom, these two differential equations are represented as a single equation. In this vector equation, the coefficients of acceleration, velocity, and displacement vectors are defined as the mass matrix, the damping matrix, and the stiffness matrix, respectively. Next, the method to compute the natural frequencies and the modal vectors, also known as mode shapes, is presented. The number of natural frequencies equals the number of degrees of freedom, which is two. Unlike in a single-degree-of-freedom (SDOF) system, there is a mode shape associated with each natural frequency. Next, free and forced vibration of both undamped and damped 2DOF systems are analyzed. Using these techniques, vibration absorbers are designed next. A vibration absorber consists of a spring, a mass, and a damper, and is attached to an SDOF main system experiencing vibration problems. After the addition of a vibration absorber to an SDOF main system, the complete system has two degrees of freedom. Last, the response is represented as a linear combination of the modal vectors, and it is shown that the response of each mode of vibration is equivalent to the response of an SDOF system.

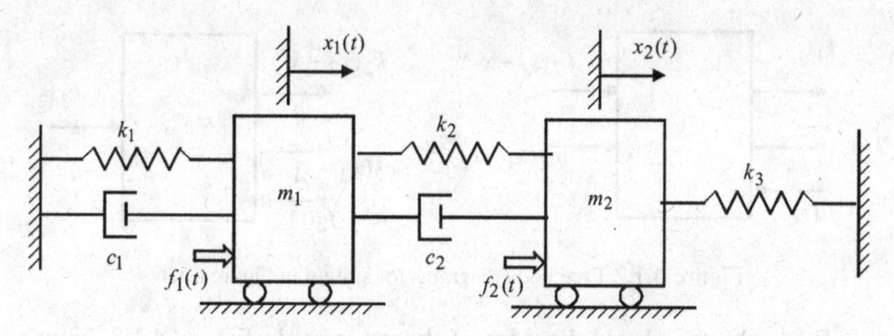

Figure 4.1.1 Two-degree-of-freedom system

4.1 MASS, STIFFNESS, AND DAMPING MATRICES

Let $x_1(t)$ and $x_2(t)$ be the displacements (linear or angular) associated with two degrees of freedom. Then, displacement $\mathbf{x}(t)$, velocity $\dot{\mathbf{x}}(t)$, and acceleration $\ddot{\mathbf{x}}(t)$ vectors are defined as follows:

$$\mathbf{x}(t) = \begin{bmatrix} x_1(t) \\ x_2(t) \end{bmatrix}; \ \dot{\mathbf{x}}(t) = \begin{bmatrix} \dot{x}_1(t) \\ \dot{x}_2(t) \end{bmatrix}; \quad \text{and} \quad \ddot{\mathbf{x}}(t) = \begin{bmatrix} \ddot{x}_1(t) \\ \ddot{x}_2(t) \end{bmatrix} \tag{4.1.1a–c}$$

Let $f_1(t)$ and $f_2(t)$ be the forces (or the torques) associated with each degree of freedom. Then, the force vector $\mathbf{f}(t)$ is described as follows:

$$\mathbf{f}(t) = \begin{bmatrix} f_1(t) \\ f_2(t) \end{bmatrix} \tag{4.1.2}$$

The dynamics of a 2DOF system is governed by a set of two coupled second-order differential equations, which is written in the matrix form as follows:

$$M\ddot{\mathbf{x}} + C\dot{\mathbf{x}} + K\mathbf{x}(t) = \mathbf{f}(t) \tag{4.1.3}$$

where the matrices M, C, and K are known as *mass*, *damping*, and *stiffness* matrices, respectively.

Example 4.1.1: Two Degree of Freedom System

Consider the 2DOF system shown in Figure 4.1.1. The free body diagram of each mass is shown in Figure 4.1.2.

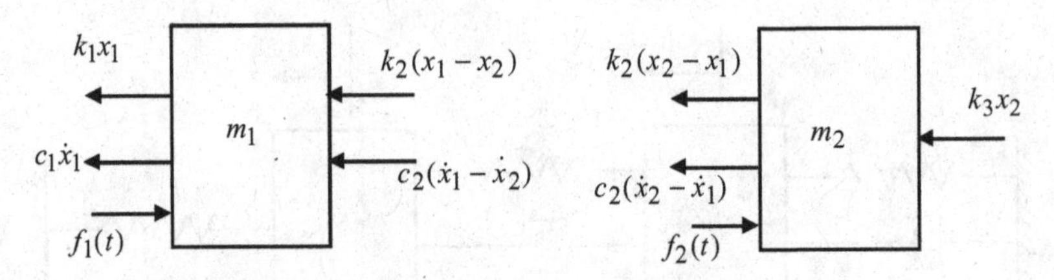

Figure 4.1.2 Free body diagrams for system in Figure 4.1.1

From the free body diagram of the mass m_1 in Figure 4.1.2, Newton's second law of motion yields

$$f_1(t) - k_1x_1 - c_1\dot{x}_1 - k_2(x_1 - x_2) - c_2(\dot{x}_1 - \dot{x}_2) = m_1\ddot{x}_1 \quad (4.1.4)$$

From the free body diagram of the mass m_2 in Figure 4.1.2, Newton's second law of motion yields

$$f_2(t) - k_3x_2 - k_2(x_2 - x_1) - c_2(\dot{x}_2 - \dot{x}_1) = m_2\ddot{x}_2 \quad (4.1.5)$$

Equations 4.1.4 and 4.1.5 are rearranged as follows:

$$m_1\ddot{x}_1 + (c_1 + c_2)\dot{x}_1 - c_2\dot{x}_2 + (k_1 + k_2)x_1 - k_2x_2 = f_1(t) \quad (4.1.6)$$

$$m_2\ddot{x}_2 - c_2\dot{x}_1 + c_2\dot{x}_2 - k_2x_1 + (k_2 + k_3)x_2 = f_2(t) \quad (4.1.7)$$

Putting Equations 4.1.6–4.1.7 in the matrix form,

$$\begin{bmatrix} m_1 & 0 \\ 0 & m_2 \end{bmatrix}\begin{bmatrix} \ddot{x}_1 \\ \ddot{x}_2 \end{bmatrix} + \begin{bmatrix} c_1 + c_2 & -c_2 \\ -c_2 & c_2 \end{bmatrix}\begin{bmatrix} \dot{x}_1 \\ \dot{x}_2 \end{bmatrix} + \begin{bmatrix} k_1 + k_2 & -k_2 \\ -k_2 & k_2 + k_3 \end{bmatrix}\begin{bmatrix} x_1 \\ x_2 \end{bmatrix} = \begin{bmatrix} f_1(t) \\ f_2(t) \end{bmatrix} \quad (4.1.8)$$

Equation 4.1.8 has the form of Equation 4.1.3,

$$M\ddot{\mathbf{x}} + C\dot{\mathbf{x}} + K\mathbf{x} = \mathbf{f}(t) \quad (4.1.9)$$

where

$$\mathbf{x}(t) = \begin{bmatrix} x_1(t) \\ x_2(t) \end{bmatrix}; \ \mathbf{f}(t) = \begin{bmatrix} f_1(t) \\ f_2(t) \end{bmatrix} \quad (4.1.10)$$

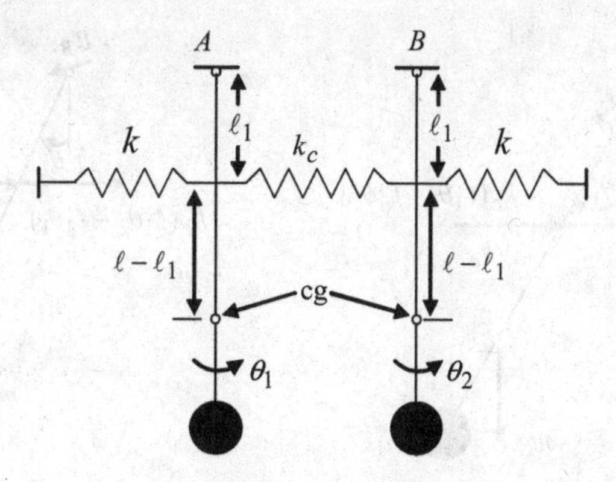

Figure 4.1.3 Double pendulum

$$M = \begin{bmatrix} m_1 & 0 \\ 0 & m_2 \end{bmatrix}, \; C = \begin{bmatrix} c_1 + c_2 & -c_2 \\ -c_2 & c_2 \end{bmatrix}, \text{ and } K = \begin{bmatrix} k_1 + k_2 & -k_2 \\ -k_2 & k_2 + k_3 \end{bmatrix} \tag{4.1.11}$$

The matrices M, C, and K are *mass*, *damping*, and *stiffness* matrices, respectively.

Example 4.1.2: Double Pendulum

Consider a double pendulum as shown in Figure 4.1.3. The mass and the mass-moment of inertia about the support point of each pendulum are m and Γ, with the center of mass located at a distance ℓ from the support joint. Derive the differential equations of motion and obtain the mass and stiffness matrices.

The free body diagram of each pendulum is shown in Figure 4.1.4, where R_A and R_B are the unknown reaction forces at the support joints A and B.

Taking moments about the support joint A,

$$-k\ell_1\theta_1\ell_1 - k_c(\ell_1\theta_1 - \ell_1\theta_2)\ell_1 - mg\ell\theta_1 = \Gamma\ddot{\theta}_1 \tag{4.1.12}$$

Taking moments about the support joint B,

$$-k\ell_1\theta_2\ell_1 - k_c(\ell_1\theta_2 - \ell_1\theta_1)\ell_1 - mg\ell\theta_2 = \Gamma\ddot{\theta}_2 \tag{4.1.13}$$

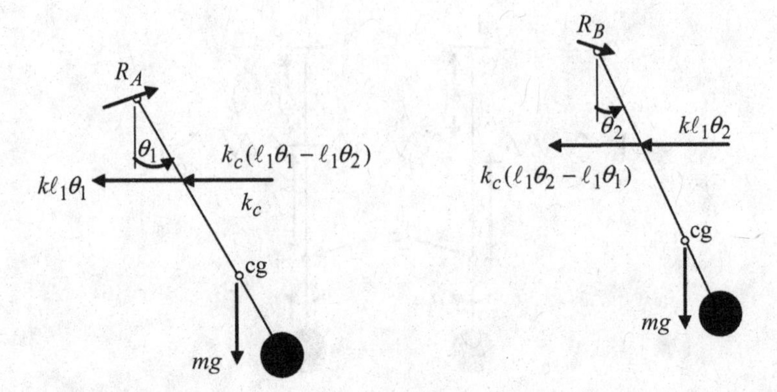

Figure 4.1.4 Free body diagram from each pendulum

Equations 4.1.12 and 4.1.13 can be put in the matrix form as

$$\begin{bmatrix} \Gamma & 0 \\ 0 & \Gamma \end{bmatrix} \begin{bmatrix} \ddot{\theta}_1 \\ \ddot{\theta}_2 \end{bmatrix} + \begin{bmatrix} (k+k_c)\ell_1^2 + mg\ell & -k_c\ell_1^2 \\ -k_c\ell_1^2 & (k+k_c)\ell_1^2 + mg\ell \end{bmatrix} \begin{bmatrix} \theta_1 \\ \theta_2 \end{bmatrix} = \begin{bmatrix} 0 \\ 0 \end{bmatrix} \tag{4.1.14}$$

Equation 4.1.14 yields mass (M) and stiffness (K) matrices as

$$M = \begin{bmatrix} \Gamma & 0 \\ 0 & \Gamma \end{bmatrix} \tag{4.1.15}$$

and

$$K = \begin{bmatrix} (k+k_c)\ell_1^2 + mg\ell & -k_c\ell_1^2 \\ -k_c\ell_1^2 & (k+k_c)\ell_1^2 + mg\ell \end{bmatrix} \tag{4.1.16}$$

Example 4.1.3: Combined Translational and Rotational Motion
Consider a rigid bar (Figure 4.1.5) with the mass m and the mass-moment of inertia I_c about the center of mass C, which is located at distances ℓ_1 and ℓ_2 from left and right ends, respectively. Further, the bar is supported by springs with stiffnesses k_1 and k_2 at its left and right ends, respectively. Derive the differential equations of motion and obtain the mass and stiffness matrices.

Note: This rigid bar and springs can be viewed as an automobile chassis and tires. To help with your imagination, dotted lines are drawn.

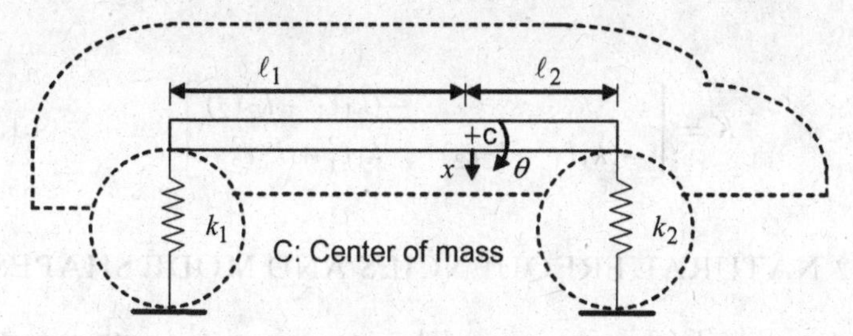

Figure 4.1.5 A rigid bar supported by springs at both ends

The free body diagram of a rigid bar is shown in Figure 4.1.6, where x is the displacement of the center of mass C and θ is the clockwise angular displacement of the bar.

Summing all forces along x-directions,

$$-k_1(x - \ell_1\theta) - k_2(x + \ell_2\theta) = m\ddot{x} \tag{4.1.17}$$

Taking moments about the center of mass C,

$$k_1(x - \ell_1\theta)\ell_1 - k_2(x + \ell_2\theta)\ell_2 = I_c\ddot{\theta} \tag{4.1.18}$$

Equations 4.1.17 and 4.1.18 are represented in the matrix form as

$$\begin{bmatrix} m & 0 \\ 0 & I_c \end{bmatrix}\begin{bmatrix} \ddot{x} \\ \ddot{\theta} \end{bmatrix} + \begin{bmatrix} k_1 + k_2 & -(k_1\ell_1 - k_2\ell_2) \\ -(k_1\ell_1 - k_2\ell_2) & k_1\ell_1^2 + k_2\ell_2^2 \end{bmatrix}\begin{bmatrix} x \\ \theta \end{bmatrix} = \begin{bmatrix} 0 \\ 0 \end{bmatrix} \tag{4.1.19}$$

Equation 4.1.19 yields mass (M) and stiffness (K) matrices as

$$M = \begin{bmatrix} m & 0 \\ 0 & I_c \end{bmatrix} \tag{4.1.20}$$

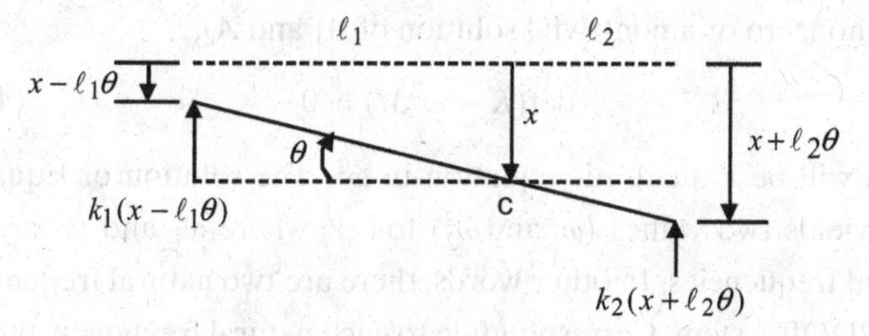

Figure 4.1.6 Free body diagram of rigid bar supported by springs

and

$$K = \begin{bmatrix} k_1 + k_2 & -(k_1\ell_1 - k_2\ell_2) \\ -(k_1\ell_1 - k_2\ell_2) & k_1\ell_1^2 + k_2\ell_2^2 \end{bmatrix} \quad (4.1.21)$$

4.2 NATURAL FREQUENCIES AND MODE SHAPES

There is a mode shape or a modal vector associated with a natural frequency. The method to compute the natural frequencies and the mode shapes is as follows.

Ignoring damping and external force terms, Equation 4.1.3 can be written as

$$M\ddot{\mathbf{x}} + K\mathbf{x} = \mathbf{0} \quad (4.2.1)$$

Let

$$\mathbf{x}(t) = \begin{bmatrix} A_1 \\ A_2 \end{bmatrix} \sin(\omega t + \phi) \quad (4.2.2)$$

where amplitudes A_1, A_2, and the phase ϕ are to be determined. Differentiating Equation 4.2.2 twice with respect to time,

$$\ddot{\mathbf{x}} = -\omega^2 \begin{bmatrix} A_1 \\ A_2 \end{bmatrix} \sin(\omega t + \phi) \quad (4.2.3)$$

Substituting Equations 4.2.2 and 4.2.3 into Equation 4.2.1,

$$(K - \omega^2 M) \begin{bmatrix} A_1 \\ A_2 \end{bmatrix} = \begin{bmatrix} 0 \\ 0 \end{bmatrix} \quad (4.2.4)$$

For a nonzero or a nontrivial solution of A_1 and A_2,

$$\det(K - \omega^2 M) = 0 \quad (4.2.5)$$

which will be a quadratic equation in ω^2. The solution of Equation 4.2.5 yields two values (ω_1^2 and ω_2^2) for ω^2 where ω_1 and ω_2 are the natural frequencies. In other words, there are two natural frequencies for a 2DOF system. Corresponding to each natural frequency, there is a mode shape which is obtained from Equation 4.2.4 after substituting $\omega = \omega_1$ or $\omega = \omega_2$, that is,

$$P_i \begin{bmatrix} A_{1,i} \\ A_{2,i} \end{bmatrix} = \begin{bmatrix} 0 \\ 0 \end{bmatrix}; \ i = 1, 2 \tag{4.2.6}$$

where

$$P_i = (K - \omega_i^2 M) = \begin{bmatrix} p_{11,i} & p_{12,i} \\ p_{21,i} & p_{22,i} \end{bmatrix}; \ i = 1, 2 \tag{4.2.7}$$

Substituting Equation 4.2.7 into Equation 4.2.6, the following two linear equations are obtained:

$$p_{11,i}A_{1,i} + p_{12,i}A_{2,i} = 0 \tag{4.2.8}$$

$$p_{21,i}A_{1,i} + p_{22,i}A_{2,i} = 0 \tag{4.2.9}$$

Because $\det P_i = 0$, the rank (Strang, 1988) of the matrix P_i cannot be two, and both the rows of the matrix P_i are not independent, or equivalently Equations 4.2.8 and 4.2.9 are essentially the same. In other words, there is only one equation in two unknowns $A_{1,i}$ and $A_{2,i}$. Therefore, only the ratio of $A_{1,i}$ and $A_{2,i}$ can be found, for example, using Equation 4.2.8,

$$\frac{A_{2,i}}{A_{1,i}} = -\frac{p_{11,i}}{p_{12,i}} \tag{4.2.10}$$

One of the variables, that is, either $A_{1,i}$ or $A_{2,i}$, can be chosen arbitrarily. For example, let $A_{1,i} = 1$. Then,

$$A_{2,i} = -\frac{p_{11,i}}{p_{12,i}} \tag{4.2.11}$$

And the modal vector $\mathbf{v}_i$ corresponding to the natural frequency ω_i can be defined as

$$\mathbf{v}_i = \begin{bmatrix} A_{1,i} \\ A_{2,i} \end{bmatrix} = \begin{bmatrix} 1 \\ -\frac{p_{11,i}}{p_{12,i}} \end{bmatrix}; i = 1, 2 \tag{4.2.12}$$

This modal vector $\mathbf{v}_i$ is also defined as the **mode shape** corresponding to the natural frequency $\omega = \omega_i$.

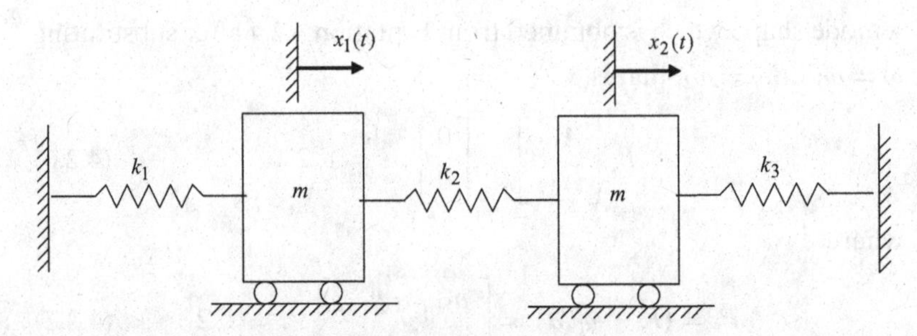

Figure 4.2.1 Undamped spring–mass system

Example 4.2.1: Natural Frequencies and Mode Shapes of the System in Figure 4.2.1

Consider the 2DOF system shown in Figure 4.1.1 with the following parameters: $c_1 = c_2 = 0$ and $m_1 = m_2 = m$. The resulting spring–mass system is shown in Figure 4.2.1. Determine the natural frequencies and the mode shapes for the two cases: $k_1 = k_2 = k_3 = k$, and $k_1 = k_2 = k$ and $k_3 = 2k$.

Case I: $k_1 = k_2 = k_3 = k$

Equation 4.1.8 yields the following mass and stiffness matrices:

$$M = \begin{bmatrix} m & 0 \\ 0 & m \end{bmatrix} \quad \text{and} \quad K = \begin{bmatrix} 2k & -k \\ -k & 2k \end{bmatrix} \tag{4.2.13}$$

Then,

$$K - \omega^2 M = \begin{bmatrix} 2k - \omega^2 m & -k \\ -k & 2k - \omega^2 m \end{bmatrix} \tag{4.2.14}$$

Therefore, from Equation 4.2.14,

$$\det(K - \omega^2 M) = (2k - \omega^2 m)^2 - k^2 = (k - \omega^2 m)(3k - \omega^2 m) = 0 \tag{4.2.15}$$

Solving Equation 4.2.15, the two natural frequencies are:

$$\omega_1 = \sqrt{\frac{k}{m}} \quad \text{and} \quad \omega_2 = \sqrt{\frac{3k}{m}} \tag{4.2.16}$$

Mode shape (modal vector) corresponding to the natural frequency ω_1
From Equation 4.2.7,

$$P_1 = (K - \omega_1^2 M) = \begin{bmatrix} p_{11,1} & p_{12,1} \\ p_{21,1} & p_{22,1} \end{bmatrix} = \begin{bmatrix} k & -k \\ -k & k \end{bmatrix} \tag{4.2.17}$$

Equation 4.2.10 yields the following eigenvector:

$$\mathbf{v}_1 = \begin{bmatrix} 1 \\ 1 \end{bmatrix} \tag{4.2.18}$$

Mode shape (modal vector) corresponding to the natural frequency ω_2
From Equation 4.2.7,

$$P_2 = (K - \omega_2^2 M) = \begin{bmatrix} p_{11,2} & p_{12,2} \\ p_{21,2} & p_{22,2} \end{bmatrix} = \begin{bmatrix} -k & -k \\ -k & -k \end{bmatrix} \tag{4.2.19}$$

Equation 4.2.10 yields the following eigenvector:

$$\mathbf{v}_2 = \begin{bmatrix} 1 \\ -1 \end{bmatrix} \tag{4.2.20}$$

Case II: $k_1 = k_2 = k$ and $k_3 = 2k$
Equation 4.1.8 yields the following mass and stiffness matrices:

$$M = \begin{bmatrix} m & 0 \\ 0 & m \end{bmatrix} \quad \text{and} \quad K = \begin{bmatrix} 2k & -k \\ -k & 3k \end{bmatrix} \tag{4.2.21}$$

Therefore, from Equation 4.2.21,

$$K - \omega^2 M = \begin{bmatrix} 2k - \omega^2 m & -k \\ -k & 3k - \omega^2 m \end{bmatrix} \tag{4.2.22}$$

From Equation 4.2.22,

$$\det(K - \omega^2 M) = (2k - \omega^2 m)(3k - \omega^2 m) - k^2 = 0 \tag{4.2.23}$$

or

$$m^2(\omega^2)^2 - 5km\omega^2 + 5k^2 = 0 \tag{4.2.24}$$

Solving Equation 4.2.24,

$$\omega^2 = \frac{5km \pm km\sqrt{5}}{2m^2} \tag{4.2.25}$$

Hence, the two natural frequencies are:

$$\omega_1 = \sqrt{\frac{1.382k}{m}} \quad \text{and} \quad \omega_2 = \sqrt{\frac{3.618k}{m}} \tag{4.2.26}$$

Mode shape (Modal vector) corresponding to the natural frequency ω_1

From Equation 4.2.7,

$$P_1 = (K - \omega_1^2 M) = \begin{bmatrix} p_{11,1} & p_{12,1} \\ p_{21,1} & p_{22,1} \end{bmatrix} = \begin{bmatrix} 0.618k & -k \\ -k & 1.618k \end{bmatrix} \tag{4.2.27}$$

Equation 4.2.10 yields the following eigenvector:

$$\mathbf{v}_1 = \begin{bmatrix} 1 \\ 0.618 \end{bmatrix} \tag{4.2.28}$$

Mode shape (Modal vector) corresponding to the natural frequency ω_2

From Equation 4.2.7,

$$P_2 = (K - \omega_2^2 M) = \begin{bmatrix} p_{11,2} & p_{12,2} \\ p_{21,2} & p_{22,2} \end{bmatrix} = \begin{bmatrix} -1.618k & -k \\ -k & -0.618k \end{bmatrix} \tag{4.2.29}$$

Equation 4.2.10 yields the following eigenvector:

$$\mathbf{v}_2 = \begin{bmatrix} 1 \\ -1.618 \end{bmatrix} \tag{4.2.30}$$

Example 4.2.2: Natural Frequencies and Mode Shapes of a Double Pendulum

Find the natural frequencies and the mode shapes of the double pendulum shown in Figure 4.1.3.

From Equations 4.1.12 and 4.1.13,

$$(K - \omega^2 M) = \begin{bmatrix} \alpha - \omega^2\Gamma & -\beta \\ -\beta & \alpha - \omega^2\Gamma \end{bmatrix} \tag{4.2.31}$$

where

$$\alpha = (k + k_c)\ell_1^2 + mg\ell \quad \text{and} \quad \beta = k_c\ell_1^2 \tag{4.2.32}$$

From Equation 4.2.31,

$$\det(K - \omega^2 M) = (\alpha - \omega^2\Gamma)^2 - \beta^2 = (\alpha + \beta - \omega^2\Gamma)(\alpha - \beta - \omega^2\Gamma) = 0 \tag{4.2.33}$$

Solving Equation 4.2.33, the two natural frequencies are

$$\omega_1^2 = \frac{\alpha - \beta}{\Gamma} \quad \text{and} \quad \omega_2^2 = \frac{\alpha + \beta}{\Gamma} \tag{4.2.34}$$

Mode shape (Modal vector) corresponding to the natural frequency ω_1
From Equation 4.2.7,

$$P_1 = (K - \omega_1^2 M) = \begin{bmatrix} p_{11,1} & p_{12,1} \\ p_{21,1} & p_{22,1} \end{bmatrix} = \begin{bmatrix} \beta & -\beta \\ -\beta & \beta \end{bmatrix} \tag{4.2.35}$$

Equation 4.2.10 yields the following modal vector:

$$\mathbf{v}_1 = \begin{bmatrix} 1 \\ 1 \end{bmatrix} \tag{4.2.36}$$

Mode shape (Modal vector) corresponding to the natural frequency ω_2
From Equation 4.2.7,

$$P_2 = (K - \omega_2^2 M) = \begin{bmatrix} p_{11,2} & p_{12,2} \\ p_{21,2} & p_{22,2} \end{bmatrix} = \begin{bmatrix} -\beta & -\beta \\ -\beta & -\beta \end{bmatrix} \tag{4.2.37}$$

Equation 4.2.10 yields the following modal vector:

$$\mathbf{v}_2 = \begin{bmatrix} 1 \\ -1 \end{bmatrix} \tag{4.2.38}$$

4.2.1 Eigenvalue/Eigenvector Interpretation

From Equation 4.2.4,

$$K \begin{bmatrix} A_1 \\ A_2 \end{bmatrix} = \omega^2 M \begin{bmatrix} A_1 \\ A_2 \end{bmatrix} \tag{4.2.39}$$

Pre-multiplying both sides of Equation 4.2.39 by M^{-1},

$$M^{-1}K\mathbf{v} = \omega^2\mathbf{v} \tag{4.2.40}$$

where

$$\mathbf{v} = \begin{bmatrix} A_1 \\ A_2 \end{bmatrix} \tag{4.2.41}$$

Therefore, the square of the natural frequency ω^2 and the corresponding modal vector $\mathbf{v}$ are the eigenvalue and the eigenvector (Boyce and DiPrima, 2005; Strang, 1988) of the matrix $M^{-1}K$, respectively. The eigenvalue/eigenvector interpretation is convenient from the computation point of view, as there are standard routines available, for example, the MATLAB routine "*eig.*"

4.3 FREE RESPONSE OF AN UNDAMPED 2DOF SYSTEM

In general, if the initial conditions are arbitrarily chosen, the free vibration will contain both the natural frequencies ω_1 and ω_2. On the basis of the assumed solution (Equation 4.2.2) and the fact that the response can be expressed as a linear combination of the modal vectors, the general solution can be written as

$$\mathbf{x}(t) = \alpha_1\mathbf{v}_1 \sin(\omega_1 t + \phi_1) + \alpha_2\mathbf{v}_2 \sin(\omega_2 t + \phi_2) \tag{4.3.1}$$

where $\mathbf{v}_1$ and $\mathbf{v}_2$ are the modal vectors (or mode shapes) corresponding to the natural frequencies ω_1 and ω_2, respectively. Constants α_1, α_2, ϕ_1, and ϕ_2 can be determined from the initial conditions $\mathbf{x}(0)$ and $\dot{\mathbf{x}}(0)$. From Equation 4.3.1,

$$\mathbf{x}(0) = \mathbf{v}_1\alpha_1 \sin\phi_1 + \mathbf{v}_2\alpha_2 \sin\phi_2 \tag{4.3.2}$$

or

$$\mathbf{x}(0) = Q \begin{bmatrix} \alpha_1 \sin\phi_1 \\ \alpha_2 \sin\phi_2 \end{bmatrix} \tag{4.3.3}$$

where Q is a 2×2 matrix defined as follows:

$$Q = \begin{bmatrix} \mathbf{v}_1 & \mathbf{v}_2 \end{bmatrix} \tag{4.3.4}$$

From Equation 4.3.3,

$$\begin{bmatrix} \alpha_1 \sin \phi_1 \\ \alpha_2 \sin \phi_2 \end{bmatrix} = Q^{-1}\mathbf{x}(0) \tag{4.3.5}$$

Differentiating Equation 4.3.1 with respect to time,

$$\dot{\mathbf{x}}(t) = \alpha_1 \mathbf{v}_1 \omega_1 \cos(\omega_1 t + \phi_1) + \alpha_2 \mathbf{v}_2 \omega_2 \cos(\omega_2 t + \phi_2) \tag{4.3.6}$$

Substituting $t = 0$ into Equation 4.3.6,

$$\dot{\mathbf{x}}(0) = \alpha_1 \mathbf{v}_1 \omega_1 \cos \phi_1 + \alpha_2 \mathbf{v}_2 \omega_2 \cos \phi_2 \tag{4.3.7}$$

Using the definition of Q given in Equation 4.3.4,

$$\dot{\mathbf{x}}(0) = Q \begin{bmatrix} \omega_1 \alpha_1 \cos \phi_1 \\ \omega_2 \alpha_2 \cos \phi_2 \end{bmatrix} \tag{4.3.8}$$

Therefore,

$$\begin{bmatrix} \omega_1 \alpha_1 \cos \phi_1 \\ \omega_2 \alpha_2 \cos \phi_2 \end{bmatrix} = Q^{-1}\dot{\mathbf{x}}(0) \tag{4.3.9}$$

In summary, $\alpha_1 \sin \phi_1$, $\alpha_2 \sin \phi_2$, $\alpha_1 \cos \phi_1$, and $\alpha_2 \cos \phi_2$ are first obtained from Equations 4.3.5 and 4.3.9. Then, α_1, α_2, ϕ_1, and ϕ_2 are determined.

As an example, let the initial conditions be as follows:

$$\mathbf{x}(0) = \chi \mathbf{v}_1 \quad \text{and} \quad \dot{\mathbf{x}}(0) = 0 \tag{4.3.10}$$

where χ is a constant.

Then, from Equations 4.3.2 and 4.3.9,

$$\begin{bmatrix} \alpha_1 \sin \phi_1 \\ \alpha_2 \sin \phi_2 \end{bmatrix} = \begin{bmatrix} \chi \\ 0 \end{bmatrix} \quad \text{and} \quad \begin{bmatrix} \alpha_1 \cos \phi_1 \\ \alpha_2 \cos \phi_2 \end{bmatrix} = \begin{bmatrix} 0 \\ 0 \end{bmatrix} \tag{4.3.11}$$

Therefore,

$$\alpha_1 = \chi,\ \phi_1 = \frac{\pi}{2}, \quad \text{and} \quad \alpha_2 = 0 \tag{4.3.12}$$

Hence, from Equations 4.3.1 and 4.3.12, the response would be

$$\mathbf{x}(t) = \chi \mathbf{v}_1 \sin\left(\omega_1 t + \frac{\pi}{2}\right) = \chi \mathbf{v}_1 \cos \omega_1 t \tag{4.3.13}$$

The response is purely sinusoidal with the frequency equal to the first natural frequency. This is due to the fact that the initial conditions match with the first modal vector and do not contain the second modal vector. If the initial conditions match the second modal vector, the free vibration will be purely sinusoidal with the frequency equal to the second natural frequency.

Example 4.3.1: Free Vibration of a Double Pendulum
Consider the double pendulum shown in Figure 4.1.3. Numerical values of the parameters are as follows: $m = 0.5\,\text{kg}$, $\Gamma = 0.6\,\text{kg} - \text{m}^2$, $k = k_c = 10,000\,\text{N/m}$, $\ell = 0.8\,\text{m}$, $\ell_1 = 0.3\,\text{m}$. Initial conditions are given as: $\theta_1(0) = 0.1\,\text{rad}$, $\theta_2(0) = 0.05\,\text{rad}$, $\dot{\theta}_1(0) = 10\,\text{rad/sec}$, $\dot{\theta}_2(0) = -15\,\text{rad/sec}$.

Determine the equation governing the free vibration.

Solution
From Equation 4.2.34,

$$\omega_1 = 38.814\,\text{rad/sec} \quad \text{and} \quad \omega_2 = 67.130\,\text{rad/sec}$$

From Equation 4.3.4,

$$Q = \begin{bmatrix} 1 & 1 \\ 1 & -1 \end{bmatrix}$$

Therefore,

$$Q^{-1} = \begin{bmatrix} 0.5 & 0.5 \\ 0.5 & -0.5 \end{bmatrix}$$

From Equation 4.3.5,

$$\begin{bmatrix} \alpha_1 \sin\phi_1 \\ \alpha_2 \sin\phi_2 \end{bmatrix} = Q^{-1}\mathbf{x}(0) = \begin{bmatrix} 0.075 \\ 0.025 \end{bmatrix} \text{rad}$$

From Equation 4.3.9,

$$\begin{bmatrix} \omega_1\alpha_1 \cos\phi_1 \\ \omega_2\alpha_2 \cos\phi_2 \end{bmatrix} = Q^{-1}\dot{\mathbf{x}}(0) = \begin{bmatrix} -2.5 \\ 12.5 \end{bmatrix} \text{rad/ sec}$$

Considering $\alpha_1 \sin\phi_1 = 0.075$ and $\omega_1\alpha_1 \cos\phi_1 = -2.5$,

$$\alpha_1 = 0.0989, \ \phi_1 = 2.2804 \,\text{rad}$$

Considering $\alpha_2 \sin\phi_2 = 0.025$ and $\omega_2\alpha_2 \cos\phi_2 = 12.5$,

$$\alpha_2 = 0.1879, \ \phi_2 = 0.1335 \,\text{rad}$$

From Equation 4.3.1,

$$\begin{bmatrix} \theta_1(t) \\ \theta_2(t) \end{bmatrix} = 0.0989 \begin{bmatrix} 1 \\ 1 \end{bmatrix} \sin(38.814t + 2.2804) + 0.1879 \begin{bmatrix} 1 \\ -1 \end{bmatrix} \sin(67.130t + 0.1335)$$

4.4 FORCED RESPONSE OF AN UNDAMPED 2DOF SYSTEM UNDER SINUSOIDAL EXCITATION

The differential equations of motion of an undamped 2DOF system can be represented as

$$M \begin{bmatrix} \ddot{x}_1 \\ \ddot{x}_2 \end{bmatrix} + K \begin{bmatrix} x_1 \\ x_2 \end{bmatrix} = \begin{bmatrix} f_{10} \\ f_{20} \end{bmatrix} \sin\omega t \tag{4.4.1}$$

where M and K are mass and stiffness matrices, respectively. The amplitudes of sinusoidal excitations are f_{10} and f_{20}.

Assume the forced response (particular integral) to be

$$\begin{bmatrix} x_1(t) \\ x_2(t) \end{bmatrix} = \begin{bmatrix} A_1 \\ A_2 \end{bmatrix} \sin\omega t \tag{4.4.2}$$

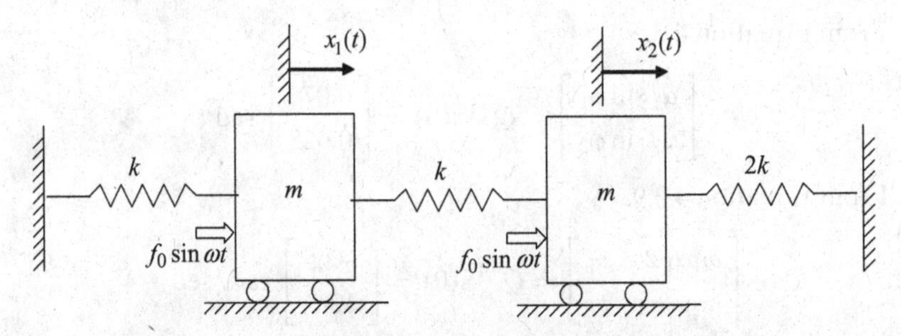

Figure 4.4.1 Two-degree-of-freedom system under sinusoidal excitation

Here, the amplitudes A_1 and A_2 are allowed to be negative. A negative value of the amplitude represents the phase angle equal to 180 degrees. Substituting Equation 4.4.2 into Equation 4.4.1, and equating the coefficients of $\sin \omega t$ on both sides,

$$(K - \omega^2 M)\begin{bmatrix} A_1 \\ A_2 \end{bmatrix} = \begin{bmatrix} f_{10} \\ f_{20} \end{bmatrix} \tag{4.4.3}$$

From Equation 4.4.3,

$$\begin{bmatrix} A_1 \\ A_2 \end{bmatrix} = (K - \omega^2 M)^{-1}\begin{bmatrix} f_{10} \\ f_{20} \end{bmatrix} \tag{4.4.4}$$

Example 4.4.1: Forced Response of a 2DOF System
Consider the 2DOF system shown in Figure 4.4.1 which is the same model as the one shown in Figure 4.1.1 with the following parameters: $c_1 = c_2 = 0$, $m_1 = m_2 = m$, $k_1 = k_2 = k$, $k_3 = 2k$, and $f_1(t) = f_2(t) = f_0 \sin \omega t$.

With respect to Equation 4.4.1,

$$f_{10} = f_{20} = f_0 \tag{4.4.5}$$

$$M = \begin{bmatrix} m & 0 \\ 0 & m \end{bmatrix} \quad \text{and} \quad K = \begin{bmatrix} 2k & -k \\ -k & 3k \end{bmatrix} \tag{4.4.6}$$

Therefore, from Equation 4.4.4,

$$\begin{bmatrix} A_1 \\ A_2 \end{bmatrix} = \begin{bmatrix} 2k - \omega^2 m & -k \\ -k & 3k - \omega^2 m \end{bmatrix}^{-1} \begin{bmatrix} f_0 \\ f_0 \end{bmatrix} \tag{4.4.7}$$

Equation 4.4.7 yields

$$\begin{bmatrix} A_1 \\ A_2 \end{bmatrix} = \frac{f_0}{(2k - \omega^2 m)(3k - \omega^2 m) - k^2} \begin{bmatrix} 4k - \omega^2 m \\ 3k - \omega^2 m \end{bmatrix} \tag{4.4.8}$$

After some algebra, the amplitudes of masses are found to be

$$\frac{A_1}{f_0/k} = \frac{0.8[1 - (\omega/\omega_{z_1})^2]}{[1 - (\omega/\omega_1)^2][1 - (\omega/\omega_2)^2]} \tag{4.4.9}$$

and

$$\frac{A_2}{f_0/k} = \frac{0.6[1 - (\omega/\omega_{z_2})^2]}{[1 - (\omega/\omega_1)^2][1 - (\omega/\omega_2)^2]} \tag{4.4.10}$$

where

$$\omega_{z_1} = \sqrt{\frac{4k}{m}}, \ \omega_{z_2} = \sqrt{\frac{3k}{m}}, \ \omega_1 = \sqrt{\frac{1.382k}{m}}, \quad \text{and} \quad \omega_2 = \sqrt{\frac{3.618k}{m}} \tag{4.4.11}$$

It should be noted that ω_1 and ω_2 are the natural frequencies (Equation 4.2.26). When $\omega = \omega_1$ or $\omega = \omega_2$, the amplitudes A_1 and A_2 are infinite (Figures 4.4.2 and 4.4.3). Furthermore, $A_1 = 0$ at $\omega = \omega_{z_1}$ and $A_2 = 0$ at $\omega = \omega_{z_2}$.

4.5 FREE VIBRATION OF A DAMPED 2DOF SYSTEM

The free vibration of a damped 2DOF system is governed by the following differential equations:

$$M\ddot{\mathbf{x}} + C\dot{\mathbf{x}} + K\mathbf{x} = 0 \tag{4.5.1}$$

where M, C, and K are mass, damping, and stiffness matrices, respectively. Here, the initial displacement $\mathbf{x}(0)$ and/or velocity $\dot{\mathbf{x}}(0)$ will be nonzero.

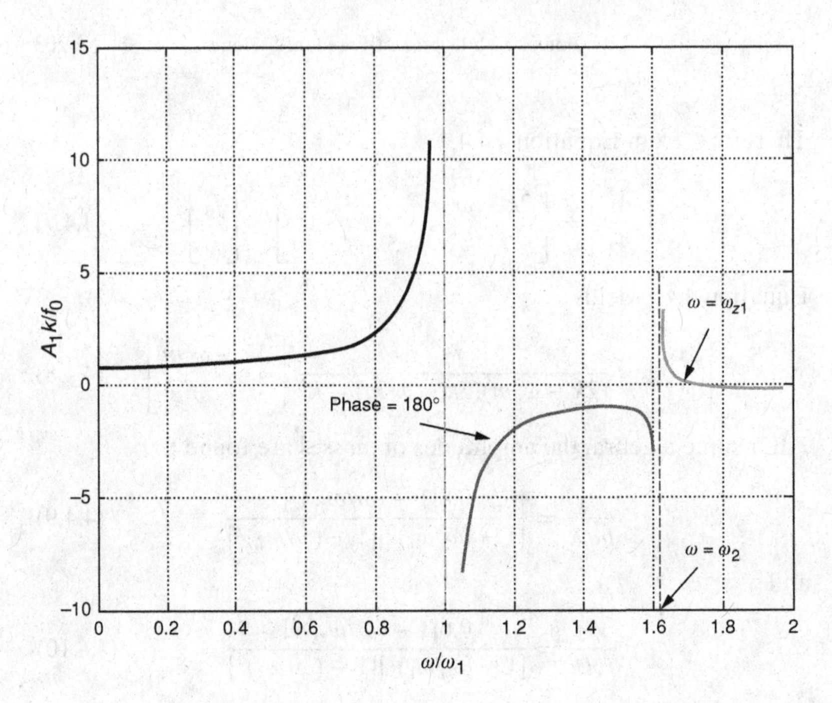

Figure 4.4.2 Amplitude of mass with displacement equal to x_1 in Figure 4.4.1

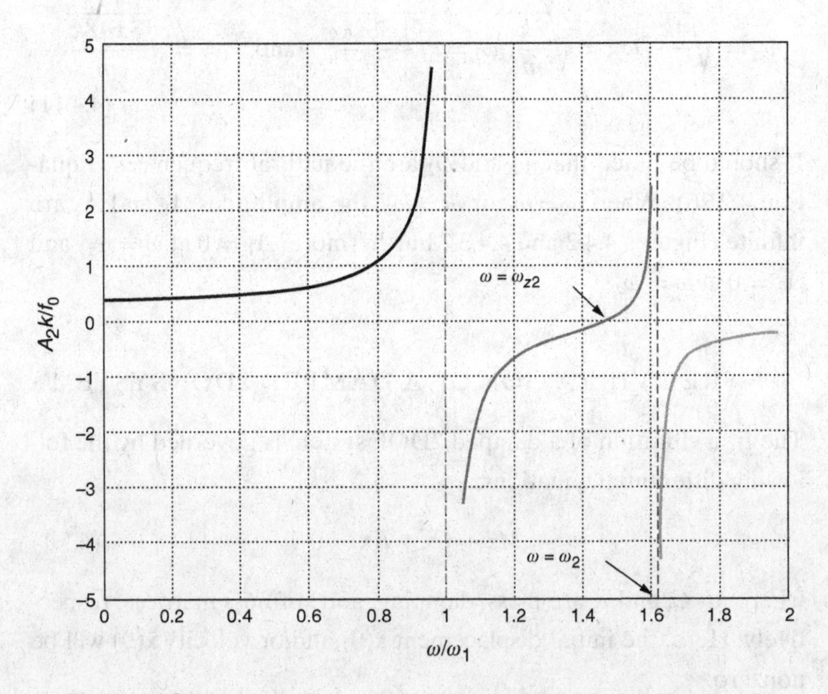

Figure 4.4.3 Amplitude of mass with displacement equal to x_2 in Figure 4.4.1

The solution can be written as

$$\mathbf{x}(t) = \begin{bmatrix} A_1 \\ A_2 \end{bmatrix} e^{st} \tag{4.5.2}$$

where s is to be determined. Substituting Equation 4.5.2 into Equation 4.5.1,

$$(Ms^2 + Cs + K) \begin{bmatrix} A_1 \\ A_2 \end{bmatrix} = \begin{bmatrix} 0 \\ 0 \end{bmatrix} \tag{4.5.3}$$

For a nontrivial solution of Equation 4.5.3,

$$\det(Ms^2 + Cs + K) = 0 \tag{4.5.4}$$

When $C = 0$,

$$\det(Ms^2 + K) = 0 \tag{4.5.5}$$

It can be shown that the roots of this equation are purely imaginary. In fact, by substituting $s = \pm j\omega$, Equation 4.5.5 reduces to

$$\det(-M\omega^2 + K) = 0 \tag{4.5.6}$$

which is the same as Equation 4.2.5. In the case of $C \neq 0$, the roots of Equation 4.5.4 can be obtained via the eigenvalue/eigenvector formulation as follows:

Define

$$\mathbf{y}_1 = \mathbf{x}(t) \quad \text{and} \quad \mathbf{y}_2 = \dot{\mathbf{x}}(t) \tag{4.5.7}$$

Then,

$$\dot{\mathbf{y}}_1 = \mathbf{y}_2 \tag{4.5.8}$$

and from Equation 4.5.1,

$$\dot{\mathbf{y}}_2 = \ddot{\mathbf{x}} = -M^{-1}K\mathbf{x}(t) - M^{-1}C\dot{\mathbf{x}}(t) = -M^{-1}K\mathbf{y}_1(t) - M^{-1}C\mathbf{y}_2(t) \tag{4.5.9}$$

Equations 4.5.8 and 4.5.9 can be written in the matrix form as

$$\dot{\mathbf{y}}(t) = R\mathbf{y}(t) \tag{4.5.10}$$

where

$$\mathbf{y}(t) = \begin{bmatrix} \mathbf{y}_1(t) \\ \mathbf{y}_2(t) \end{bmatrix} \tag{4.5.11}$$

and

$$R = \begin{bmatrix} \mathbf{0} & I \\ -M^{-1}K & -M^{-1}C \end{bmatrix} \tag{4.5.12}$$

It can be shown that the eigenvalues of the matrix R are the same as the roots of Equation 4.5.4. These eigenvalues can be complex or real. A set of complex conjugate eigenvalues $(-\sigma \pm j\omega)$ correspond to an underdamped SDOF system and can be written as

$$-\sigma \pm j\omega = -\xi\omega_n \pm \omega_n\sqrt{1-\xi^2 j} \tag{4.5.13}$$

where ξ and ω_n are the damping ratio and the undamped natural frequency respectively. Therefore,

$$\omega_n = \sqrt{\sigma^2 + \omega^2} \tag{4.5.14}$$

and

$$\xi = \frac{\sigma}{\sqrt{\sigma^2 + \omega^2}} \tag{4.5.15}$$

The undamped natural frequency (Equation 4.5.14) will be one of the roots of $\det(K - \omega^2 M)$.

The solution of Equation 4.5.10 can be described as

$$\mathbf{y}(t) = e^{Rt}\mathbf{y}(0) \tag{4.5.16}$$

where e^{Rt} is the matrix exponential described as

$$e^{Rt} = I + Rt + R^2\frac{t^2}{2!} + R^3\frac{t^3}{3!} + \cdots \tag{4.5.17}$$

and

$$R^2 = RR,\ R^3 = R^2 R, \ldots \tag{4.5.18}$$

In MATLAB, the matrix exponential can be calculated by using the command "***expm***." It should also be noted that

$$\mathbf{y}(0) = \begin{bmatrix} \mathbf{y}_1(0) \\ \mathbf{y}_2(0) \end{bmatrix} = \begin{bmatrix} \mathbf{x}(0) \\ \dot{\mathbf{x}}(0) \end{bmatrix} \tag{4.5.19}$$

That is, $\mathbf{y}(0)$ is composed of the initial displacement $\mathbf{x}(0)$ and the velocity $\dot{\mathbf{x}}(0)$ vectors.

Example 4.5.1: Free Vibration of a Damped 2DOF Freedom System
Consider the spring–mass–damper model in Figure 4.1.1 with $f_1(t) = f_2(t) = 0$. Let $m_1 = m_2 = 0.5\,\text{kg}$, $k_1 = k_2 = k_3 = 10,000\,\text{N/m}$, $c_1 = 10\,\text{N} - \text{sec/m}$, and $c_2 = 0$. Obtain $x_1(t)$ and $x_2(t)$ when the initial conditions are as follows: $x_1(0) = 1.5$, $x_2(0) = 0.5$, $\dot{x}_1(0) = 0$, and $\dot{x}_2(0) = 0$.

From Equation 4.1.8,

$$M = \begin{bmatrix} m_1 & 0 \\ 0 & m_2 \end{bmatrix} = \begin{bmatrix} 0.5 & 0 \\ 0 & 0.5 \end{bmatrix}$$

$$C = \begin{bmatrix} c_1 + c_2 & -c_2 \\ -c_2 & c_2 \end{bmatrix} = \begin{bmatrix} 10 & 0 \\ 0 & 0 \end{bmatrix}$$

and

$$K = \begin{bmatrix} 20,000 & -10,000 \\ -10,000 & 20,000 \end{bmatrix}$$

The eigenvalues of the matrix R (Equation 4.5.12) are presented in Table 4.1.

The response is obtained from the MATLAB program 4.5.1 and is presented in Figure 4.5.1.

Table 4.1 Eigenvalues of matrix R

Characteristic roots $(-\sigma \pm j\omega)$	ξ	ω_n (rad/sec)
$-5.01 \pm 141.51j$	0.0354	141.5987
$-4.99 \pm 244.59i$	0.0204	244.6422

MATLAB Program 4.5.1: Free Vibration of a Damped 2DOF System

```
%
clear all
close all
%
M=[0.5 0;0 0.5];
C=[10 0;0 0];
K=[20000 -10000;-10000 20000];
%
MI=inv(M);
%
R=[zeros(2,2) eye(2);-MI*K -MI*C]
%
ada=eig(R);
ogn1=abs(ada(1));
xi1=-real(ada(1))/ogn1;
ogn2=abs(ada(3));
xi2=-real(ada(3))/ogn2;
%
%Initial conditions
y0=[1 0.5 0 0]';
%
T=2*pi/ogn1;
delt=T/40;
t=-delt;
```

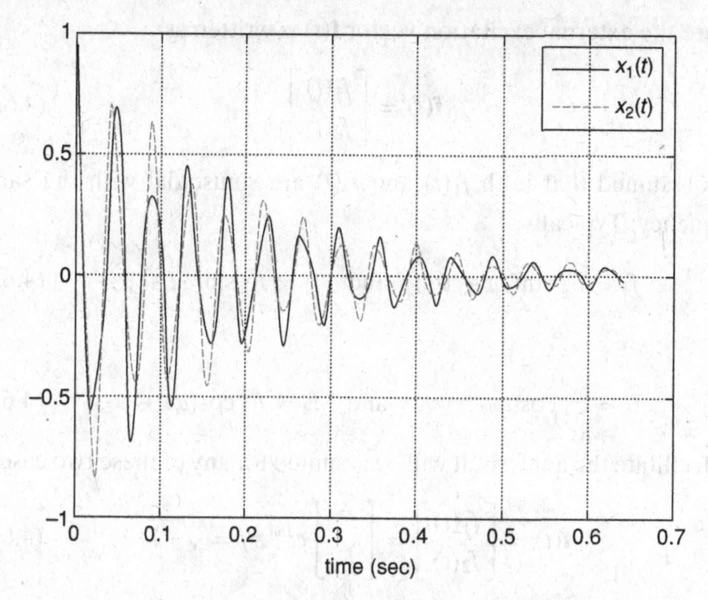

Figure 4.5.1 Free vibration of a 2DOF system

```
for i=1:1000
        t=t+delt;
        tv(i)=t;
        y(:,i)=expm(R*t)*y0;
end
plot(tv,y(1,:),tv,y(2,:),'--')
xlabel('time(sec.)')
legend('x_1(t)','x_2(t)')
grid
```

4.6 STEADY-STATE RESPONSE OF A DAMPED 2DOF SYSTEM UNDER SINUSOIDAL EXCITATION

The vibration of a damped 2DOF system is governed by the differential Equation 4.1.3 that is rewritten here:

$$M\ddot{\mathbf{x}} + C\dot{\mathbf{x}} + K\mathbf{x} = \mathbf{f}(t) \tag{4.6.1}$$

where the external excitation vector $\mathbf{f}(t)$ is written as

$$\mathbf{f}(t) = \begin{bmatrix} f_1(t) \\ f_2(t) \end{bmatrix} \tag{4.6.2}$$

It is assumed that both $f_1(t)$ and $f_2(t)$ are sinusoidal with the same frequency. Typically,

$$f_1 = \hat{F}_1 \sin(\omega t + \psi_1) \quad \text{and} \quad f_2 = \hat{F}_2 \sin(\omega t + \psi_2) \tag{4.6.3}$$

or

$$f_1 = \hat{F}_1 \cos(\omega t + \psi_1) \quad \text{and} \quad f_2 = \hat{F}_2 \cos(\omega t + \psi_2) \tag{4.6.4}$$

To facilitate the analysis, it will be assumed for any of these two cases,

$$\mathbf{f}(t) = \begin{bmatrix} f_1(t) \\ f_2(t) \end{bmatrix} = \begin{bmatrix} F_1 \\ F_2 \end{bmatrix} e^{j\omega t}; \; j = \sqrt{-1} \tag{4.6.5}$$

where

$$\begin{bmatrix} F_1 \\ F_2 \end{bmatrix} = \begin{bmatrix} \hat{F}_1 e^{j\psi_1} \\ \hat{F}_2 e^{j\psi_2} \end{bmatrix} \tag{4.6.6}$$

The steady-state solution of Equation 4.6.1 is assumed to be

$$\begin{bmatrix} x_1(t) \\ x_2(t) \end{bmatrix} = \begin{bmatrix} X_1 \\ X_2 \end{bmatrix} e^{j\omega t} \tag{4.6.7}$$

Therefore,

$$\dot{\mathbf{x}}(t) = \begin{bmatrix} \dot{x}_1(t) \\ \dot{x}_2(t) \end{bmatrix} = \begin{bmatrix} X_1 \\ X_2 \end{bmatrix} j\omega e^{j\omega t} \tag{4.6.8}$$

and

$$\ddot{\mathbf{x}}(t) = \begin{bmatrix} \ddot{x}_1(t) \\ \ddot{x}_2(t) \end{bmatrix} = \begin{bmatrix} X_1 \\ X_2 \end{bmatrix} (-\omega^2) e^{j\omega t} \tag{4.6.9}$$

Substituting Equations 4.6.5, 4.6.7–4.6.9 into Equation 4.6.1,

$$[(K - \omega^2 M) + jC\omega] \begin{bmatrix} X_1 \\ X_2 \end{bmatrix} e^{j\omega t} = \begin{bmatrix} F_1 \\ F_2 \end{bmatrix} e^{j\omega t} \tag{4.6.10}$$

Equating the coefficients of $e^{j\omega t}$ on both sides,

$$[(K - \omega^2 M) + jC\omega] \begin{bmatrix} X_1 \\ X_2 \end{bmatrix} = \begin{bmatrix} F_1 \\ F_2 \end{bmatrix} \tag{4.6.11}$$

Therefore,

$$\begin{bmatrix} X_1 \\ X_2 \end{bmatrix} = [(K - \omega^2 M) + jC\omega]^{-1} \begin{bmatrix} F_1 \\ F_2 \end{bmatrix} \tag{4.6.12}$$

Corresponding to Equation 4.6.3,

$$\begin{bmatrix} x_1(t) \\ x_2(t) \end{bmatrix} = \begin{bmatrix} A_1 \sin(\omega t + \phi_1) \\ A_2 \sin(\omega t + \phi_2) \end{bmatrix} \tag{4.6.13}$$

where

$$A_1 = |X_1|, \; A_2 = |X_2|, \; \phi_1 = \angle X_1, \quad \text{and} \quad \phi_2 = \angle X_2 \tag{4.6.14}$$

If excitations were in the form of Equation 4.6.4,

$$\begin{bmatrix} x_1(t) \\ x_2(t) \end{bmatrix} = \begin{bmatrix} A_1 \cos(\omega t + \phi_1) \\ A_2 \cos(\omega t + \phi_2) \end{bmatrix} \tag{4.6.15}$$

where A_1, A_2, ϕ_1, and ϕ_2 are given by Equation 4.6.14.

Example 4.6.1: Steady-State Response of a Damped 2DOF System
Find the steady-state response of system in Figure 4.1.1 for which $f_1(t) = 10\sin(240t)$ N and $f_2(t) = 100\cos(240t)$ N. As in Example 4.5.1, let $m_1 = m_2 = 0.5$ kg, $k_1 = k_2 = k_3 = 10,000$ N/m, $c_1 = 10$ N – sec/m, and $c_2 = 0$.

Using K, M, and C in Example 4.5.1,

$$(K - \omega^2 M) + j\omega C = 10^4 \begin{bmatrix} -0.88 + j0.24 & -1 \\ -1 & -0.88 \end{bmatrix}$$

Therefore,

$$[(K - \omega^2 M) + j\omega C]^{-1} = 10^{-3} \begin{bmatrix} 0.2079 - j0.1946 & -0.2362 + j0.2211 \\ -0.2362 + j0.2211 & 0.1548 - j0.2513 \end{bmatrix}$$

With respect to Equation 4.6.3,

$$\hat{F}_1 = 10,\ \psi_1 = 0,\ \hat{F}_2 = 100, \quad \text{and} \quad \psi_2 = \frac{\pi}{2}$$

Therefore,

$$\begin{bmatrix} F_1 \\ F_2 \end{bmatrix} = \begin{bmatrix} \hat{F}_1 e^{j\psi_1} \\ \hat{F}_2 e^{j\psi_2} \end{bmatrix} = \begin{bmatrix} 10 \\ j100 \end{bmatrix}$$

From Equation 4.6.12,

$$\begin{bmatrix} X_1 \\ X_2 \end{bmatrix} = [(K - \omega^2 M) + jC\omega]^{-1} \begin{bmatrix} F_1 \\ F_2 \end{bmatrix} = \begin{bmatrix} 0.0242 + j0.0217 \\ -0.0275 - j0.0133 \end{bmatrix}$$

Next,

$$A_1 = |X_1| = 0.0325,\ A_2 = |X_2| = 0.0305,$$
$$\phi_1 = \angle X_1 = 0.7306\,\text{rad}, \quad \text{and} \quad \phi_2 = \angle X_2 = -2.6919\,\text{rad}$$

Last, Equation 4.6.13 yields the following steady-state response:

$$\begin{bmatrix} x_1(t) \\ x_2(t) \end{bmatrix} = \begin{bmatrix} 0.0325 \sin(240t + 0.7306) \\ 0.0305 \sin(240t - 2.6919) \end{bmatrix} \text{m}$$

4.7 VIBRATION ABSORBER

In this section, design techniques for two types (undamped and damped) of vibration absorbers are presented.

4.7.1 Undamped Vibration Absorber

Consider an undamped spring–mass system with the stiffness k_1 and the mass m_1 subjected to a sinusoidal excitation $f_0 \sin \omega t$ (Figure 4.7.1). When $\omega = \sqrt{k_1/m_1}$, there will be resonance and a large amount of vibration. In many applications, it is not possible to add damping, or to change the natural frequency, or the excitation frequency. In these situations, another spring–mass system with the stiffness k_2 and the

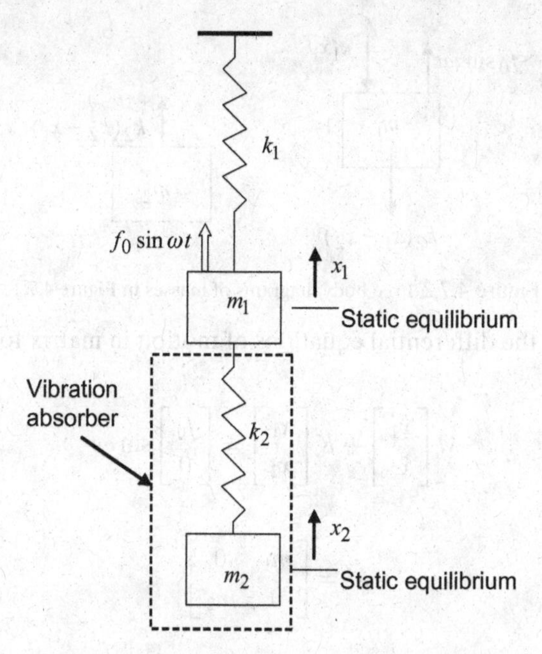

Figure 4.7.1 An undamped vibration absorber

mass m_2 is added to suppress this vibration. The spring–mass system with the stiffness k_2 and the mass m_2 is called the undamped vibration absorber. The objectives are to select the absorber stiffness k_2 and the absorber mass m_2 for the suppression of vibration of the main mass m_1 in a reliable manner.

Applying Newton's second law of motion to free body diagrams in Figure 4.7.2,

$$f_0 \sin \omega t - k_1 x_1 - k_2(x_1 - x_2) = m_1 \ddot{x}_1 \tag{4.7.1}$$

$$-k_2(x_2 - x_1) = m_2 \ddot{x}_2 \tag{4.7.2}$$

Rearranging the differential equations of motion (Equations 4.7.1 and 4.7,2),

$$m_1 \ddot{x}_1 + (k_1 + k_2)x_1 - k_2 x_2 = f_0 \sin \omega t \tag{4.7.3}$$

$$m_2 \ddot{x}_2 - k_2 x_1 + k_2 x_2 = 0 \tag{4.7.4}$$

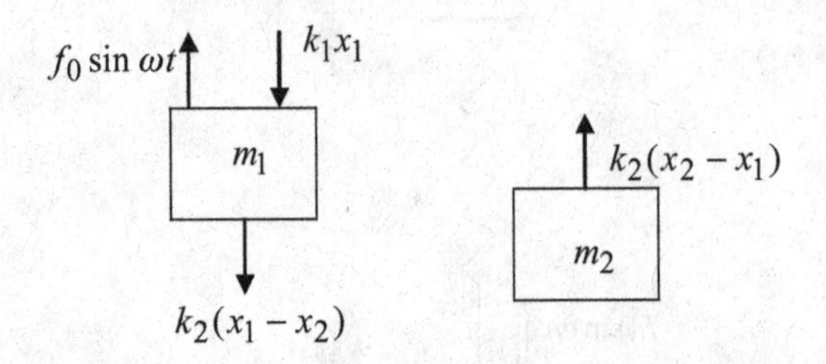

Figure 4.7.2 Free body diagrams of masses in Figure 4.7.1

Therefore, the differential equations of motion in matrix form can be written as

$$M \begin{bmatrix} \ddot{x}_1 \\ \ddot{x}_2 \end{bmatrix} + K \begin{bmatrix} x_1 \\ x_2 \end{bmatrix} = \begin{bmatrix} f_0 \\ 0 \end{bmatrix} \sin \omega t \tag{4.7.5}$$

where

$$M = \begin{bmatrix} m_1 & 0 \\ 0 & m_2 \end{bmatrix} \tag{4.7.6}$$

and

$$K = \begin{bmatrix} k_1 + k_2 & -k_2 \\ -k_2 & k_2 \end{bmatrix} \tag{4.7.7}$$

From Equations 4.7.6 and 4.7.7,

$$K - \omega^2 M = \begin{bmatrix} k_1 + k_2 - \omega^2 m_1 & -k_2 \\ -k_2 & k_2 - \omega^2 m_2 \end{bmatrix} \tag{4.7.8}$$

From Equation 4.7.8,

$$(K - \omega^2 M)^{-1} = \frac{1}{\Delta} \begin{bmatrix} k_2 - \omega^2 m_2 & +k_2 \\ +k_2 & k_1 + k_2 - \omega^2 m_1 \end{bmatrix} \tag{4.7.9}$$

where

$$\Delta = \det(K - \omega^2 M) = (k_1 + k_2 - \omega^2 m_1)(k_2 - \omega^2 m_2) - k_2^2 \tag{4.7.10}$$

From Equation 4.4.4,

$$\begin{aligned} \begin{bmatrix} A_1 \\ A_2 \end{bmatrix} &= \frac{1}{\Delta} \begin{bmatrix} k_2 - \omega^2 m_2 & k_2 \\ k_2 & k_1 + k_2 - \omega^2 m_1 \end{bmatrix} \begin{bmatrix} f_0 \\ 0 \end{bmatrix} \\ &= \frac{1}{\Delta} \begin{bmatrix} (k_2 - \omega^2 m_2) f_0 \\ k_2 f_0 \end{bmatrix} \end{aligned} \tag{4.7.11}$$

where A_1 and A_2 are the amplitudes of x_1 and x_2, respectively. From Equations 4.7.11 and 4.7.10,

$$A_1 = \frac{(k_2 - \omega^2 m_2) f_0}{(k_1 + k_2 - \omega^2 m_1)(k_2 - \omega^2 m_2) - k_2^2} \tag{4.7.12}$$

and

$$A_2 = \frac{k_2 f_0}{(k_1 + k_2 - \omega^2 m_1)(k_2 - \omega^2 m_2) - k_2^2} \tag{4.7.13}$$

The amplitude A_1 of the main mass m_1 will be zero when the absorber parameters k_2 and m_2 are chosen such that

$$\frac{k_2}{m_2} = \omega^2 \tag{4.7.14}$$

Define

$$\sqrt{\frac{k_1}{m_1}} = \omega_{11} \tag{4.7.15}$$

and

$$\sqrt{\frac{k_2}{m_2}} = \omega_{22} \tag{4.7.16}$$

The condition in Equation 4.7.14 can then be expressed as

$$\omega = \omega_{22} \tag{4.7.17}$$

In other words, whenever ω_{22} is chosen such that it equals the excitation frequency ω, $A_1 = 0$, that is, the vibration of the main mass will be completely suppressed. The parameter ω_{22} is designed such that the vibration of the main mass will be suppressed at the resonance condition of the original SDOF system, that is, $\omega = \omega_{11}$. As a result, the design condition is

$$\omega_{22} = \omega_{11} \tag{4.7.18}$$

From the definitions in Equations 4.7.15 and 4.7.16, the condition in Equation 4.7.18 is also expressed as

$$\frac{k_2}{m_2} = \frac{k_1}{m_1} \tag{4.7.19}$$

In summary, when k_2 and m_2 are selected to satisfy Equation 4.7.19, $A_1 = 0$ when $\omega = \omega_{11}$. And the amplitude of the absorber mass at this condition $\omega = \omega_{11} = \omega_{22}$ will be

$$A_2 = -\frac{f_0}{k_2} \tag{4.7.20}$$

Define the mass ratio μ as

$$\mu = \frac{m_2}{m_1} \tag{4.7.21}$$

Then, from Equation 4.7.19,

$$\frac{k_2}{k_1} = \frac{m_2}{m_1} = \mu \tag{4.7.22}$$

Using Equation 4.7.22, Equations 4.7.12 and 4.7.13 are represented as follows:

$$\frac{A_1 k_1}{f_0} = \frac{(1 - r^2)}{(1 + \mu - r^2)(1 - r^2) - \mu} \tag{4.7.23}$$

and

$$\frac{A_2 k_1}{f_0} = \frac{1}{(1 + \mu - r^2)(1 - r^2) - \mu} \tag{4.7.24}$$

where

$$r = \frac{\omega}{\omega_{11}} = \frac{\omega}{\omega_{22}} \tag{4.7.25}$$

In Figure 4.7.3, the magnitude of the amplitude $|A_1|$ is plotted as a function of ω with the condition in Equation 4.7.18. It should be noted that $A_1 \neq 0$ when $\omega \neq \omega_{11}$. In fact, the amplitude A_1 is infinite when ω equals one of the natural frequencies of the 2DOF system, which are the roots of Equation 4.2.5 as follows:

$$\Delta = \det(K - \omega^2 M) = (k_1 + k_2 - \omega^2 m_1)(k_2 - \omega^2 m_2) - k_2^2 = 0 \tag{4.7.26}$$

In Figure 4.7.4, the magnitude of the amplitude $|A_2|$ is plotted as a function of ω with the condition in Equation 4.7.18. Dividing

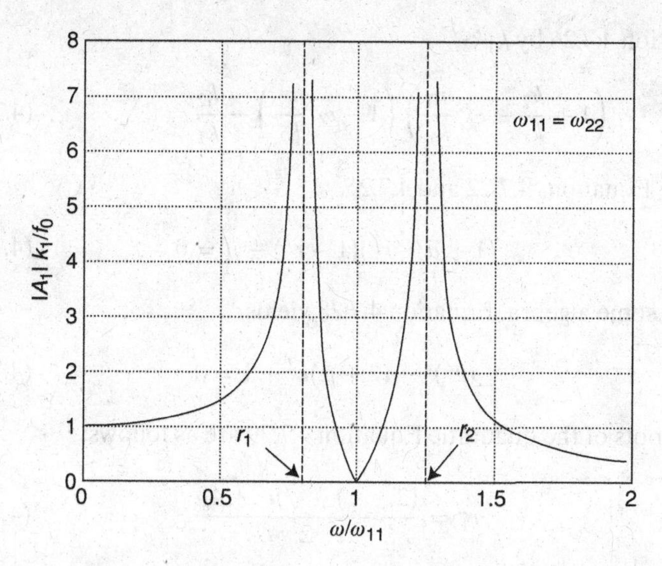

Figure 4.7.3 Amplitude of the main mass versus excitation frequencies

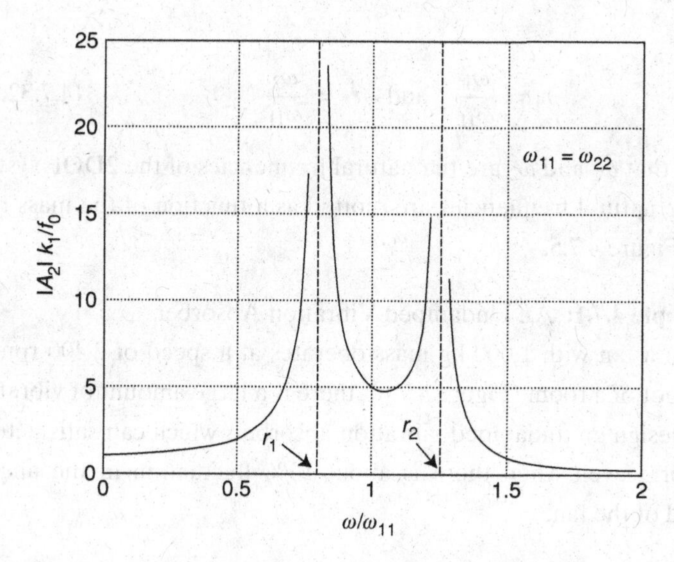

Figure 4.7.4 Amplitude of the absorber mass versus excitation frequencies

Equation 4.7.26 by k_1k_2,

$$\left(1+\frac{k_2}{k_1}-\omega^2\frac{m_1}{k_1}\right)\left(1-\omega^2\frac{m_2}{k_2}\right)-\frac{k_2}{k_1}=0 \tag{4.7.27}$$

Using Equations 4.7.22 and 4.7.25,

$$(1+\mu-r^2)(1-r^2)-\mu=0 \tag{4.7.28}$$

After some algebra, Equation 4.7.28 yields

$$(r^2)^2-(2+\mu)r^2+1=0 \tag{4.7.29}$$

The roots of the quadratic Equation 4.7.29 are as follows:

$$r_1^2=\frac{(2+\mu)-\sqrt{\mu^2+4\mu}}{2} \tag{4.7.30}$$

$$r_2^2=\frac{(2+\mu)+\sqrt{\mu^2+4\mu}}{2} \tag{4.7.31}$$

where

$$r_1=\frac{\omega_1}{\omega_{11}} \quad \text{and} \quad r_2=\frac{\omega_2}{\omega_{11}} \tag{4.7.32a, b}$$

Note that ω_1 and ω_2 are the natural frequencies of the 2DOF system. These natural frequencies are plotted as a function of the mass ratio μ in Figure 4.7.5.

Example 4.7.1: An Undamped Vibration Absorber

When a fan with 1,000 kg mass operates at a speed of 2,400 rpm on the roof of a room (Figure 4.7.6), there is a large amount of vibration.

Design an undamped vibration absorber, which can satisfactorily perform even when there is about 10% fluctuation in the angular speed of the fan.

Solution

$$\omega_{11}=\omega=\frac{2,400\times 2\pi}{60}=80\pi \text{ rad/sec}$$

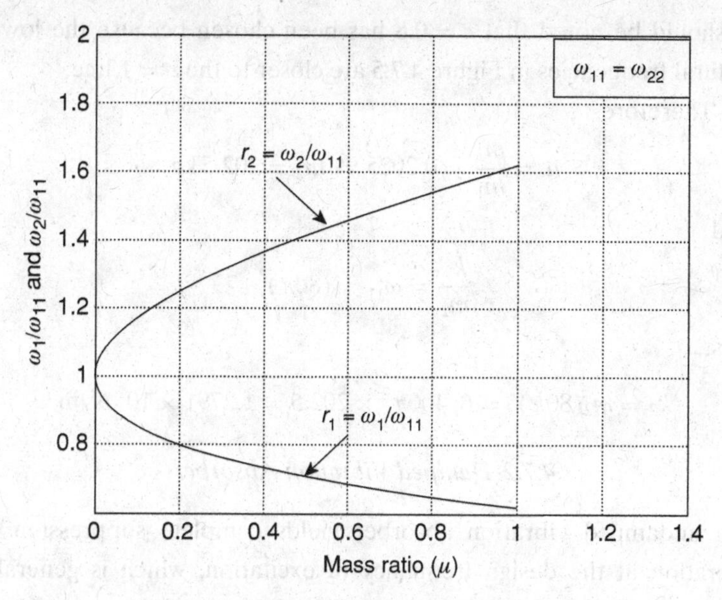

Figure 4.7.5 Natural frequencies versus mass ratio

To ensure a good margin of safety, let us keep the natural frequencies of the 2DOF system to be at least 20% away from ω_{11} (Figure 4.7.5). In this case, $r = 0.8$ in the following equation:

$$(r^2)^2 - (2 + \mu)r^2 + 1 = 0$$

or

$$\mu = \frac{r^4 + 1}{r^2} - 2 = \frac{(0.8)^4 + 1}{(0.8)^2} - 2 = 0.2025$$

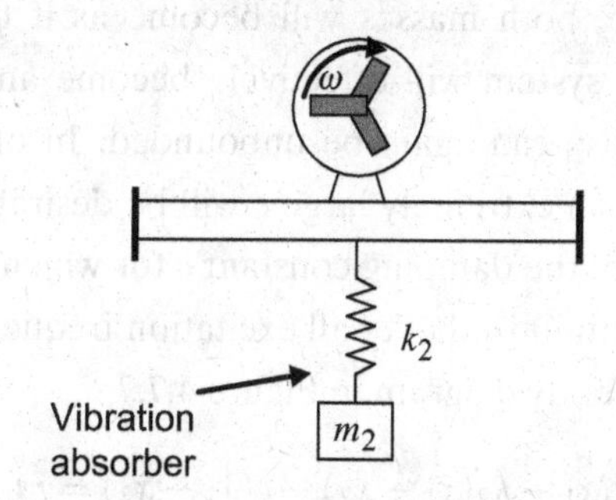

Figure 4.7.6 Undamped vibration absorber attached to the ceiling of room

It should be noted that $r = 0.8$ has been chosen because the lower natural frequencies in Figure 4.7.5 are closer to the $r = 1$ line.

Therefore,

$$\mu = \frac{m_2}{m_1} = 0.2025 \Rightarrow m_2 = 202.5\,\text{kg}$$

and

$$\frac{k_2}{m_2} = \omega_{11}^2 = (80\pi)^2$$

Last,

$$k_2 = m_2(80\pi)^2 = 6,400\pi^2 \times 202.5 = 1.2791 \times 10^7\,\text{N/m}$$

4.7.2 Damped Vibration Absorber

An undamped vibration absorber yields complete suppression of vibration at the design frequency of excitation, which is generally selected to be the resonance condition of the original system. However, if the excitation frequency is a variable, unbounded response is obtained when the excitation frequency matches either of the two natural frequencies of the 2DOF system. In order to alleviate this situation, a damper is included in the absorber design as shown in Figure 4.7.7, and the objective is to choose damper parameters such that the vibration of the main mass is minimized over all excitation frequencies. When the damping constant $c = 0$, the amplitude of the main mass can be unbounded. Similarly, when the damping constant is extremely large, both masses will become as if they are welded together, and the system will effectively become an SDOF system, that is, the response can again be unbounded. In other words, neither a small c nor an extremely large c will be desirable. There exists an optimal value of the damping constant c for which the vibration of the main mass is minimized over all excitation frequencies.

From the free body diagram in Figure 4.7.7,

$$f_0 \sin \omega t - k_1 x_1 - k_2(x_1 - x_2) - c(\dot{x}_1 - \dot{x}_2) = m_1 \ddot{x}_1 \tag{4.7.33}$$

$$-k_2(x_2 - x_1) - c(\dot{x}_2 - \dot{x}_1) = m_2 \ddot{x}_2 \tag{4.7.34}$$

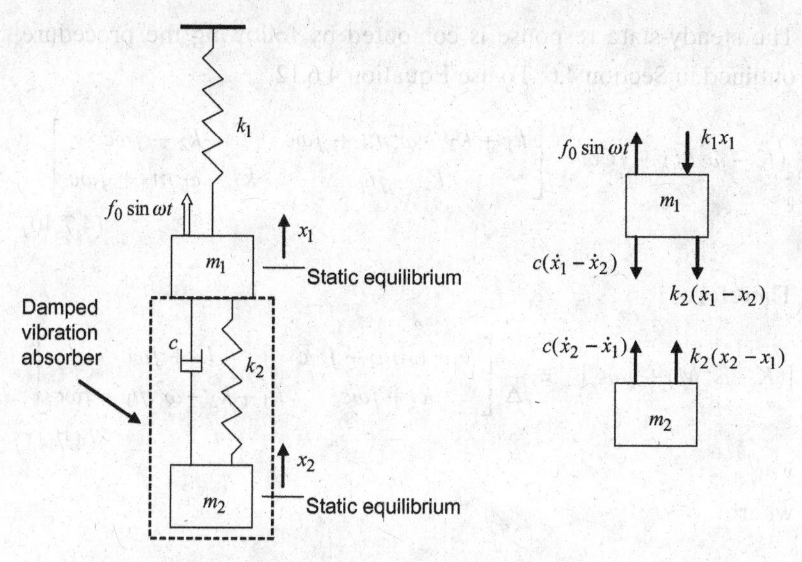

Figure 4.7.7 Damped vibration absorber and associated free body diagrams

or

$$m_1 \ddot{x}_1 + c\dot{x}_1 - c\dot{x}_2 + (k_1 + k_2)x_1 - k_2 x_2 = f_0 \sin \omega t \tag{4.7.35}$$

$$m_2 \ddot{x}_2 - c\dot{x}_1 + c\dot{x}_2 - k_2 x_1 + k_2 x_2 = 0 \tag{4.7.36}$$

Equations 4.7.35–4.7.36 are put in the matrix form as

$$M\ddot{\mathbf{x}} + C\dot{\mathbf{x}} + K\mathbf{x}(t) = \mathbf{f}(t) \tag{4.7.37}$$

where

$$\mathbf{x}(t) = \begin{bmatrix} x_1(t) \\ x_2(t) \end{bmatrix}; \ \mathbf{f}(t) = \begin{bmatrix} f_0 \\ 0 \end{bmatrix} \sin \omega t \tag{4.7.38}$$

$$M = \begin{bmatrix} m_1 & 0 \\ 0 & m_2 \end{bmatrix}, \ C = \begin{bmatrix} c & -c \\ -c & c \end{bmatrix}, \ \text{and} \ K = \begin{bmatrix} k_1 + k_2 & -k_2 \\ -k_2 & k_2 \end{bmatrix} \tag{4.7.39}$$

The steady-state response is computed by following the procedure outlined in Section 4.6. To use Equation 4.6.12,

$$(K-\omega^2 M)+jC\omega=\begin{bmatrix} k_1+k_2-\omega^2 m_1+j\omega c & -k_2-j\omega c \\ -k_2-j\omega c & k_2-\omega^2 m_2+j\omega c \end{bmatrix} \tag{4.7.40}$$

Therefore,

$$[(K-\omega^2 M)+j\omega C]^{-1}=\frac{1}{\Delta}\begin{bmatrix} k_2-\omega^2 m_2+j\omega c & k_2+j\omega c \\ k_2+j\omega c & k_1+k_2-\omega^2 m_1+j\omega c \end{bmatrix} \tag{4.7.41}$$

where

$$\begin{aligned}\Delta &= \det(K-\omega^2 M+j\omega C)\\ &= (k_1+k_2-\omega^2 m_1+j\omega c)(k_2-\omega^2 m_2+j\omega c)-(k_2+j\omega c)^2\end{aligned} \tag{4.7.42}$$

After some algebra,

$$\Delta=[(k_1-\omega^2 m_1)(k_2-\omega^2 m_2)-\omega^2 k_2 m_2]+j\omega c(k_1-\omega^2(m_1+m_2)) \tag{4.7.43}$$

Furthermore

$$\begin{bmatrix} F_1 \\ F_2 \end{bmatrix}=\begin{bmatrix} f_0 \\ 0 \end{bmatrix} \tag{4.7.44}$$

From Equation 4.6.12,

$$\begin{bmatrix} X_1 \\ X_2 \end{bmatrix}=\frac{1}{\Delta}\begin{bmatrix} (k_2-\omega^2 m_2+j\omega c)f_0 \\ (k_2+j\omega c)f_0 \end{bmatrix} \tag{4.7.45}$$

The amplitude of the main mass is given by

$$A_1=|X_1|=\frac{|(k_2-\omega^2 m_2+j\omega c)f_0|}{|\Delta|} \tag{4.7.46}$$

Using Equation 4.7.43,

$$A_1 = \frac{f_0\sqrt{(k_2-\omega^2 m_2)^2+(c\omega)^2}}{\sqrt{[(k_1-\omega^2 m_1)(k_2-\omega^2 m_2)-\omega^2 k_2 m_2]^2+[\omega c(k_1-\omega^2(m_1+m_2)]^2}} \tag{4.7.47}$$

or

$$\frac{A_1 k_1}{f_0} = \frac{\sqrt{\left(\frac{k_2}{k_1}-\frac{\omega^2 m_2}{k_1}\right)^2+\left(\frac{c\omega}{k_1}\right)^2}}{\sqrt{\left[\frac{(k_1-\omega^2 m_1)(k_2-\omega^2 m_2)}{k_1^2}-\frac{\omega^2 k_2 m_2}{k_1^2}\right]^2+\left[\frac{\omega c(k_1-\omega^2(m_1+m_2))}{k_1^2}\right]^2}} \tag{4.7.48}$$

Define the following nondimensional variables:

$$\mu = \frac{m_2}{m_1},\ \omega_{11} = \sqrt{\frac{k_1}{m_1}},\ \omega_{22} = \sqrt{\frac{k_2}{m_2}},\ f = \frac{\omega_{22}}{\omega_{11}} \tag{4.7.49}$$

$$g = \frac{\omega}{\omega_{11}} \quad \text{and} \quad \xi = \frac{c}{2m_2\omega_{11}} \tag{4.7.50}$$

Then,

$$\frac{k_2}{k_1} = \frac{k_2}{m_2}\frac{m_1}{k_1}\frac{m_2}{m_1} = \left(\frac{\omega_{22}}{\omega_{11}}\right)^2 \mu = f^2\mu \tag{4.7.51}$$

$$\frac{\omega^2 m_2}{k_1} = \frac{\omega^2 m_1}{k_1}\frac{m_2}{m_1} = g^2\mu \tag{4.7.52}$$

$$\frac{c\omega}{k_1} = \frac{2c\omega}{2m_2\omega_{11}}\frac{\omega_{11}m_2}{k_1} = 2\xi\frac{\omega}{\omega_{11}} = 2\xi g\mu \tag{4.7.53}$$

$$\frac{(\omega c)(k_1-\omega^2(m_1+m_2))}{k_1^2} = \left(\frac{\omega c}{k_1}\right)\left(1-\frac{\omega^2}{\omega_{11}^2}-\frac{\omega^2}{\omega_{11}^2}\mu\right)$$
$$= (2\xi g\mu)(1-g^2-\mu g^2) \tag{4.7.54}$$

$$\frac{(k_1-\omega^2 m_1)(k_2-\omega^2 m_2)}{k_1^2} = (1-g^2)\left(\frac{k_2}{k_1}-\frac{\omega^2 m_2}{k_1}\right)$$
$$= (1-g^2)(f^2-g^2)\mu \tag{4.7.55}$$

$$\frac{\omega^2 k_2 m_2}{k_1^2} = \frac{\omega^2 k_2}{k_1}\frac{m_1}{k_1}\frac{m_2}{m_1} = g^2 f^2 \mu^2 \tag{4.7.56}$$

Substituting Equations 4.7.49–4.7.56 into Equation 4.7.48,

$$\frac{A_1 k_1}{f_0} = \frac{\sqrt{(f^2 - g^2)^2 + (2\xi g)^2}}{\sqrt{[(1-g^2)(f^2-g^2) - g^2 f^2 \mu]^2 + (2\xi g)^2 (1 - g^2 - \mu g^2)^2}} \tag{4.7.57}$$

Den Hartog (1956) has obtained optimal absorber parameters to minimize the following function:

$$I = \max_{g} A_1(g) \tag{4.7.58}$$

Here, $A_1(g)$ is a representation of the fact that A_1 is a function of frequency ratio g. The objective function I is the maximum value of A_1 with respect to the variation in g.

Case I: Tuned Case ($f = 1$ or $\omega_{22} = \omega_{11}$)

For the minimum value of I, the optimal value of ξ is

$$\xi^2 = \frac{\mu(\mu+3)(1+\sqrt{\mu/(\mu+2)}}{8(1+\mu)} \tag{4.7.59}$$

And the objective function I is

$$I = \max_{g} A_1(g) = \frac{f_0}{k_1}\frac{1}{(-\mu + (1+\mu)\sqrt{\mu/(\mu+2)})} \tag{4.7.60}$$

The amplitude of this main mass due to this optimal absorber is shown in Figure 4.7.8.

Case II: No restriction on f (Absorber not tuned to main system)

For the minimum value of I,

$$f = \frac{1}{1+\mu} \tag{4.7.61}$$

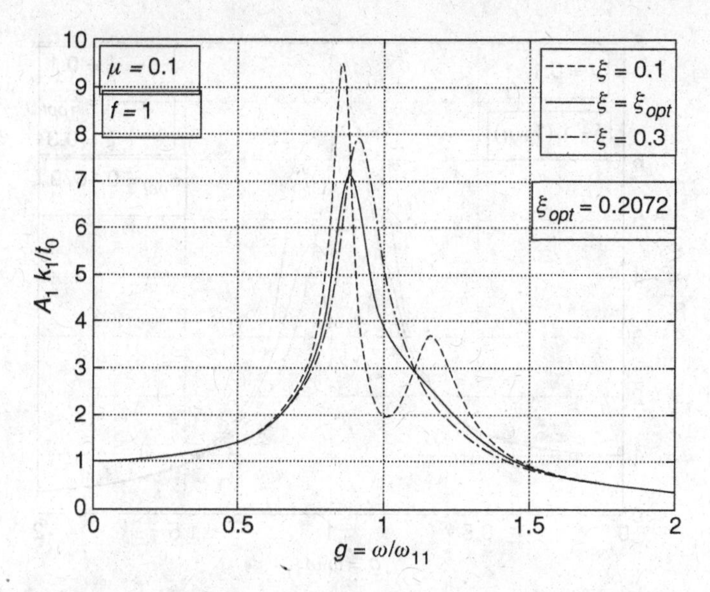

Figure 4.7.8 Amplitude of the main mass for various damping ratios (Absorber tuned to main system, $f = 1$)

and

$$\xi^2 = \frac{3\mu}{8(1+\mu)^3} \tag{4.7.62}$$

The objective function I is

$$I = \max_{g} A_1(g) = \frac{f_0\sqrt{1+2/\mu}}{k_1} \tag{4.7.63}$$

The amplitude of this main mass due to this optimal absorber is shown in Figure 4.7.9.

Example 4.7.2: A Damped Vibration Absorber

When a fan with 1,000 kg mass operates at a speed of 2,400 rpm on the roof of a room (Figure 4.7.10), there is a large amount of vibration.

Design an optimally damped vibration absorber with the mass ratio equal to 0.2025, which is same as that in Example 4.7.1.

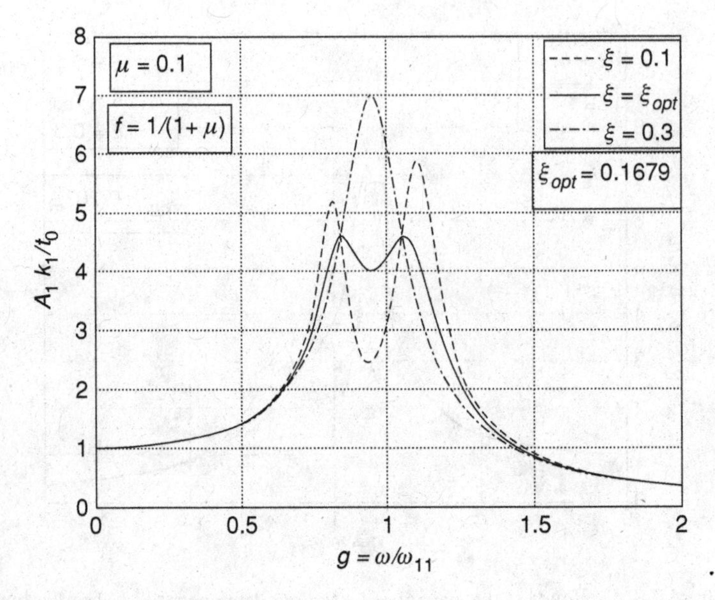

Figure 4.7.9 Amplitude of the main mass for various damping ratios (Absorber not tuned to main system)

From Equation (4.7.61),

$$f = \frac{1}{1+\mu} = \frac{1}{1.2025} = 0.8316$$

and

$$\frac{k_2}{m_2} = f^2 \omega_{11}^2 = (0.8316 \times 80\pi)^2$$

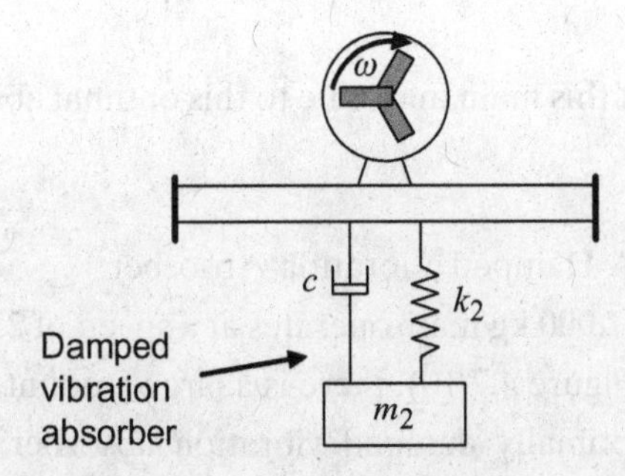

Figure 4.7.10 Damped vibration absorber attached to the ceiling of room

Therefore,

$$k_2 = m_2(f \times 80\pi)^2 = 6,400\pi^2 f^2 \times 202.5 = 8.8457 \times 10^6 \text{ N/m}$$

From Equation (4.7.62),

$$\xi = \sqrt{\frac{3\mu}{8(1+\mu)^3}} = 0.209$$

$$c = 2\xi m_2 \omega_{11} = 2.1274 \times 10^4 \text{ N-sec/m}$$

4.8 MODAL DECOMPOSITION OF RESPONSE

The general differential Equation 4.1.3 is rewritten

$$M\ddot{\mathbf{x}} + C\dot{\mathbf{x}} + K\mathbf{x} = \mathbf{f}(t) \tag{4.8.1}$$

Pre-multiplying both sides of Equation 4.8.1 by M^{-1},

$$\ddot{\mathbf{x}} + M^{-1}C\dot{\mathbf{x}} + M^{-1}K\mathbf{x} = M^{-1}\mathbf{f}(t) \tag{4.8.2}$$

In general, the response $\mathbf{x}(t)$ is a linear combination of the modal vectors $\mathbf{v}_1$ and $\mathbf{v}_2$, that is,

$$\mathbf{x}(t) = \mathbf{v}_1 y_1(t) + \mathbf{v}_2 y_2(t) \tag{4.8.3}$$

where $y_1(t)$ and $y_2(t)$ are the coefficients of modal vectors $\mathbf{v}_1$ and $\mathbf{v}_2$, respectively. Equation 4.8.3 can be represented in a compact form as follows:

$$\mathbf{x}(t) = V\mathbf{y}(t) \tag{4.8.4}$$

where

$$V = [\mathbf{v}_1 \;\; \mathbf{v}_2] \tag{4.8.5}$$

and

$$\mathbf{y}(t) = \begin{bmatrix} y_1(t) \\ y_2(t) \end{bmatrix} \tag{4.8.6}$$

Substituting Equation 4.8.4 into Equation 4.8.2 and pre-multiplying by V^{-1},

$$\ddot{\mathbf{y}} + C_d\dot{\mathbf{y}} + K_d\mathbf{y} = \mathbf{f}_v(t) \tag{4.8.7}$$

where

$$C_d = V^{-1}M^{-1}CV,\ K_d = V^{-1}M^{-1}KV, \quad \text{and} \quad \mathbf{f}_v(t) = V^{-1}M^{-1}\mathbf{f}(t) \tag{4.8.8}$$

From Equation 4.8.8 and Equation 4.2.40,

$$K_d = V^{-1}M^{-1}KV = \Lambda = \begin{bmatrix} \omega_1^2 & 0 \\ 0 & \omega_2^2 \end{bmatrix} \tag{4.8.9}$$

Let

$$\mathbf{f}_v(t) = \begin{bmatrix} f_{v1}(t) \\ f_{v2}(t) \end{bmatrix} \tag{4.8.10}$$

Case I: Undamped System ($C = 0$)
Equations 4.8.7 and 4.8.10 yield

$$\ddot{y}_1 + \omega_1^2 y_1 = f_{v1}(t) \tag{4.8.11}$$

$$\ddot{y}_2 + \omega_2^2 y_2 = f_{v2}(t) \tag{4.8.12}$$

Equations 4.8.11 and 4.8.12 can be solved by techniques developed for an SDOF system. After that, the response can be obtained from Equation 4.8.3.

Case II: Damped System ($C \neq 0$)
For a general damping matrix, the matrix C_d will not be diagonal, and the differential equations for $y_1(t)$ and $y_2(t)$ will be coupled. However, the modal equations are decoupled for special cases, for example,

$$C = \alpha M + \beta K \tag{4.8.13}$$

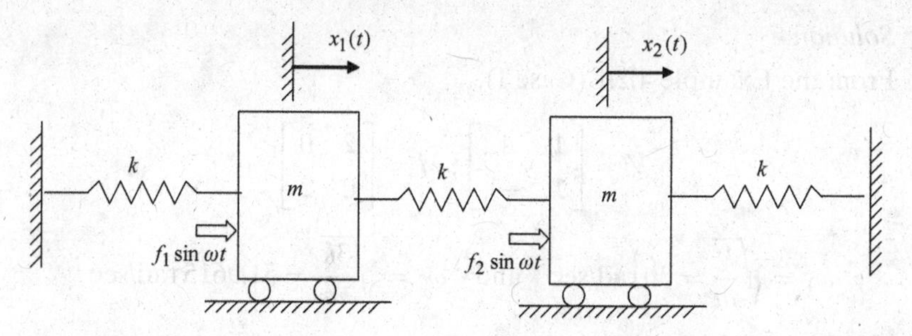

Figure 4.8.1 Two-degree-of-freedom system with sinusoidal excitation and modal damping

where α and β are constants. This form of damping is known as Proportional or Rayleigh damping. In this case, from Equation 4.8.8,

$$C_d = V^{-1}M^{-1}CV = \alpha I + \beta \Lambda \tag{4.8.14}$$

Equation 4.8.7 yields

$$\ddot{y}_1 + (\alpha + \beta\omega_1^2)\dot{y}_1 + \omega_1^2 y_1 = f_{v1}(t) \tag{4.8.15}$$

$$\ddot{y}_2 + (\alpha + \beta\omega_2^2)\dot{y}_2 + \omega_2^2 y_2 = f_{v2}(t) \tag{4.8.16}$$

Equations 4.8.15 and 4.8.16 can be solved by techniques developed for an SDOF system. After that, the response can be obtained from Equation 4.8.3.

On the basis of Equations 4.8.15 and 4.8.16, the modal damping in each mode ξ_i; $i = 1, 2$, can be described as

$$\xi_i = \frac{\alpha + \beta\omega_i^2}{2\omega_i}; i = 1, 2 \tag{4.8.17}$$

Example 4.8.1: Application of Modal Decomposition

Consider the system shown in Figure 4.8.1 and assume that the damping ratios in modes 1 and 2 are 0.05 and 0.1 respectively. Determine the steady-state response via modal decomposition. Following parameters are provided: $m = 2$ kg, $k = 1,800$ N/m, $f_1 = 10$ N, $f_2 = 20$ N, and $\omega = 45$ rad/sec.

Solution

From the Example 4.2.1 (Case I),

$$V = \begin{bmatrix} 1 & 1 \\ 1 & -1 \end{bmatrix}; M = \begin{bmatrix} 2 & 0 \\ 0 & 2 \end{bmatrix}$$

$$\omega_1 = \sqrt{\frac{k}{m}} = 30\,\text{rad/sec} \quad \text{and} \quad \omega_2 = \sqrt{\frac{3k}{m}} = 51.9615\,\text{rad/sec}$$

Note that

$$\xi_1 = 0.05,\ r_1 = \frac{\omega}{\omega_1} = \frac{45}{30} = 1.5$$

$$\xi_2 = 0.1,\ r_2 = \frac{\omega}{\omega_2} = \frac{45}{51.9615} = 0.866$$

$$\mathbf{f}(t) = \begin{bmatrix} 10 \\ 20 \end{bmatrix} \sin 45t$$

From Equation 4.8.8,

$$\mathbf{f}_v(t) = V^{-1}M^{-1}\mathbf{f}(t) = \begin{bmatrix} f_{v1}(t) \\ f_{v2}(t) \end{bmatrix} = \begin{bmatrix} 7.5 \\ -2.5 \end{bmatrix} \sin \omega t$$

In steady state, the responses of Equations 4.8.15 and 4.8.16 are given as

$$y_1(t) = Y_1 \sin(45t + \phi_1)$$

$$y_2(t) = Y_2 \sin(45t + \phi_2)$$

where

$$Y_1 = \frac{7.5}{\omega_1^2[(1 - r_1^2)^2 + (2\xi_1 r_1)^2]^{0.5}} = 0.0066,$$

$$\phi_1 = \tan^{-1} \frac{2\xi_1 r_1}{(1 - r_1^2)} = 3.022\,\text{rad}$$

$$Y_2 = \frac{-2.5}{\omega_2^2[(1 - r_2^2)^2 + (2\xi_2 r_2)^2]^{0.5}} = -0.0030,$$

$$\phi_2 = \tan^{-1} \frac{2\xi_2 r_2}{(1 - r_2^2)} = 0.6058\,\text{rad}$$

Last, from Equation 4.8.4,

$$x_1(t) = y_1(t) + y_2(t)$$

$$x_2(t) = y_1(t) - y_2(t)$$

EXERCISE PROBLEMS

P4.1 Consider the system in Figure P4.1.

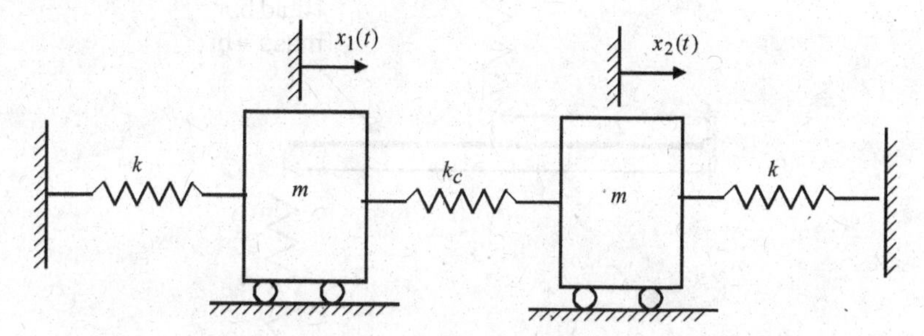

Figure P4.1 Undamped 2DOF system with coupling stiffness k_c

a. Derive the differential equations of motion and obtain the mass and stiffness matrices.
b. Calculate the natural frequencies and the mode shapes.
c. Find the initial conditions such that the free vibration is sinusoidal with each natural frequency.

P4.2 Consider the system in Figure P4.2.

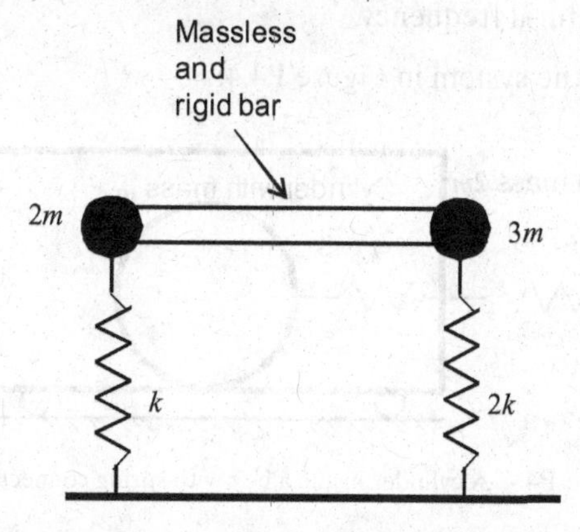

Figure P4.2 Massless rigid bar with masses lumped at ends and supported on springs

a. Derive the differential equations of motion and obtain the mass and stiffness matrices.
b. Calculate the natural frequencies and the mode shapes.
c. Find the initial conditions such that the free vibration is sinusoidal with each natural frequency.

P4.3 Consider the system in Figure P4.3.

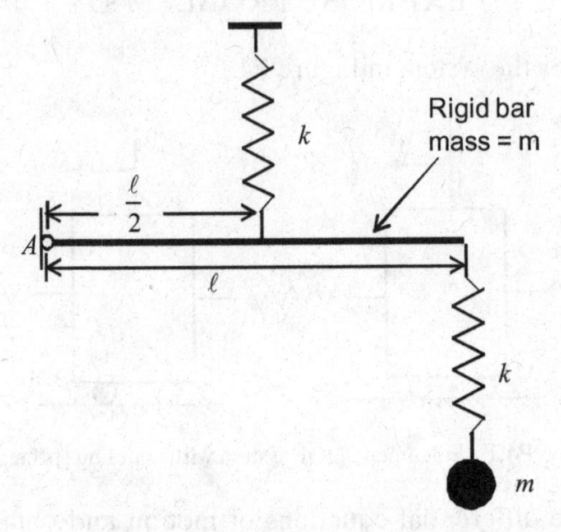

Figure P4.3 A rigid bar connected to another mass via a spring

a. Derive the differential equations of motion and obtain the mass and stiffness matrices.
b. Calculate the natural frequencies and the mode shapes.
c. Find the initial conditions such that the free vibration is sinusoidal with each natural frequency.

P4.4 Consider the system in Figure P4.4.

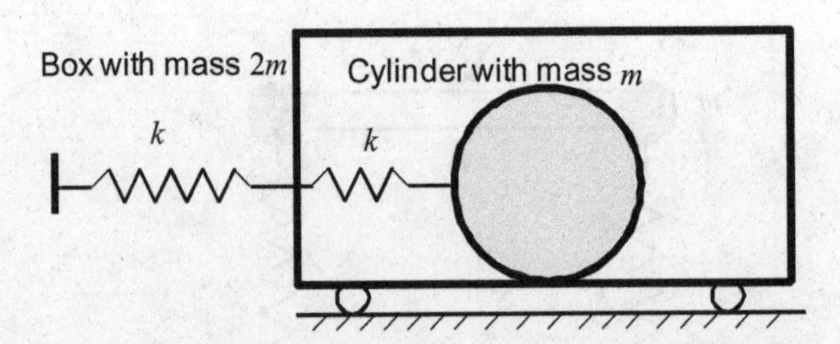

Figure P4.4 A cylinder inside a box with spring connections

a. Derive the differential equations of motion and obtain the mass and stiffness matrices.
b. Calculate the natural frequencies and the mode shapes.
c. Find the initial conditions such that the free vibration is sinusoidal with each natural frequency.

P4.5 Obtain and plot free vibration response for the system shown in Figure P4.1 when $m = 1\,\text{kg}$, $k = 530\,\text{N/m}$, $k_c = 130\,\text{N/m}$. Assume that $x_1(0) = 0.01\,\text{m}$, $\dot{x}_1(0) = 1\,\text{m/sec}$, $x_2(0) = 0$, and $\dot{x}_2(0) = 0$.

P4.6 Consider the system in Figure P4.6.

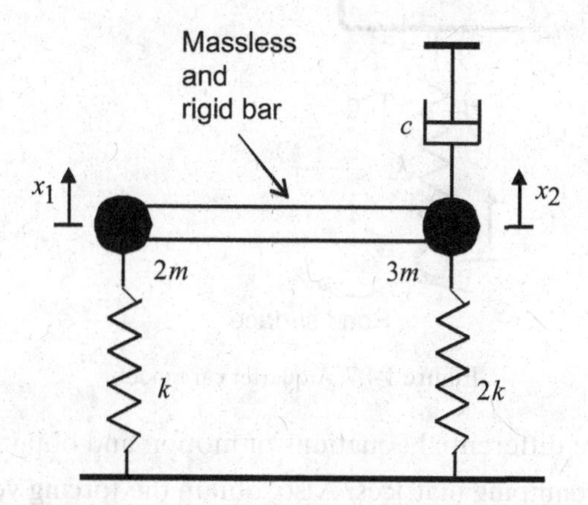

Figure P4.6 Massless rigid bar with masses lumped at ends and supported by springs and a damper

a. Derive the differential equations of motion and obtain mass, stiffness, and damping matrices.
b. Assume that $m = 11\,\text{kg}$, $k = 4,511\,\text{N/m}$, and $c = 20\,\text{N} - \text{sec/m}$. Determine the damping ratio and the undamped natural frequency for each mode.
c. Obtain and plot response when $x_1(0) = 0.01\,\text{m}$, $\dot{x}_1(0) = 1\,\text{m/sec}$, $x_2(0) = -0.02\,\text{m}$, and $\dot{x}_2(0) = 0$.

P4.7 A quarter car model of an automobile is shown in Figure P4.7. The vehicle is traveling with a velocity V on a sinusoidal road surface with amplitude = 0.011 m and a wavelength of 5.3 m.

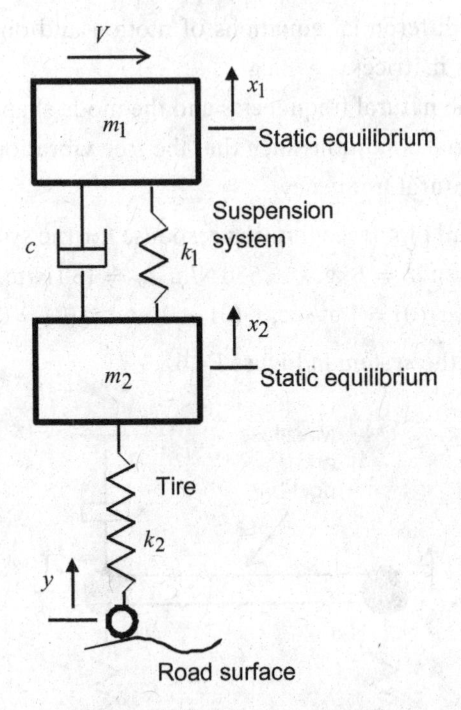

Figure P4.7 A quarter car model

a. Derive the differential equations of motion and obtain mass, stiffness, and damping matrices. Also, obtain the forcing vector.
b. Assume that $m_1 = 1,010\,\text{kg}$, $m_2 = 76\,\text{kg}$, $k_1 = 31,110\,\text{N/m}$, $k_2 = 321,100\,\text{N/m}$, and $c = 4,980\,\text{N} - \text{sec/m}$. Determine the damping ratio and the undamped natural frequency for each mode.
c. Compute the amplitudes and the phases of steady-state responses when the velocity $V = 100\,\text{km/h}$ with and without the damper.
d. Plot amplitudes of both masses as a function of the velocity V in the presence of a damper.

P4.8 A rotor–shaft system (Figure P4.8) consisting of torsional stiffness k_1 and mass-moment of inertia J_1 is subjected to a sinusoidal torque with magnitude 1.3 kN-meter. When the excitation frequency equals 80 Hz, there is a large amount of vibration.

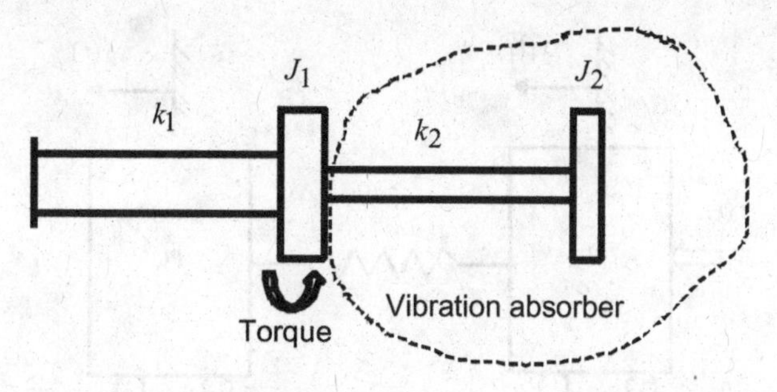

Figure P4.8 Undamped torsional vibration absorber

Design an undamped vibration absorber with the requirement that the system will be safe for 20% fluctuation in excitation frequency around 80 Hz. Determine the amplitude of the absorber rotor J_2 at 80 Hz.

P4.9 A rotor–shaft system consisting (Figure P4.9) of torsional stiffness k_1 and mass-moment of inertia J_1 is subjected to a sinusoidal torque with magnitude 1.3 kN-meter. When the excitation frequency equals 80 Hz, there is a large amount of vibration.

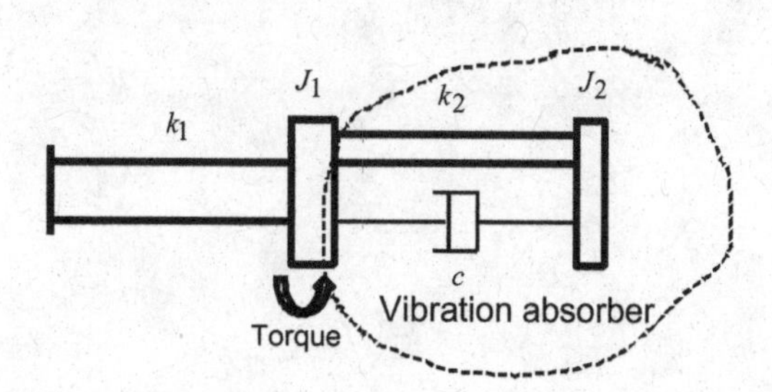

Figure P4.9 Damped torsional vibration absorber

Design an optimally damped vibration absorber. Determine the amplitudes of both rotors as a function of excitation frequencies.

P4.10 Determine the response of the two-mass system in Figure P4.10 via modal decomposition when the force $f(t)$ is a step function of magnitude 5 N and the modal damping ratio in the vibratory mode

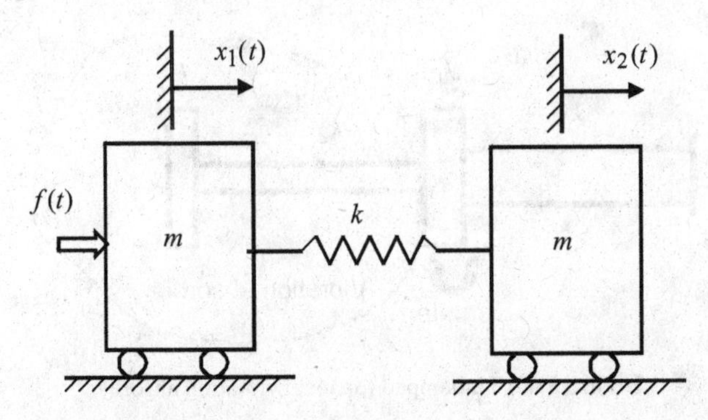

Figure P4.10 Two masses connected by a spring

is 0.05. Assume that $m = 1.5\,\text{kg}$, $k = 1,250\,\text{N/m}$, $x_1(0) = 0.02\,\text{m}$, $\dot{x}_1(0) = 1\,\text{m/sec}$, $x_2(0) = -0.01\,\text{m}$, and $\dot{x}_2(0) = 0$.

5

FINITE AND INFINITE (CONTINUOUS) DIMENSIONAL SYSTEMS

This chapter begins with the computation of the natural frequencies and the mode shapes of a discrete multi-degree-of-freedom (MDOF) system. It is shown that the natural frequencies and the modal vectors (mode shapes) are computed as the eigenvalues and the eigenvectors of a matrix dependent on mass and stiffness matrices. The orthogonal properties of modal vectors are derived. These orthogonal principles are the foundation of the modal decomposition technique, which leads to a significant reduction in the computational effort required to compute the response. Next, the following cases of continuous systems are considered: transverse vibration of a string, longitudinal vibration of a bar, torsional vibration of a circular shaft, and transverse vibration of a beam. These continuous systems have mass continuously distributed, are infinitely dimensional, and are governed by partial differential equations. The method of separation of variables is used and the natural frequencies and the modal vectors are calculated. Again, modal decomposition is used to compute the response. Last, the finite element method is introduced via examples of the longitudinal vibration of a bar and the transverse vibration of a beam.

5.1 MULTI-DEGREE-OF-FREEDOM SYSTEMS

The differential equations of an MDOF system is written as

$$M\ddot{\mathbf{x}} + C\dot{\mathbf{x}} + K\mathbf{x} = \mathbf{f}(t) \tag{5.1.1}$$

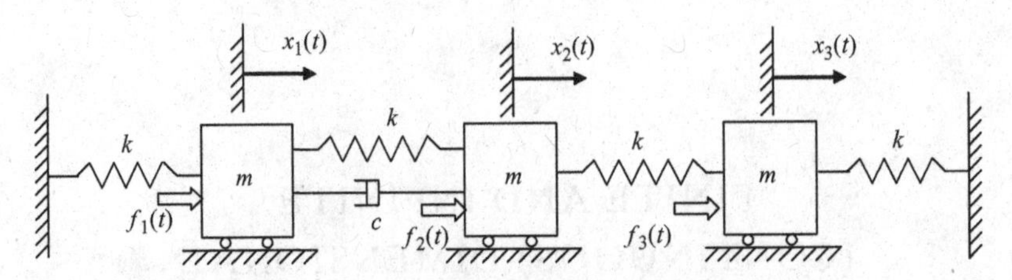

Figure 5.1.1 Three-degree-of-freedom system

The mass matrix M, the damping matrix C, and the stiffness matrix K are $n \times n$ matrices where n is the number of degrees of freedom. The force vector $\mathbf{f}(t)$ is n-dimensional.

Example 5.1.1: Consider the three-degree-of-freedom system shown in Figure 5.1.1.

The free body diagram of each mass is shown in Figure 5.1.2. Applying Newton's law of motion to each mass, three second-order differential equations are obtained as follows:

$$f_1(t) - kx_1 - k(x_1 - x_2) - c(\dot{x}_1 - \dot{x}_2) = m\ddot{x}_1 \quad (5.1.2)$$

$$f_2(t) - k(x_2 - x_1) - c(\dot{x}_2 - \dot{x}_1) - k(x_2 - x_3) = m\ddot{x}_2 \quad (5.1.3)$$

$$f_3(t) - k(x_3 - x_2) - kx_3 = m\ddot{x}_3 \quad (5.1.4)$$

With the number of degrees of freedom n equal to 3, Equations 5.1.2–5.1.4 can be put in the matrix form (Equation 5.1.1).

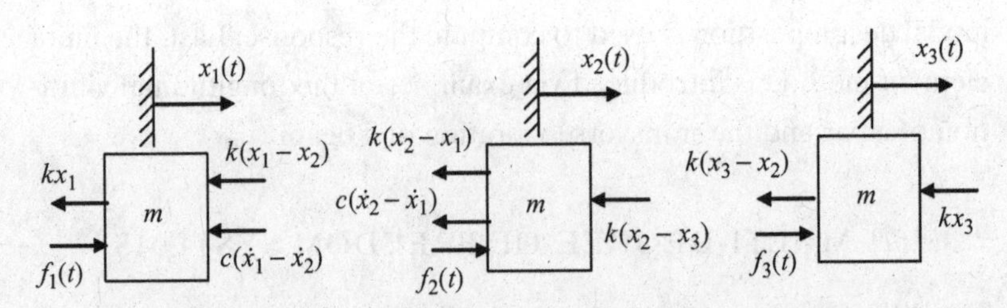

Figure 5.1.2 Free body diagram for each mass in Figure 5.1.1

where

$$\mathbf{x}(t) = \begin{bmatrix} x_1(t) \\ x_2(t) \\ x_3(t) \end{bmatrix}; \quad \mathbf{f}(t) = \begin{bmatrix} f_1(t) \\ f_2(t) \\ f_3(t) \end{bmatrix} \tag{5.1.5}$$

$$M = \begin{bmatrix} m & 0 & 0 \\ 0 & m & 0 \\ 0 & 0 & m \end{bmatrix}, K = \begin{bmatrix} 2k & -k & 0 \\ -k & 2k & -k \\ 0 & -k & 2k \end{bmatrix}, \text{ and}$$

$$C = \begin{bmatrix} c & -c & 0 \\ -c & c & 0 \\ 0 & 0 & 0 \end{bmatrix} \tag{5.1.6}$$

5.1.1 Natural Frequencies and Modal Vectors (Mode Shapes)

There is a mode shape or a modal vector associated with a natural frequency. A general method to compute the natural frequencies and the mode shapes is as follows.

Ignoring damping and external force terms, Equation 5.1.1 can be written as

$$M\ddot{\mathbf{x}} + K\mathbf{x} = \mathbf{0} \tag{5.1.7}$$

Let

$$\mathbf{x}(t) = \mathbf{a}\sin(\omega t + \phi) \tag{5.1.8}$$

where $n \times 1$ vector $\mathbf{a}$, the frequency ω, and the phase ϕ are to be determined.

Differentiating Equation 5.1.8 twice with respect to time,

$$\ddot{\mathbf{x}} = -\omega^2\mathbf{a}\sin(\omega t + \phi) \tag{5.1.9}$$

Substituting Equations 5.1.8 and 5.1.9 into Equation 5.1.7,

$$(K - \omega^2 M)\mathbf{a} = \mathbf{0} \tag{5.1.10}$$

For a nonzero or a nontrivial solution of **a**,

$$\det(K - \omega^2 M) = 0 \tag{5.1.11}$$

which will be a polynomial equation of degree n in ω^2. Equation 5.1.10 can also be written as

$$K\mathbf{a} = \omega^2 M\mathbf{a} \tag{5.1.12}$$

or

$$M^{-1}K\mathbf{a} = \omega^2\mathbf{a} \tag{5.1.13}$$

Equation 5.1.13 clearly indicates that ω^2 and **a** are an eigenvalue and an eigenvector (Strang, 1988) of the matrix $M^{-1}K$. In addition, Equation 5.1.12 suggests that ω^2 and **a** are generalized eigenvalues and eigenvectors of the stiffness matrix K with respect to the mass matrix M. The MATLAB command for computation of generalized eigenvalues and eigenvectors is $eig(K, M)$. The formulation of the generalized eigenvalue/eigenvector problem is convenient for a large number of degrees of freedom because the inverse of the mass matrix is not required.

Example 5.1.2: Eigenvalues and Eigenvectors of Three-Mass Chain
From Example 5.1.1, the mass and stiffness matrices are as follows:

$$M = \begin{bmatrix} m & 0 & 0 \\ 0 & m & 0 \\ 0 & 0 & m \end{bmatrix}; \quad K = \begin{bmatrix} 2k & -k & 0 \\ -k & 2k & -k \\ 0 & -k & 2k \end{bmatrix} \tag{5.1.14a, b}$$

Therefore,

$$K - \omega^2 M = \begin{bmatrix} 2k - \omega^2 m & -k & 0 \\ -k & 2k - \omega^2 m & -k \\ 0 & -k & 2k - \omega^2 m \end{bmatrix} \tag{5.1.15}$$

From Equation 5.1.15,

$$\det(K - \omega^2 M) = (2k - m\omega^2)[(2k - m\omega^2)^2 - 2k^2] \tag{5.1.16}$$

From Equation 5.1.16, three natural frequencies are as follows:

$$\omega_1 = \sqrt{\frac{(2-\sqrt{2})k}{m}}, \quad \omega_2 = \sqrt{\frac{2k}{m}}, \quad \text{and} \quad \omega_3 = \sqrt{\frac{(2+\sqrt{2})k}{m}} \tag{5.1.17}$$

Using Equation 5.1.10,

$$(K-\omega^2 M)\mathbf{a} = \begin{bmatrix} 2k-\omega^2 m & -k & 0 \\ -k & 2k-\omega^2 m & -k \\ 0 & -k & 2k-\omega^2 m \end{bmatrix} \begin{bmatrix} a_1 \\ a_2 \\ a_3 \end{bmatrix} = \begin{bmatrix} 0 \\ 0 \\ 0 \end{bmatrix} \tag{5.1.18}$$

I. Modal Vector for $\omega^2 = \frac{(2-\sqrt{2})k}{m}$

From Equation 5.1.18,

$$\begin{bmatrix} \sqrt{2}k & -k & 0 \\ -k & \sqrt{2}k & -k \\ 0 & -k & \sqrt{2}k \end{bmatrix} \begin{bmatrix} a_1 \\ a_2 \\ a_3 \end{bmatrix} = \begin{bmatrix} 0 \\ 0 \\ 0 \end{bmatrix} \tag{5.1.19}$$

Two independent equations are

$$\sqrt{2}ka_1 - ka_2 = 0 \quad \text{and} \quad -ka_2 + \sqrt{2}ka_3 = 0 \qquad \text{(5.1.20a, b)}$$

Since there are two equations in three unknowns, a_1 is arbitrarily chosen to be 1. Then, the solutions of Equations 5.1.20a,b yield $a_2 = \sqrt{2}$ and $a_3 = 1$. In other words, the modal vector is

$$\mathbf{a} = [1 \quad \sqrt{2} \quad 1]^T \tag{5.1.21}$$

II. Modal Vector for $\omega^2 = 2k/m$

From Equation 5.1.18,

$$\begin{bmatrix} 0 & -k & 0 \\ -k & 0 & -k \\ 0 & -k & 0 \end{bmatrix} \begin{bmatrix} a_1 \\ a_2 \\ a_3 \end{bmatrix} = \begin{bmatrix} 0 \\ 0 \\ 0 \end{bmatrix} \tag{5.1.22}$$

Two independent equations are

$$ka_2 = 0 \quad \text{and} \quad -ka_1 - ka_3 = 0 \tag{5.1.23a, b}$$

Again, a_1 is arbitrarily chosen to be 1. Then, the solutions of Equations 5.1.23a,b yield $a_2 = 0$ and $a_3 = -1$. In other words, the modal vector is

$$\mathbf{a} = [1 \quad 0 \quad -1]^T \tag{5.1.24}$$

III. Modal Vector for $\omega^2 = \frac{(2+\sqrt{2})k}{m}$

From Equation 5.1.18,

$$\begin{bmatrix} -\sqrt{2}k & -k & 0 \\ -k & -\sqrt{2}k & -k \\ 0 & -k & -\sqrt{2}k \end{bmatrix} \begin{bmatrix} a_1 \\ a_2 \\ a_3 \end{bmatrix} = \begin{bmatrix} 0 \\ 0 \\ 0 \end{bmatrix} \tag{5.1.25}$$

Two independent equations are

$$-\sqrt{2}ka_1 - ka_2 = 0 \quad \text{and} \quad -ka_2 - \sqrt{2}ka_3 = 0 \tag{5.1.26a, b}$$

Again, a_1 is arbitrarily chosen to be 1. Then, the solutions of Equations 5.1.26a,b yield $a_2 = -\sqrt{2}$ and $a_3 = 1$. In other words, the modal vector is

$$\mathbf{a} = [1 \quad -\sqrt{2} \quad 1]^T \tag{5.1.27}$$

5.1.2 Orthogonality of Eigenvectors for Symmetric Mass and Symmetric Stiffness Matrices

The orthogonality of eigenvectors is an important property for the vibration analysis of an MDOF system. The derivation of this property is as follows.

Let ω_i^2 and $\mathbf{v}_i$ be the eigenvalue and eigenvector pair where $i = 1, 2, \ldots, n$. Then,

$$K\mathbf{v}_i = \omega_i^2 M\mathbf{v}_i \tag{5.1.28}$$

and

$$K\mathbf{v}_j = \omega_j^2 M\mathbf{v}_j \tag{5.1.29}$$

Pre-multiplying both sides of Equation 5.1.28 by $\mathbf{v}_j^T$,

$$\mathbf{v}_j^T K\mathbf{v}_i = \omega_i^2 \mathbf{v}_j^T M\mathbf{v}_i \tag{5.1.30}$$

Pre-multiplying both sides of Equation 5.1.29 by $\mathbf{v}_i^T$ and then taking the transpose,

$$\left(\mathbf{v}_i^T K\mathbf{v}_j\right)^T = \omega_j^2\left(\mathbf{v}_i^T M\mathbf{v}_j\right)^T \tag{5.1.31}$$

or

$$\mathbf{v}_j^T K^T\mathbf{v}_i = \omega_j^2 \mathbf{v}_j^T M^T\mathbf{v}_i \tag{5.1.32}$$

For symmetric mass and symmetric stiffness matrices,

$$K = K^T \tag{5.1.33}$$

and

$$M = M^T \tag{5.1.34}$$

Using Equations 5.1.33 and 5.1.34, Equation 5.1.32 yields

$$\mathbf{v}_j^T K\mathbf{v}_i = \omega_j^2 \mathbf{v}_j^T M\mathbf{v}_i \tag{5.1.35}$$

Substituting Equation 5.1.30 into Equation 5.1.35,

$$\left(\omega_i^2 - \omega_j^2\right)\mathbf{v}_j^T M\mathbf{v}_i = 0 \tag{5.1.36}$$

As a result,

$$\mathbf{v}_j^T M\mathbf{v}_i = 0 \quad \text{for } \omega_i \neq \omega_j \tag{5.1.37}$$

From Equations 5.1.30 and 5.1.37,

$$\mathbf{v}_j^T K\mathbf{v}_i = 0 \quad \text{for } \omega_i \neq \omega_j \tag{5.1.38}$$

Usually, each eigenvector is scaled such that

$$\mathbf{v}_i^T M\mathbf{v}_i = 1; \quad i = 1, 2, \ldots, n \tag{5.1.39}$$

In this case, from Equation 5.1.30,

$$\mathbf{v}_i^T K \mathbf{v}_i = \omega_i^2; \quad i = 1, 2, \ldots, n \tag{5.1.40}$$

Define a modal matrix V as follows:

$$V = [\mathbf{v}_1 \quad \mathbf{v}_2 \quad \cdots \quad \mathbf{v}_{n-1} \quad \mathbf{v}_n] \tag{5.1.41}$$

Then, Equations 5.1.37–5.1.40 are expressed as

$$V^T M V = I_n \tag{5.1.42}$$

and

$$V^T K V = \Lambda \tag{5.1.43}$$

where

$$\Lambda = \begin{bmatrix} \omega_1^2 & 0 & \cdots & 0 & 0 \\ 0 & \omega_2^2 & \cdots & 0 & 0 \\ \vdots & \vdots & \ddots & \vdots & \vdots \\ 0 & 0 & \cdots & \omega_{n-1}^2 & 0 \\ 0 & 0 & \cdots & 0 & \omega_n^2 \end{bmatrix} \tag{5.1.44}$$

Note: The derivations (Equations 5.1.42 and 5.1.43) are shown only for nonrepeated natural frequencies. However, it may be possible to diagonalize when some of the natural frequencies are repeated.

Example 5.1.3: Orthogonality of Modal Vectors in Three-Mass Chain

Modal vectors in Example 5.1.2 are normalized to satisfy Equation 5.1.42 as follows:

$$\mathbf{v}_1 = \frac{1}{2\sqrt{m}} \begin{bmatrix} 1 \\ \sqrt{2} \\ 1 \end{bmatrix}, \quad \mathbf{v}_2 = \frac{1}{\sqrt{2m}} \begin{bmatrix} 1 \\ 0 \\ -1 \end{bmatrix}, \quad \text{and} \quad \mathbf{v}_3 = \frac{1}{2\sqrt{m}} \begin{bmatrix} 1 \\ -\sqrt{2} \\ 1 \end{bmatrix} \tag{5.1.45}$$

Therefore, the modal matrix is

$$V = [\mathbf{v}_1 \quad \mathbf{v}_2 \quad \mathbf{v}_3] = \frac{1}{2\sqrt{2m}} \begin{bmatrix} \sqrt{2} & 2 & \sqrt{2} \\ 2 & 0 & -2 \\ \sqrt{2} & -2 & \sqrt{2} \end{bmatrix} \quad (5.1.46)$$

Hence,

$$V^T K V = \frac{1}{8m} \begin{bmatrix} \sqrt{2} & 2 & \sqrt{2} \\ 2 & 0 & -2 \\ \sqrt{2} & -2 & \sqrt{2} \end{bmatrix} \begin{bmatrix} 2k & -k & 0 \\ -k & 2k & -k \\ 0 & -k & 2k \end{bmatrix} \begin{bmatrix} \sqrt{2} & 2 & \sqrt{2} \\ 2 & 0 & -2 \\ \sqrt{2} & -2 & \sqrt{2} \end{bmatrix} \quad (5.1.47)$$

It can be verified that

$$V^T K V = \frac{k}{m} \begin{bmatrix} 2-\sqrt{2} & 0 & 0 \\ 0 & 2 & 0 \\ 0 & 0 & 2+\sqrt{2} \end{bmatrix} = \begin{bmatrix} \omega_1^2 & 0 & 0 \\ 0 & \omega_2^2 & 0 \\ 0 & 0 & \omega_3^2 \end{bmatrix} \quad (5.1.48)$$

5.1.3 Modal Decomposition

In general, the response $\mathbf{x}(t)$ is a linear combination of the modal vectors $\mathbf{v}_i$; $i = 1, 2, \ldots, n$, that is,

$$\mathbf{x}(t) = \mathbf{v}_1 y_1(t) + \mathbf{v}_2 y_2(t) + \cdots + \mathbf{v}_n y_n(t) \quad (5.1.49)$$

where $y_i(t)$ is the coefficient of the modal vectors $\mathbf{v}_i$; $i = 1, 2, \ldots, n$. Equation 5.1.49 can be represented in a compact form as follows:

$$\mathbf{x}(t) = V\mathbf{y}(t) \quad (5.1.50)$$

where the matrix V is defined by Equation 5.1.41 and the vector $\mathbf{y}(t)$ is defined as

$$\mathbf{y}(t) = [\, y_1 \quad y_2 \quad \cdots \quad y_{n-1} \quad y_n \,]^T \quad (5.1.51)$$

Substituting Equation 5.1.50 into Equation 5.1.1, and pre-multiplying by V^T,

$$V^T M V \ddot{\mathbf{y}} + V^T C V \dot{\mathbf{y}} + V^T K V \mathbf{y} = V^T \mathbf{f}(t) \quad (5.1.52)$$

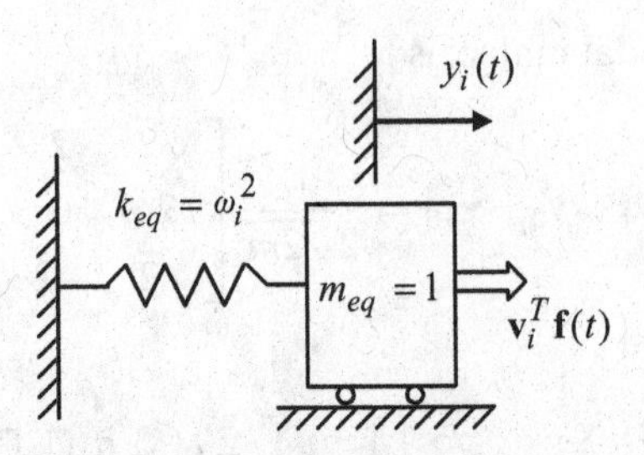

Figure 5.1.3 An equivalent undamped SDOF system for each mode

Equations 5.1.52 are often called **modal equations** as they are in terms of modal components y_i; $i = 1, 2, \ldots, n$. Matrices $V^T MV$ and $V^T KV$ are diagonal, but there is no guarantee that $V^T CV$ is diagonal. Two special cases of damping resulting in decoupled modal equations are considered as follows.

Case I: Undamped System $(C = 0)$
Substituting Equations 5.1.42 and 5.1.43 into Equation 5.1.52,

$$\ddot{\mathbf{y}} + \Lambda \mathbf{y} = V^T \mathbf{f}(t) \tag{5.1.53}$$

or

$$\ddot{y}_i + \omega_i^2 y_i = \mathbf{v}_i^T \mathbf{f}(t); \quad i = 1, 2, \ldots, n \tag{5.1.54}$$

Here, modal equations are decoupled and each modal equation in Equation 5.1.54 can be viewed as an equivalent undamped single-degree-of-freedom system subjected to the force $\mathbf{v}_i^T \mathbf{f}(t)$ (Figure 5.1.3). The quantity $\mathbf{v}_i^T \mathbf{f}(t)$ is also known as the **modal force.**

Example 5.1.4: Consider the system in Figure 5.1.1 with zero damping.
Let $m = 2$ kg and $k = 1{,}000$ kg, and

$$\mathbf{f}(t) = \begin{bmatrix} 1 \\ 0.5 \\ 2 \end{bmatrix} \sin 30t \text{ N} \tag{5.1.55}$$

Let the initial displacement and velocity vectors be $\mathbf{x}(0) = [0.03\ 0.02\ 0.04]^T$ m and $\dot{\mathbf{x}}(0) = [3\ 5\ 8]^T$ m/sec.

The MATLAB program is listed in Program 5.1. This yields

$$V = \begin{bmatrix} 0.3536 & -0.5 & -0.3536 \\ 0.5 & 0 & 0.5 \\ 0.3536 & 0.5 & -0.3536 \end{bmatrix} \tag{5.1.56}$$

and

$$\Lambda = \begin{bmatrix} \omega_1^2 & 0 & 0 \\ 0 & \omega_2^2 & 0 \\ 0 & 0 & \omega_3^2 \end{bmatrix} = \begin{bmatrix} (17.114)^2 & 0 & 0 \\ 0 & (31.6228)^2 & 0 \\ 0 & 0 & (41.3171)^2 \end{bmatrix} \tag{5.1.57}$$

The conditions in Equations 5.1.42 and 5.1.43 have been verified. Next,

$$V^T\mathbf{f}(t) = \begin{bmatrix} \mathbf{v}_1^T\mathbf{f}(t) \\ \mathbf{v}_2^T\mathbf{f}(t) \\ \mathbf{v}_3^T\mathbf{f}(t) \end{bmatrix} = \begin{bmatrix} 1.3107 \\ 0.5 \\ -0.8107 \end{bmatrix} \sin 30t \tag{5.1.58}$$

Initial conditions are

$$\begin{bmatrix} y_1(0) \\ y_2(0) \\ y_3(0) \end{bmatrix} = V^{-1}\mathbf{x}(0) = \begin{bmatrix} 0.0695 \\ 0.01 \\ -0.0295 \end{bmatrix} \tag{5.1.59}$$

and

$$\begin{bmatrix} \dot{y}_1(0) \\ \dot{y}_2(0) \\ \dot{y}_3(0) \end{bmatrix} = V^{-1}\dot{\mathbf{x}}(0) = \begin{bmatrix} 12.7782 \\ 5 \\ -2.7782 \end{bmatrix} \tag{5.1.60}$$

Equation 5.1.54 yields

$$\ddot{y}_1 + (17.114)^2 y_1 = 1.3107 \sin 30t; \quad y_1(0) = 0.0695, \quad \dot{y}_1(0) = 12.7782 \tag{5.1.61a}$$

$$\ddot{y}_2 + (31.6228)^2 y_2 = 0.5 \sin 30t; \quad y_2(0) = 0.01, \quad \dot{y}_2(0) = 5 \tag{5.1.61b}$$

$$\ddot{y}_3 + (41.3171)^2 y_3 = -0.8107 \sin 30t; \quad y_3(0) = -0.0295, \quad \dot{y}_3(0) = -2.7782 \tag{5.1.61c}$$

These equations can be easily solved using techniques presented in Chapter 2, and the response of the MDOF system is then given by

$$\begin{bmatrix} x_1(t) \\ x_2(t) \\ x_3(t) \end{bmatrix} = V \begin{bmatrix} y_1(t) \\ y_2(t) \\ y_3(t) \end{bmatrix} \tag{5.1.62}$$

MATLAB Program 5.1: Modal Vectors and Modal Initial Conditions

```
%
clear all
close all
%
m=2;
k=1000;
M=[2 0 0;0 2 0;0 0 2];
K=[2*k -k 0;-k 2*k -k;0 -k 2*k];
f=[1 0.5 2]';
%
[V,D]=eig(K,M);
V'*f
y0=inv(V)*[0.03 0.02 0.04]'%initial value of y
dy0=inv(V)*[3 5 8]'%initial value of dy/dt
```

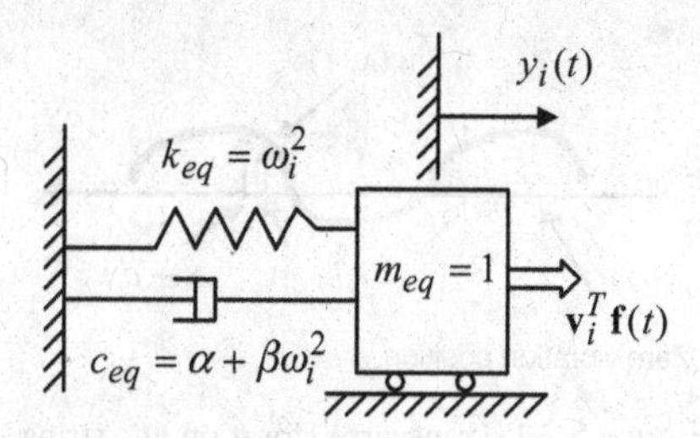

Figure 5.1.4 An equivalent damped SDOF system for each mode

Case II: Proportional or Rayleigh Damping

Assume that the damping matrix has the following form:

$$C = \alpha M + \beta K \tag{5.1.63}$$

where α and β are the constants. This form of damping is known as **Proportional** and **Rayleigh** damping. Substituting Equations 5.1.42, 5.1.43, and 5.1.63 into Equation 5.1.52,

$$\ddot{\mathbf{y}} + (\alpha I + \beta\Lambda)\dot{\mathbf{y}} + \Lambda\mathbf{y} = V^T\mathbf{f}(t) \tag{5.1.64}$$

or

$$\ddot{y}_i + \left(\alpha + \beta\omega_i^2\right)\dot{y}_i + \omega_i^2 y_i = \mathbf{v}_i^T\mathbf{f}(t); \quad i = 1, 2, \ldots, n \tag{5.1.65}$$

Again, the modal equations are decoupled and each modal equation in Equation 5.1.65 can be viewed as an equivalent damped single-degree-of-freedom system subjected to the modal force $\mathbf{v}_i^T\mathbf{f}(t)$ (Figure 5.1.4).

Example 5.1.5: Nonproportional Damping

Consider the Example 5.1.1 with $c = 5$ N-sec/meter. The mass and stiffness matrices are the same as those in Example 5.1.4. In this case,

$$V^T C V = \begin{bmatrix} 0.1072 & 0.3661 & 0.6250 \\ 0.3661 & 1.25 & 2.1339 \\ 0.6250 & 2.1339 & 3.6428 \end{bmatrix} \tag{5.1.66}$$

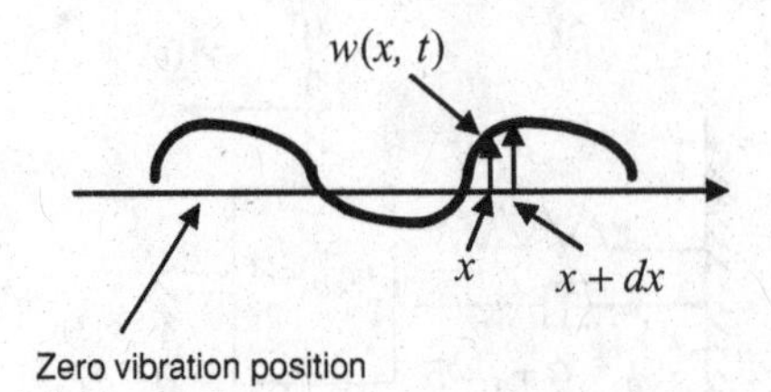

Figure 5.2.1 Transverse vibration of a string

Because $V^T CV$ is not diagonal, the damping in the system is not proportional.

5.2 CONTINUOUS SYSTEMS GOVERNED BY WAVE EQUATIONS

This section deals with the vibration of a string, the longitudinal vibration of a bar, and the torsional vibration of a circular shaft.

5.2.1 Transverse Vibration of a String

Consider a string (Figure 5.2.1) with tension P, mass per unit length μ, and external force per unit length $f_\ell(x, t)$. Let $w(x, t)$ be the transverse deflection of the string at a position x and time t. The free body diagram of a string section with the length dx is shown in Figure 5.2.2.

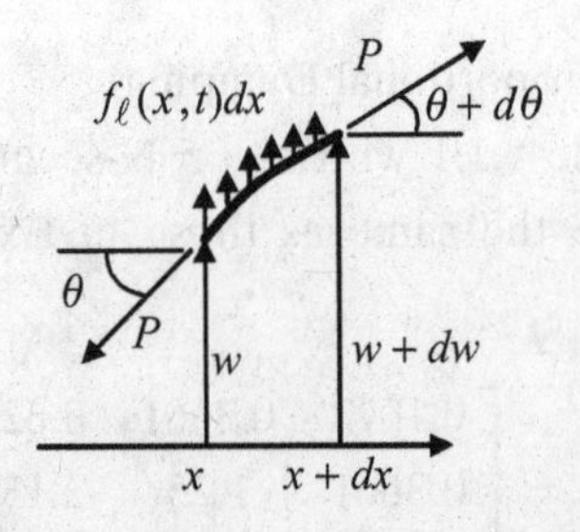

Figure 5.2.2 Free body diagram of a string element

Applying Newton's second law of motion to the free body diagram in Figure 5.2.2,

$$f_\ell(x,t)dx + P\sin(\theta + d\theta) - P\sin\theta = \mu dx \frac{\partial^2 w}{\partial t^2} \tag{5.2.1}$$

where θ is the slope of the function $w(x,t)$, that is,

$$\tan\theta = \frac{\partial w}{\partial x} \tag{5.2.2}$$

For a small θ,

$$\sin\theta \approx \tan\theta = \frac{\partial w}{\partial x} \tag{5.2.3}$$

and

$$\sin(\theta + d\theta) \approx \tan(\theta + d\theta) = \frac{\partial w}{\partial x} + \frac{\partial^2 w}{\partial x^2}dx \tag{5.2.4}$$

Substituting Equations 5.2.3 and 5.2.4 into Equation 5.2.1,

$$\frac{1}{\mu}f_\ell(x,t) + \chi^2 \frac{\partial^2 w}{\partial x^2} = \frac{\partial^2 w}{\partial t^2} \tag{5.2.5}$$

where

$$\chi = \sqrt{\frac{P}{\mu}} \tag{5.2.6}$$

Equation 5.2.5 is the governing partial differential equation of motion.

Natural Frequencies and Mode Shapes

Natural frequencies and mode shapes are the characteristics of free vibration. Therefore, the external force term in Equation 5.2.5 is ignored to obtain

$$\chi^2 \frac{\partial^2 w}{\partial x^2} = \frac{\partial^2 w}{\partial t^2} \tag{5.2.7}$$

The partial differential Equation 5.2.7 is known as the **wave equation**.

The solution of the partial differential Equation 5.2.7 is represented as follows:

$$w(x,t) = X(x)T(t) \tag{5.2.8}$$

where $X(x)$ is a function of position x only and $T(t)$ is a function of time t only. This technique is known as the **separation of variables** (Boyce and DiPrima, 2005). Differentiating Equation 5.2.8,

$$\frac{\partial^2 w}{\partial x^2} = \frac{d^2 X}{dx^2} T \tag{5.2.9}$$

and

$$\frac{\partial^2 w}{\partial t^2} = X \frac{d^2 T}{dt^2} \tag{5.2.10}$$

Substituting Equations 5.2.9 and 5.2.10 into Equation 5.2.7,

$$\chi^2 \frac{1}{X} \frac{d^2 X}{dx^2} = \frac{1}{T} \frac{d^2 T}{dt^2} \tag{5.2.11}$$

Since the left-hand side of Equation 5.2.11 is only a function of x and the right-hand side of Equation 5.2.11 is only a function of t, they can be equal only by being a constant. Also, it has been found that this constant must be a negative number for a physically meaningful solution. Denoting this negative constant as $-\omega^2$,

$$\chi^2 \frac{1}{X} \frac{d^2 X}{dx^2} = \frac{1}{T} \frac{d^2 T}{dt^2} = -\omega^2 \tag{5.2.12}$$

Equation 5.2.12 represent the following two ordinary differential equations:

$$\frac{d^2 T}{dt^2} + \omega^2 T = 0 \tag{5.2.13}$$

and

$$\frac{d^2 X}{dx^2} + \frac{\omega^2}{\chi^2} X = 0 \tag{5.2.14}$$

Both equations have the same form as the differential equation governing the undamped free vibration of an equivalent single-degree-of-freedom system in Chapter 1. Therefore, the solution of Equation 5.2.13 is as follows:

$$T(t) = A_1 \sin \omega t + B_1 \cos \omega t \tag{5.2.15}$$

where the constants A_1 and B_1 depend on the initial position and the velocity of the string. The form of the solution Equation 5.2.15 clearly indicates that the constant ω is the natural frequency of vibration.

The solution of Equation 5.2.14 is as follows:

$$X(x) = A_2 \sin\left(\frac{\omega}{\chi}x\right) + B_2 \cos\left(\frac{\omega}{\chi}x\right) \tag{5.2.16}$$

where the constants A_2 and B_2 depend on the boundary conditions which also yield natural frequencies and associated mode shapes. As an example, a string fixed at both ends is considered as follows.

String Fixed at Both Ends Let a string of length ℓ be fixed at both ends. In this case, the boundary conditions are described as

$$w(0, t) = 0 \quad \text{for all } t \tag{5.2.17}$$

and

$$w(\ell, t) = 0 \quad \text{for all } t \tag{5.2.18}$$

From Equations 5.2.8 and 5.2.17,

$$X(0) = 0 \tag{5.2.19}$$

From Equations 5.2.8 and 5.2.18,

$$X(\ell) = 0 \tag{5.2.20}$$

Imposing the conditions (Equations 5.2.19 and 5.2.20) to Equation 5.2.16,

$$B_2 = 0 \tag{5.2.21}$$

and

$$X(\ell) = A_2 \sin\left(\frac{\omega}{\chi}\ell\right) = 0 \tag{5.2.22}$$

For a nontrivial solution ($A_2 \neq 0$),

$$\sin\left(\frac{\omega}{\chi}\ell\right) = 0 \tag{5.2.23}$$

The number of solutions of Equation 5.2.23 is infinite as follows:

$$\frac{\omega}{\chi}\ell = n\pi; \quad n = 1, 2, 3, \ldots, \tag{5.2.24}$$

Equation 5.2.24 leads to the natural frequencies for the transverse vibration of the string:

$$\omega_n = \frac{n\pi\chi}{\ell} = \frac{n\pi}{\ell}\sqrt{\frac{P}{\mu}}; \quad n = 1, 2, 3, \ldots, \tag{5.2.25}$$

The number of natural frequencies is infinite as the number of degrees of freedom of a continuous structure is infinite. Substituting Equations 5.2.21 and 5.2.25 into Equation 5.2.16,

$$X(x) = A_2 \sin\left(\frac{\omega_n}{\chi}x\right) = A_2 \sin\left(n\pi\frac{x}{\ell}\right) \tag{5.2.26}$$

Setting A_2 arbitrarily equal to one, the mode shape associated with the frequency ω_n is written as

$$\phi_n(x) = \sin\left(n\pi\frac{x}{\ell}\right); \quad n = 1, 2, 3, \ldots, \tag{5.2.27}$$

These mode shapes are shown in Figure 5.2.3.

The mode shapes (Equation 5.2.27) are **orthogonal** to each other in the following sense:

$$\int_0^\ell \phi_i(x)\phi_j(x)dx = 0; \quad i \neq j \tag{5.2.28}$$

It should also be noted that

$$\int_0^\ell \phi_i^2(x)dx = \frac{\ell}{2} \tag{5.2.29}$$

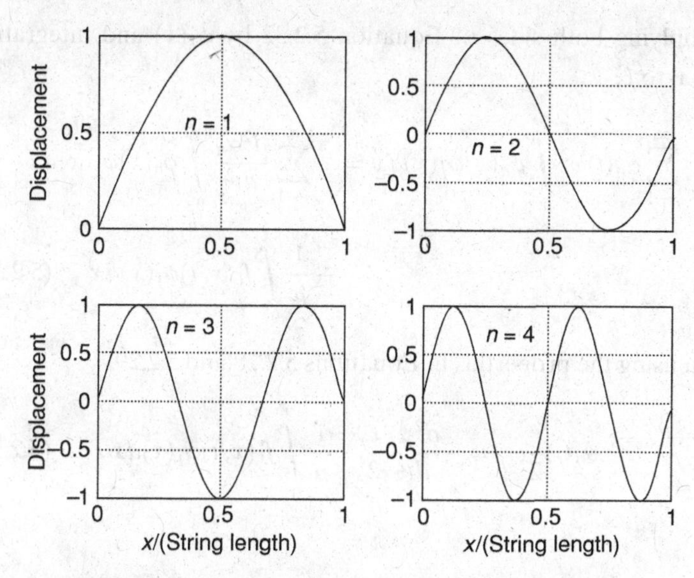

Figure 5.2.3 Mode shapes $\phi_n(x)$ of strings fixed at both ends

Computation of Response

In general, the response can be expressed as a linear combination of mode shapes, that is,

$$w(x,t) = \sum_{n=1}^{\infty} \alpha_n(t)\phi_n(x) \tag{5.2.30}$$

where time-dependent coefficients $\alpha_n(t)$ are to be determined.

Differentiating Equation 5.2.30 twice and using Equation 5.2.14,

$$\chi^2 \frac{\partial^2 w}{\partial x^2} = \sum_{n=1}^{\infty} \alpha_n(t)\chi^2 \frac{d^2\phi_n}{dx^2} = -\sum_{n=1}^{\infty} \alpha_n(t)\omega_n^2\phi_n(x) \tag{5.2.31}$$

Substituting Equation 5.2.31 into the partial differential equation of motion (Equation 5.2.5),

$$\sum_{n=1}^{\infty} \alpha_n(t)\omega_n^2\phi_n(x) = -\sum_{n=1}^{\infty} \frac{d^2\alpha_n}{dt^2}\phi_n(x) + \frac{1}{\mu} f_\ell(x,t) \tag{5.2.32}$$

Multiplying both sides of Equation 5.2.32 by $\phi_j(x)$ and integrating from 0 to ℓ,

$$\sum_{n=1}^{\infty} \alpha_n(t)\omega_n^2 \int_0^{\ell} \phi_n(x)\phi_j(x)dx = -\sum_{n=1}^{\infty} \frac{d^2\alpha_n}{dt^2} \int_0^{\ell} \phi_n(x)\phi_j(x)dx + \frac{1}{\mu}\int_0^{\ell} f_\ell(x,t)\phi_j(x)dx \qquad (5.2.33)$$

Then using the properties in Equations 5.2.28 and 5.2.29,

$$\alpha_j(t)\frac{\ell}{2}\omega_j^2 = -\frac{d^2\alpha_j}{dt^2}\frac{\ell}{2} + \frac{1}{\mu}\int_0^{\ell} f_\ell(x,t)\phi_j(x)dx \qquad (5.2.34)$$

or

$$\frac{d^2\alpha_j}{dt^2} + \omega_j^2\alpha_j(t) = \frac{2}{\mu\ell}\int_0^{\ell} f_\ell(x,t)\phi_j(x)dx \qquad (5.2.35)$$

Initial conditions are obtained from $w(x,0)$ and $\dot{w}(x,0)$. From Equation 5.2.30,

$$w(x,0) = \sum_{n=1}^{\infty} \alpha_n(0)\phi_n(x) \qquad (5.2.36)$$

$$\dot{w}(x,0) = \sum_{n=1}^{\infty} \dot{\alpha}_n(0)\phi_n(x) \qquad (5.2.37)$$

Using the properties in Equations 5.2.28 and 5.2.29,

$$\alpha_j(0) = \frac{2}{\ell}\int_0^{\ell} w(x,0)\phi_j(x)dx \qquad (5.2.38)$$

and

$$\dot{\alpha}_j(0) = \frac{2}{\ell}\int_0^{\ell} \dot{w}(x,0)\phi_j(x)dx \qquad (5.2.39)$$

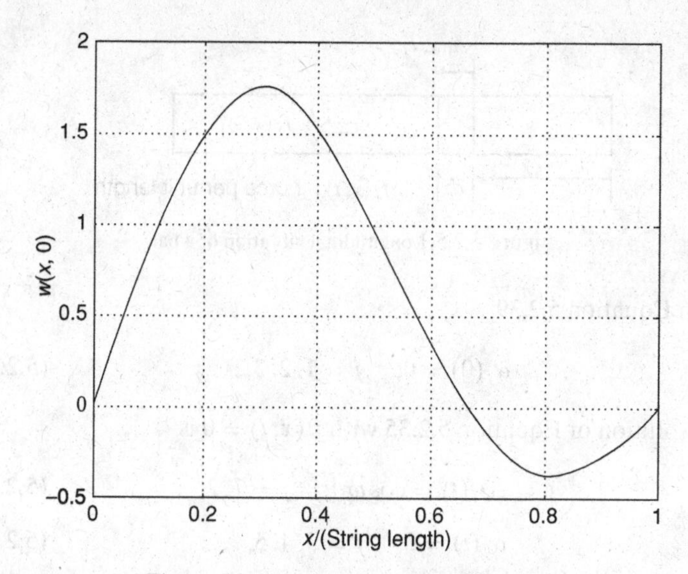

Figure 5.2.4 Initial displacement of a string

Example 5.2.1: Free Vibration of a String
Consider a string for which the initial displacement is shown in Figure 5.2.4, which can be analytically expressed as

$$w(x,0) = \sin\left(\frac{\pi x}{\ell}\right) + \sin\left(\frac{2\pi x}{\ell}\right) \tag{5.2.40}$$

Assuming that the initial velocity of string $\dot{w}(x,0) = 0$, determine the free response of the string.

From Equation 5.2.38,

$$\alpha_1(0) = \frac{2}{\ell}\int_0^\ell \left[\sin\left(\frac{\pi x}{\ell}\right) + \sin\left(\frac{2\pi x}{\ell}\right)\right]\sin\left(\frac{\pi x}{\ell}\right)dx = 1 \tag{5.2.41}$$

$$\alpha_2(0) = \frac{2}{\ell}\int_0^\ell \left[\sin\left(\frac{\pi x}{\ell}\right) + \sin\left(\frac{2\pi x}{\ell}\right)\right]\sin\left(\frac{2\pi x}{\ell}\right)dx = 1 \tag{5.2.42}$$

and

$$\alpha_j(0) = 0; \quad j = 3, 4, 5, \ldots, \tag{5.2.43}$$

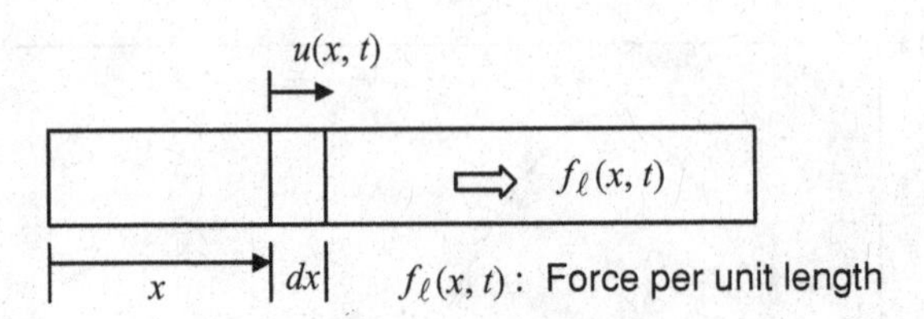

Figure 5.2.5 Longitudinal vibration of a bar

From Equation 5.2.39,

$$\dot{\alpha}_j(0) = 0; \quad j = 1, 2, 3, \ldots, \tag{5.2.44}$$

The solution of Equation 5.2.35 with $f_\ell(x, t) = 0$ is

$$\alpha_j(t) = \cos \omega_j t; \quad j = 1, 2 \tag{5.2.45}$$

$$\alpha_j(t) = 0; \quad j = 3, 4, 5, \ldots, \tag{5.2.46}$$

Therefore, from Equation 5.2.30, the response is

$$w(x, t) = \cos(\omega_1 t) \sin\left(\frac{\pi x}{\ell}\right) + \cos(\omega_2 t) \sin\left(\frac{2\pi x}{\ell}\right) \tag{5.2.47}$$

5.2.2 Longitudinal Vibration of a Bar

Consider a longitudinal bar shown in Figure 5.2.5 for which $u(x, t)$ is the axial displacement at a distance x from the left end and at any time t. The force per unit length along the axial direction is $f_\ell(x, t)$. The free body diagram of an element of the length dx is shown in Figure 5.2.6, where P is the force on the element from the part of the bar on the left of the element. Similarly, $P + dP$ is the force on the element from the part of the bar that is on the right side of the element.

Applying Newton's second law of motion to the element in Figure 5.2.6,

$$f_\ell(x, t)dx + P + dP - P = \rho A dx \frac{\partial^2 u}{\partial t^2} \tag{5.2.48}$$

where ρ and A are the mass density and the cross-sectional area, respectively. Note that the mass of the element of length dx is $\rho A dx$.

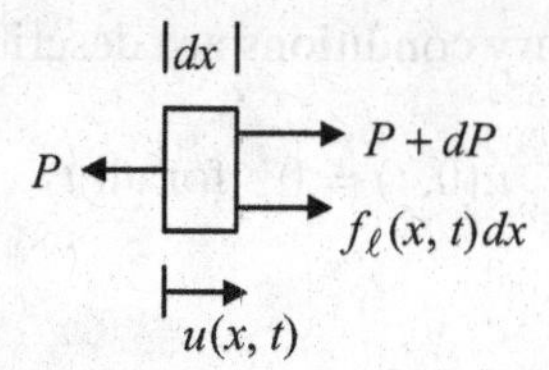

Figure 5.2.6 Free body diagram of an element of length dx

The strain ε (Crandall et al., 1999) at a position x in Figure 5.2.5 is

$$\varepsilon = \frac{\partial u}{\partial x} \tag{5.2.49}$$

Therefore, the stress σ at a position x in Figure 5.2.5 is

$$\sigma = E\frac{\partial u}{\partial x} \tag{5.2.50}$$

where E is the Young's modulus of elasticity. Using Equation 5.2.50, the internal force P at a position x in Figure 5.2.5 is

$$P = \sigma A = EA\frac{\partial u}{\partial x} \tag{5.2.51}$$

Differentiating Equation 5.2.51,

$$dP = EA\frac{\partial^2 u}{\partial x^2}dx \tag{5.2.52}$$

Substituting Equation 5.2.52 into Equation 5.2.48,

$$c^2\frac{\partial^2 u}{\partial x^2} + \frac{f_\ell(x,t)}{\rho A} = \frac{\partial^2 u}{\partial t^2} \tag{5.2.53}$$

where

$$c = \sqrt{\frac{E}{\rho}} \tag{5.2.54}$$

The governing partial differential equation of motion (Equation 5.2.53) is also a **wave equation**.

Example 5.2.2: Find the natural frequencies and the mode shapes of a fixed–free longitudinal bar.

In this case, the boundary conditions are described as

$$u(0,t) = 0 \quad \text{for all } t \tag{5.2.55}$$

and

$$\frac{du}{dx}(\ell,t) = 0 \quad \text{for all } t \tag{5.2.56}$$

Similar to Equation 5.2.8,

$$u(x,t) = X(x)T(t) \tag{5.2.57}$$

The solutions of $T(t)$ and $X(x)$ are given by Equations 5.2.15 and 5.2.16 where χ is replaced by c.

From Equations 5.2.55 and 5.2.57

$$X(0) = 0 \tag{5.2.58}$$

From Equations 5.2.56 and 5.2.57,

$$\frac{dX}{dx}(\ell) = 0 \tag{5.2.59}$$

Imposing the conditions in Equations 5.2.58 and 5.2.59 to Equation 5.2.16,

$$B_2 = 0 \tag{5.2.60}$$

and

$$\frac{dX}{dx}(\ell) = A_2\frac{\omega}{c}\cos\left(\frac{\omega}{c}\ell\right) = 0 \tag{5.2.61}$$

For a nontrivial solution ($A_2 \neq 0$),

$$\cos\left(\frac{\omega}{c}\ell\right) = 0 \tag{5.2.62}$$

The number of solutions of Equation 5.2.62 is infinite as follows:

$$\frac{\omega}{c}\ell = \frac{2n-1}{2}\pi; \quad n = 1, 2, 3, \ldots, \tag{5.2.63}$$

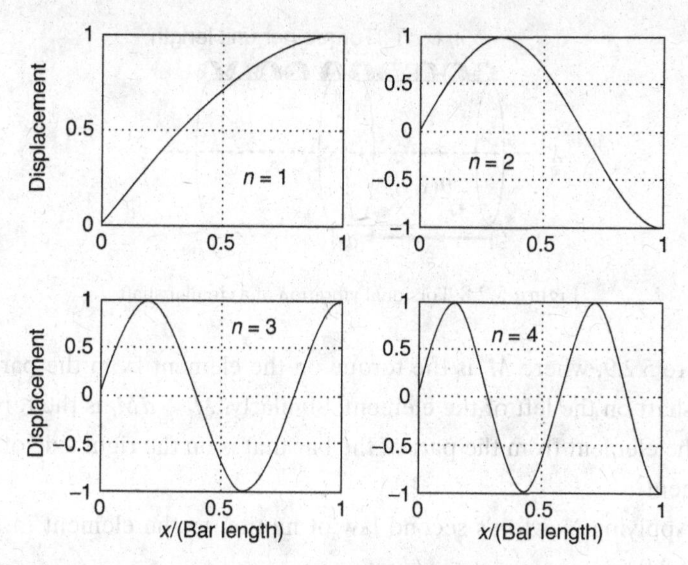

Figure 5.2.7 Mode shapes $\phi_n(x)$ for longitudinal vibration of a bar (fixed-free)

Equation 5.2.63 leads to the natural frequencies for the longitudinal vibration of a bar:

$$\omega_n = \frac{(2n-1)\pi c}{2\ell} = \frac{(2n-1)\pi}{2\ell}\sqrt{\frac{E}{\rho}}; \quad n = 1, 2, 3, \ldots, \tag{5.2.64}$$

The mode shapes of the longitudinal bar vibration are

$$\phi_n(x) = \sin\left(\frac{\omega_n}{c}x\right) = \sin\left(\frac{(2n-1)\pi}{2\ell}x\right) \tag{5.2.65}$$

These mode shapes are plotted in Figure 5.2.7.

5.2.3 *Torsional Vibration of a Circular Shaft*

Consider a shaft with a circular cross section shown in Figure 5.2.8 for which $\theta(x, t)$ is the angle of twist of a section at a distance x from the left end and at any time t.

The torque per unit length along the axial direction is $n_\ell(x, t)$. The free body diagram of an element of the length dx is shown in

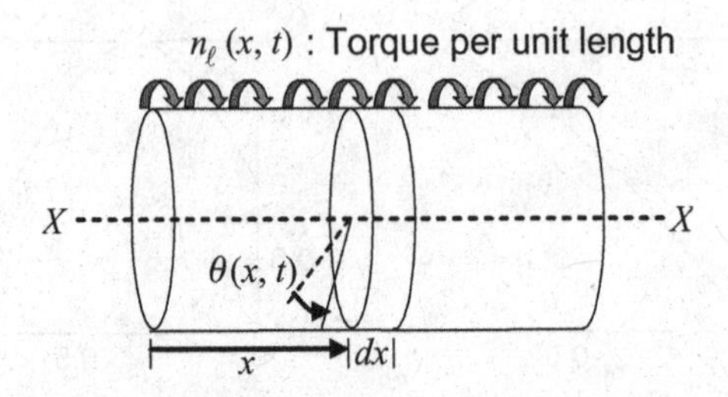

Figure 5.2.8 Torsional vibration of a circular shaft

Figure 5.2.9, where M_t is the torque on the element from the part of the shaft on the left of the element. Similarly, $M_t + dM_t$ is the torque on the element from the part of the bar that is on the right side of the element.

Applying Newton's second law of motion to the element in Figure 5.2.9,

$$M_t + dM_t + n_\ell(x, t)dx - M_t = I_0 dx \frac{\partial^2 \theta}{\partial t^2} \tag{5.2.66}$$

where I_0 is the mass-moment of inertia per unit length about the axis XX. It is known (Crandall et al., 1999) that

$$M_t = GJ \frac{\partial \theta}{\partial x} \tag{5.2.67}$$

where G and J are the Shear modulus of elasticity and the area moment of inertia of the circular cross section about the axis XX,

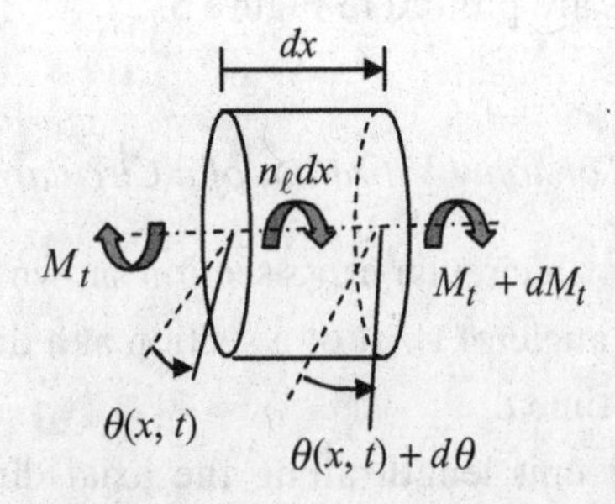

Figure 5.2.9 Free body diagram of an element of length dx

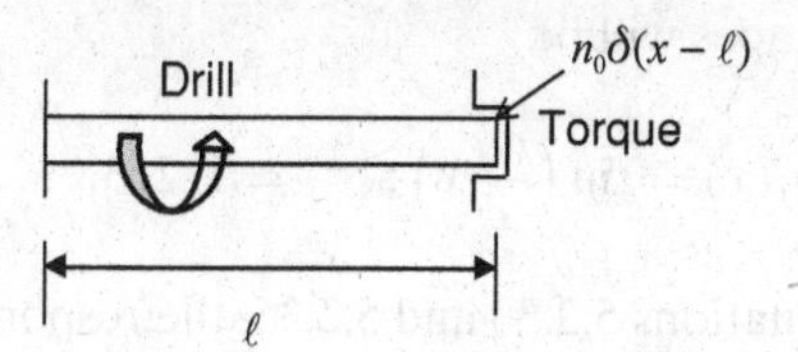

Figure 5.2.10 A simple model of drill

respectively. From Equation 5.2.67,

$$dM_t = GJ\frac{\partial^2\theta}{\partial x^2}dx \tag{5.2.68}$$

Substituting Equation 5.2.68 into Equation 5.2.66,

$$c^2\frac{\partial^2\theta}{\partial x^2} + \frac{n_\ell(x,t)}{I_0} = \frac{\partial^2\theta}{\partial t^2} \tag{5.2.69}$$

where

$$c = \sqrt{\frac{GJ}{I_0}} \tag{5.2.70}$$

The governing differential equation of motion (Equation 5.2.69) is also a **wave equation.**

Example 5.2.3: Forced Response of a Circular Drill
Consider a circular drill of length ℓ and diameter d. When the drill makes a hole on a work surface, it experiences a torque n_0 at $x = \ell$ (Figure 5.2.10). Assuming that $\theta(x, 0) = 0$, $\dot{\theta}(x, 0) - 0$, and treating the shaft as fixed–free, determine the response $\theta(x, t)$, the angle of twist of a section at a distance x from the left.

Since the shaft is fixed–free, the natural frequencies and the mode shapes can be shown to be given by Equations 5.2.64 and 5.2.65, that is,

$$\omega_j = \frac{(2j-1)\pi c}{2\ell} = \frac{(2j-1)\pi}{2\ell}\sqrt{\frac{GJ}{I_0}}; \quad j = 1, 2, 3, \ldots, \tag{5.2.71}$$

And the mode shapes will be

$$\phi_j(x) = \sin\left(\frac{\omega_j}{c}x\right); \quad j = 1, 2, 3, \ldots, \tag{5.2.72}$$

Following Equations 5.2.30 and 5.2.35, the response is given by

$$\theta(x,t) = \sum_{j=1}^{\infty} \alpha_j(t)\phi_j(x) \tag{5.2.73}$$

where

$$\frac{d^2\alpha_j}{dt^2} + \omega_j^2\alpha_j(t) = \frac{2}{I_0\ell}\int_0^{\ell} n_\ell(x,t)\phi_j(x)dx \tag{5.2.74}$$

Here,

$$n_\ell(x,t) = n_0\delta(x-\ell) \tag{5.2.75}$$

where $\delta(x-\ell)$ is the Dirac delta or the unit impulse function. Substituting Equation 5.2.75 into Equation 5.2.74, and using Equation 3.2.5,

$$\frac{d^2\alpha_j}{dt^2} + \omega_j^2\alpha_j(t) = \frac{2n_0\phi_j(\ell)}{I_0\ell} \tag{5.2.76}$$

Because $\theta(x,0) = 0$ and $\dot{\theta}(x,0) = 0$,

$$\alpha_j(0) = 0 \quad \text{and} \quad \dot{\alpha}_j(0) = 0; \quad j = 1, 2, 3, \ldots, \tag{5.2.77}$$

Following the solution procedure in Chapter 2 (Equation 2.1.21),

$$\alpha_j(t) = \frac{2n_0\phi_j(\ell)}{I_0\ell\omega_j^2}(1-\cos\omega_j t); \quad j = 1, 2, 3, \ldots, \tag{5.2.78}$$

From Equations 5.2.73, 5.2.72, and 5.2.78,

$$\theta(x,t) = \sum_{j=1}^{\infty} \frac{2n_0\phi_j(\ell)}{I_0\ell\omega_j^2}(1-\cos\omega_j t)\sin\left(\frac{\omega_j}{c}x\right) \tag{5.2.79}$$

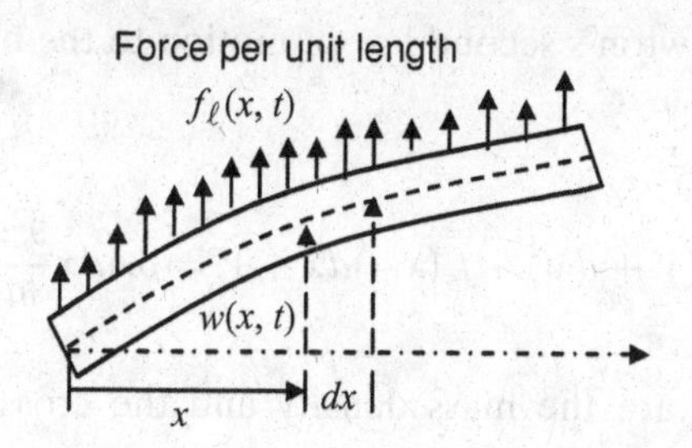

Figure 5.3.1 Transverse vibration of a beam

5.3 CONTINUOUS SYSTEMS: TRANSVERSE VIBRATION OF A BEAM

5.3.1 Governing Partial Differential Equation of Motion

Consider a beam shown in Figure 5.3.1 for which $w(x, t)$ is the transverse displacement at a distance x from the left end and at any time t. The force per unit length along the lateral direction is $f_\ell(x, t)$. The free body diagram of an element of the length dx is shown in Figure 5.3.2, where $V(x, t)$ and $M(x, t)$ are the shear force and the bending moment, respectively, on the element from the part of the beam on the left of the element. Similarly, $V(x, t) + dV(x, t)$ and $M(x, t) + dM(x, t)$ are the shear force and the bending moment, respectively, on the element from the part of the beam which is on the right side of the element.

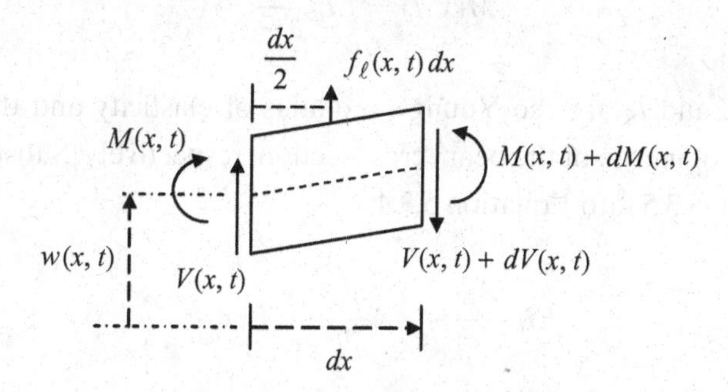

Figure 5.3.2 Free body diagram of a beam element

Applying Newton's second law of motion to the beam element in Figure 5.3.2,

$$-(V+dV)+f_\ell(x,t)dx+V=\rho A dx\frac{\partial^2 w}{\partial t^2} \tag{5.3.1}$$

where ρ and A are the mass density and the cross-sectional area, respectively.

It is well known (Crandall et al., 1999) that

$$V=\frac{\partial M}{\partial x} \tag{5.3.2}$$

Therefore,

$$dV=\frac{\partial^2 M}{\partial x^2}dx \tag{5.3.3}$$

Substituting Equation 5.3.3 into Equation 5.3.1,

$$-\frac{\partial^2 M}{\partial x^2}+f_\ell(x,t)=\rho A\frac{\partial^2 w}{\partial t^2} \tag{5.3.4}$$

From the elementary beam theory (Crandall et al., 1999),

$$M(x,t)=EI_a\frac{\partial^2 w}{\partial x^2} \tag{5.3.5}$$

where E and I_a are the Young's modulus of elasticity and the area moment of inertia of the beam cross section, respectively. Substituting Equation 5.3.5 into Equation 5.3.4,

$$EI_a\frac{\partial^4 w}{\partial x^4}+\rho A\frac{\partial^2 w}{\partial t^2}=f_\ell(x,t) \tag{5.3.6}$$

Equation 5.3.6 is the governing partial differential equation of motion.

5.3.2 *Natural Frequencies and Mode Shapes*

Setting the external force $f_\ell(x, t) = 0$, Equation 5.3.6 can be written as

$$\beta^2 \frac{\partial^4 w}{\partial x^4} = -\frac{\partial^2 w}{\partial t^2} \tag{5.3.7}$$

where

$$\beta^2 = \frac{EI_a}{\rho A} \tag{5.3.8}$$

Following the method of separation of variables, assume that

$$w(x, t) = X(x)T(t) \tag{5.3.9}$$

where $X(x)$ is a function of x only and $T(t)$ is a function of time t only. From Equation 5.3.9,

$$\frac{\partial^4 w}{\partial x^4} = \frac{d^4 X}{dx^4} T \tag{5.3.10}$$

and

$$\frac{\partial^2 w}{\partial t^2} = X \frac{d^2 T}{dt^2} \tag{5.3.11}$$

Substituting Equation 5.3.10 and 5.3.11 into Equation 5.3.7

$$-\beta^2 \frac{1}{X} \frac{d^4 X}{dx^4} = \frac{1}{T} \frac{d^2 T}{dt^2} \tag{5.3.12}$$

Since the left-hand side of Equation 5.3.12 is only a function of x and the right-hand side of Equation 5.3.12 is only a function of t, they can be equal only by being a constant. Also, it has been found that this constant must be a negative number for a physically meaningful solution. Denoting this negative constant as $-\omega^2$,

$$-\beta^2 \frac{1}{X} \frac{d^4 X}{dx^4} = \frac{1}{T} \frac{d^2 T}{dt^2} = -\omega^2 \tag{5.3.13}$$

From Equation 5.3.13,

$$\frac{d^2 T}{dt^2} + \omega^2 T = 0 \tag{5.3.14}$$

The solution of Equation 5.3.14 is

$$T(t) = A_1 \sin \omega t + B_1 \cos \omega t \tag{5.3.15}$$

where A_1 and B_1 are the constants. Equation 5.3.13 also yields

$$\frac{d^4X}{dx^4} - \gamma^4 X = 0 \tag{5.3.16}$$

where

$$\gamma^4 = \frac{\omega^2}{\beta^2} \tag{5.3.17}$$

The solution of Equation 5.3.16 is assumed as

$$X(x) = Ae^{sx} \tag{5.3.18}$$

Substituting Equation 5.3.18 into Equation 5.3.16,

$$(s^4 - \gamma^4)Ae^{sx} = 0 \tag{5.3.19}$$

For a nontrivial solution ($A \neq 0$),

$$s^4 - \gamma^4 = 0 \tag{5.3.20}$$

or

$$(s + \gamma)(s - \gamma)(s + i\gamma)(s - i\gamma) = 0 \tag{5.3.21}$$

where $i = \sqrt{-1}$. The four roots of Equation 5.3.21 are as follows:

$$s_1 = -\gamma, \quad s_2 = \gamma, \quad s_3 = -i\gamma, \quad \text{and} \quad s_4 = i\gamma \tag{5.3.22}$$

Therefore, the general solution of Equation 5.3.16 is

$$X(x) = A_2 e^{-\gamma x} + B_2 e^{\gamma x} + C_2 e^{-i\gamma x} + D_2 e^{i\gamma x} \tag{5.3.23}$$

Equation 5.3.23 can also be expressed as

$$X(x) = A_3 \cosh \gamma x + B_3 \sinh \gamma x + C_3 \cos \gamma x + D_3 \sin \gamma x \tag{5.3.24}$$

Here, the constants A_3 and B_3 are related to A_2 and B_2. Also, the constants C_3 and D_3 are related to C_2 and D_2.

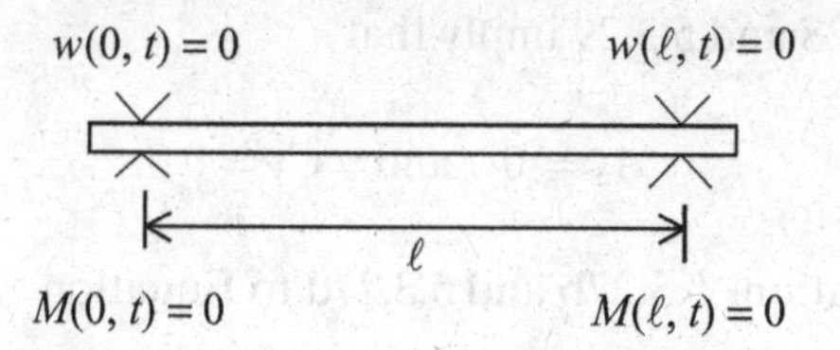

Figure 5.3.3 A simply supported beam

Simply Supported Beam

For a simply supported beam (Figure 5.3.3) of length ℓ, deflections at both ends are zero, that is,

$$w(0, t) = 0 \quad \text{and} \quad w(\ell, t) = 0 \tag{5.3.25}$$

Also, the bending moments at both ends of a simply supported beam are zero, that is,

$$M(0, t) = 0 \quad \text{and} \quad M(\ell, t) = 0 \tag{5.3.26}$$

The boundary conditions in Equations 5.3.25 and 5.3.26 along with the assumed form of the solution (Equation 5.3.9) and Equation 5.3.5 lead to the following four conditions on $X(x)$:

$$X(0) = 0 \tag{5.3.27a}$$

$$X(\ell) = 0 \tag{5.3.27b}$$

$$\frac{d^2X}{dx^2}(0) = 0 \tag{5.3.27c}$$

and

$$\frac{d^2X}{dx^2}(\ell) = 0 \tag{5.3.27d}$$

From Equations 5.3.27a, 5.3.27c, and 5.3.24,

$$A_3 + C_3 = 0 \tag{5.3.28}$$

$$A_3 - C_3 = 0 \tag{5.3.29}$$

Equations 5.3.28 and 5.3.29 imply that

$$A_3 = 0 \quad \text{and} \quad C_3 = 0 \tag{5.3.30a, b}$$

Applying Equations 5.3.27b and 5.3.27d to Equation 5.3.24,

$$\begin{bmatrix} \sinh \gamma \ell & \sin \gamma \ell \\ \sinh \gamma \ell & -\sin \gamma \ell \end{bmatrix} \begin{bmatrix} B_3 \\ D_3 \end{bmatrix} = \begin{bmatrix} 0 \\ 0 \end{bmatrix} \tag{5.3.31}$$

For a nontrivial solution of Equation 5.3.31,

$$\det \begin{bmatrix} \sinh \gamma \ell & \sin \gamma \ell \\ \sinh \gamma \ell & -\sin \gamma \ell \end{bmatrix} = 0 \tag{5.3.32}$$

The condition in Equation 5.3.32 yields

$$\sin \gamma \ell = 0 \tag{5.3.33}$$

Substituting Equation 5.3.33 into Equation 5.3.31 yields

$$B_3 = 0 \tag{5.3.34}$$

Because of Equations 5.3.30a,b and 5.3.34, Equation 5.3.24 yields

$$X(x) = D_3 \sin \gamma x \tag{5.3.35}$$

Equations 5.3.33 and 5.3.35 lead to the natural frequencies and the mode shapes, respectively. From Equation 5.3.33,

$$\gamma \ell = n\pi; \quad n = 1, 2, 3, \ldots, \tag{5.3.36}$$

From Equations 5.3.17 and 5.3.36, the natural frequencies of a simply supported beam are

$$\omega_n = \frac{n^2 \pi^2 \beta}{\ell^2}; \quad n = 1, 2, 3, \ldots, \tag{5.3.37}$$

There are an infinite number of natural frequencies as the number of degrees of freedom of a continuous structure is infinite. The

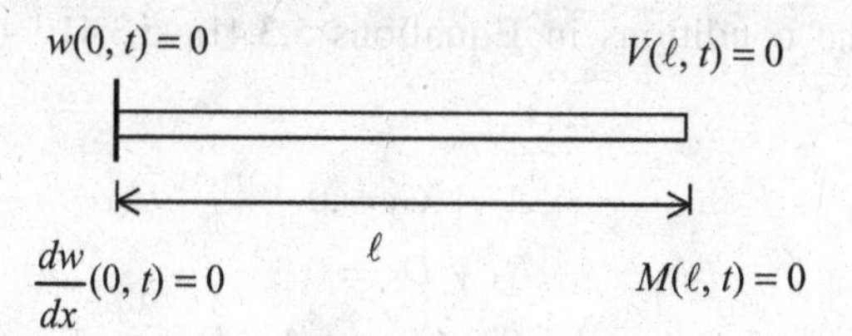

Figure 5.3.4 A cantilever beam

associated mode shapes are obtained from the Equation 5.3.35 by arbitrarily setting $D_3 = 1$:

$$\phi_n(x) = \sin\left(\frac{n\pi x}{\ell}\right); \quad n = 1, 2, 3, \ldots, \tag{5.3.38}$$

Cantilever Beam

For a cantilever beam (Figure 5.3.4) of length ℓ, the deflection and the slope at the left end is zero, that is,

$$w(0, t) = 0 \quad \text{and} \quad \frac{dw}{dx}(0, t) = 0 \tag{5.3.39}$$

Also, the bending moment and the shear force at the right end of a cantilever beam are zero, that is,

$$M(\ell, t) = 0 \quad \text{and} \quad V(\ell, t) = 0 \tag{5.3.40}$$

The boundary conditions in Equations 5.3.39 and 5.3.40 along with the assumed form of solution (Equations 5.3.9) and Equation 5.3.5 lead to the following four conditions on $X(x)$:

$$X(0) = 0 \tag{5.3.41a}$$

$$\frac{dX}{dx}(0) = 0 \tag{5.3.41b}$$

$$\frac{d^2X}{dx^2}(\ell) = 0 \tag{5.3.41c}$$

and

$$\frac{d^3X}{dx^3}(\ell) = 0 \tag{5.3.41d}$$

Applying the conditions in Equations 5.3.41a–5.3.41d to Equation 5.3.24,

$$A_3 + C_3 = 0 \quad (5.3.42a)$$

$$B_3 + D_3 = 0 \quad (5.3.42b)$$

$$A_3 \cosh \gamma\ell + B_3 \sinh \gamma\ell - C_3 \cos \gamma\ell - D_3 \sin \gamma\ell = 0 \quad (5.3.42c)$$

$$A_3 \sinh \gamma\ell + B_3 \cosh \gamma\ell + C_3 \sin \gamma\ell - D_3 \cos \gamma\ell = 0 \quad (5.3.42d)$$

Because of Equations 5.3.42a and 5.3.42b, Equations 5.3.42c and 5.3.42d can be written as

$$\begin{bmatrix} (\cosh \gamma\ell + \cos \gamma\ell) & (\sinh \gamma\ell + \sin \gamma\ell) \\ (\sinh \gamma\ell - \sin \gamma\ell) & (\cosh \gamma\ell + \cos \gamma\ell) \end{bmatrix} \begin{bmatrix} A_3 \\ B_3 \end{bmatrix} = \begin{bmatrix} 0 \\ 0 \end{bmatrix} \quad (5.3.43)$$

For a nontrivial solution of Equation 5.3.43,

$$\det \begin{bmatrix} (\cosh \gamma\ell + \cos \gamma\ell) & (\sinh \gamma\ell + \sin \gamma\ell) \\ (\sinh \gamma\ell - \sin \gamma\ell) & (\cosh \gamma\ell + \cos \gamma\ell) \end{bmatrix} = 0 \quad (5.3.44)$$

The condition in Equation 5.3.44 yields

$$\cosh \gamma\ell \cos \gamma\ell = -1 \quad (5.3.45)$$

And from Equation 5.3.43,

$$\frac{B_3}{A_3} = -\frac{(\cosh \gamma\ell + \cos \gamma\ell)}{(\sinh \gamma\ell + \sin \gamma\ell)} \quad (5.3.46)$$

From Equation 5.3.24,

$$X(x) = A_3[(\cosh \gamma x - \cos \gamma x) + \Lambda(\sinh \gamma x - \sin \gamma x)] \quad (5.3.47)$$

where

$$\Lambda = \frac{B_3}{A_3} = -\frac{(\cosh \gamma\ell + \cos \gamma\ell)}{(\sinh \gamma\ell + \sin \gamma\ell)} \quad (5.3.48)$$

There are an infinite number of roots of Equation 5.3.45. The first four roots of Equation 5.3.45 are

$$\gamma_1\ell = 1.875104, \quad \gamma_2\ell = 4.694091,$$

$$\gamma_3\ell = 7.854757, \quad \text{and} \quad \gamma_4\ell = 10.995541 \quad (5.3.49)$$

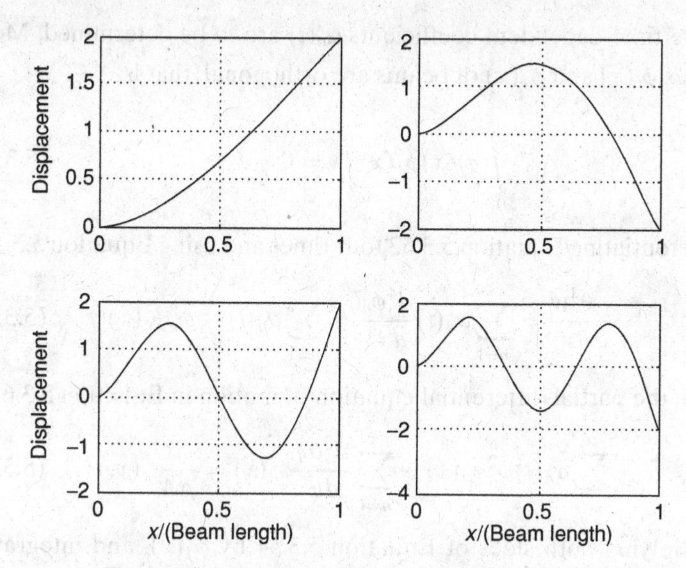

Figure 5.3.5 Mode shapes $\phi_n(x)$ of a cantilever beam

From Equations 5.3.17 and 5.3.49, the natural frequencies of a cantilever beam are

$$\omega_n = \frac{\beta}{\ell^2}(\gamma_n \ell)^2; \quad n = 1, 2, 3, \ldots, \tag{5.3.50}$$

The associated mode shapes are obtained from Equation 5.3.47 by arbitrarily setting $A_3 = 1$:

$$\phi_n(x) = (\cosh \gamma_n x - \cos \gamma_n x) + \Lambda_n(\sinh \gamma_n x - \sin \gamma_n x) \tag{5.3.51}$$

where Λ_n is given by Equation 5.3.48 with $\gamma = \gamma_n$. These mode shapes are plotted in Figure 5.3.5.

5.3.3 Computation of Response

In general, the response can be expressed as a linear combination of mode shapes, that is,

$$w(x, t) = \sum_{n=1}^{\infty} \alpha_n(t)\phi_n(x) \tag{5.3.52}$$

where time-dependent coefficients $\alpha_n(t)$ are to be determined. Mode shapes $\phi_i(x)$ and $\phi_j(x)$ of beams are orthogonal, that is,

$$\int_0^\ell \phi_i(x)\phi_j(x)dx = 0; \quad i \neq j \tag{5.3.53}$$

Differentiating Equation 5.3.52 four times and using Equation 5.3.16,

$$\frac{\partial^4 w}{\partial x^4} = \sum_{n=1}^{\infty} \alpha_n(t)\frac{d^4\phi_n}{dx^4} = \sum_{n=1}^{\infty} \alpha_n(t)\frac{1}{\beta^2}\omega_n^2\phi_n(x) \tag{5.3.54}$$

From the partial differential equation of motion in Equation 5.3.6,

$$\sum_{n=1}^{\infty} \alpha_n(t)\omega_n^2\phi_n(x) + \sum_{n=1}^{\infty} \frac{d^2\alpha_n}{dt^2}\phi_n(x) = \frac{1}{\rho A} f_\ell(x,t) \tag{5.3.55}$$

Multiplying both sides of Equation 5.3.54 by $\phi_j(x)$, and integrating from 0 to ℓ,

$$\sum_{n=1}^{\infty} \alpha_n(t)\omega_n^2 \int_0^\ell \phi_n(x)\phi_j(x)dx + \sum_{n=1}^{\infty} \frac{d^2\alpha_n}{dt^2} \int_0^\ell \phi_n(x)\phi_j(x)dx$$

$$= \frac{1}{\rho A}\int_0^\ell f_\ell(x,t)\phi_j(x)dx \tag{5.3.56}$$

Then using the property in Equation 5.3.53,

$$\alpha_j(t)\eta_j\omega_j^2 + \frac{d^2\alpha_j}{dt^2}\eta_j = \frac{1}{\rho A}\int_0^\ell f_\ell(x,t)\phi_j(x)dx \tag{5.3.57}$$

or

$$\frac{d^2\alpha_j}{dt^2} + \omega_j^2\alpha_j(t) = \frac{1}{\rho A\eta_j}\int_0^\ell f_\ell(x,t)\phi_j(x)dx \tag{5.3.58}$$

where

$$\eta_j = \int_0^\ell \phi_j^2(x)dx \tag{5.3.59}$$

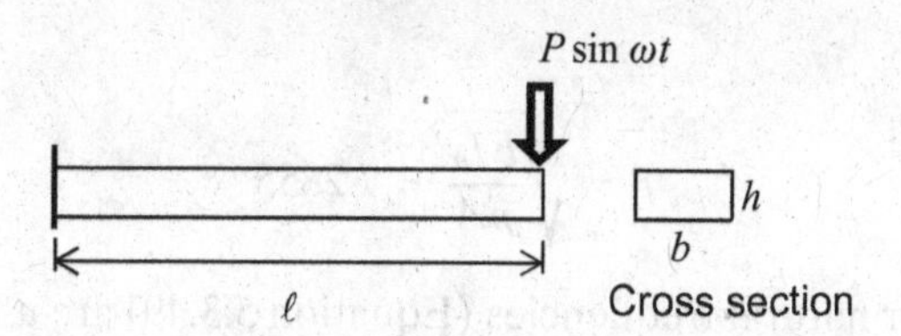

Figure 5.3.6 A cantilever beam subjected to sinusoidal excitation

Initial conditions are obtained from $w(x,0)$ and $\dot{w}(x,0)$. From Equation 5.3.52,

$$w(x,0) = \sum_{n=1}^{\infty} \alpha_n(0)\phi_n(x) \tag{5.3.60}$$

$$\dot{w}(x,0) = \sum_{n=1}^{\infty} \dot{\alpha}_n(0)\phi_n(x) \tag{5.3.61}$$

Using the properties in Equations 5.3.53 and 5.3.59,

$$\alpha_j(0) = \frac{1}{\eta_j}\int_0^{\ell} w(x,0)\phi_j(x)dx \tag{5.3.62}$$

and

$$\dot{\alpha}_j(0) = \frac{1}{\eta_j}\int_0^{\ell} \dot{w}(x,0)\phi_j(x)dx \tag{5.3.63}$$

Example 5.3.1: Response of a Cantilever Beam
Consider a **steel** cantilever beam of rectangular cross section (Figure 5.3.6) with the width $b = 0.01$ m and the thickness $h = 0.005$ m. The length ℓ of the beam is 0.8 m. At the tip of the beam, a sinusoidal force with magnitude $P = 10$ N and frequency $\omega = 100$ rad/sec is applied. Determine the steady-state response by assuming that the damping ratio in each mode is 0.02.

For steel, $E = 2\times10^{11}$ N/m^2 and $\rho = 7,850$ kg/m^3.

$$A = bh = 5\times10^{-5}\text{N/m}^2 \quad \text{and} \quad I_a = \frac{1}{12}bh^3$$

Then,

$$\beta = \sqrt{\frac{EI_a}{\rho A}} = 7.2855$$

The first four natural frequencies (Equation 5.3.49) are as follows:

$$\omega_1 = 40\,\text{rad/sec}, \quad \omega_2 = 250.8\,\text{rad/sec},$$
$$\omega_3 = 702.3\,\text{rad/sec}, \quad \text{and} \quad \omega_4 = 1376.3\,\text{rad/sec}$$

Here,

$$f_\ell(x,t) = P \sin \omega t \delta(x - \ell) \tag{5.3.64}$$

where $\delta(x - \ell)$ is the Dirac delta or the unit impulse function.

After introducing the damping ratio ξ in each mode, Equation 5.3.58 becomes

$$\frac{d^2\alpha_j}{dt^2} + 2\xi\omega_j\dot{\alpha}_j + \omega_j^2\alpha_j(t) = f_{eq}(j)\sin\omega t \tag{5.3.65}$$

where

$$f_{eq}(j) = \frac{P\phi_j(\ell)}{\rho A \eta_j} \tag{5.3.66}$$

Using Equation 5.3.59,

$$\eta_j = \int_0^\ell \phi_j^2(x)dx = \ell \int_0^1 \phi_j^2(\bar{x})d\bar{x} \tag{5.3.67}$$

where

$$\bar{x} = \frac{x}{\ell} \tag{5.3.68}$$

and

$$\phi_j(\bar{x}) = [\cosh(\gamma_j\ell\bar{x}) - \cos(\gamma_j\ell\bar{x}) + \Lambda_j(\sinh(\gamma_j\ell\bar{x}) - \sin(\gamma_j\ell\bar{x})] \tag{5.3.69}$$

It has been verified numerically that

$$\eta_j = \ell \tag{5.3.70}$$

Using results in Chapter 2, the steady-state response of Equation 5.3.65 can be written as

$$\alpha_j(t) = \Delta_j \sin(\omega t - \theta_j) \tag{5.3.71}$$

where

$$\Delta_j = \frac{f_{eq}(j)}{\sqrt{\left(\omega_j^2 - \omega^2\right)^2 + (2\xi\omega_j\omega)^2}} \tag{5.3.72}$$

and

$$\theta_j = \frac{2\xi\omega_j\omega}{\omega_j^2 - \omega^2} \tag{5.3.73}$$

From Equation 5.3.52, the steady-state displacement at $x = \ell$ is given by

$$w(\ell, t) = \sum_{j=1}^{n} \phi_j(\ell)\alpha_j(t) = \sum_{j=1}^{n} \phi_j(\ell)\Delta_j \sin(\omega t - \theta_j) \tag{5.3.74}$$

After some algebra, Equation 5.3.74 is written as

$$w(\ell, t) = B_1 \sin \omega t - B_2 \cos \omega t \tag{5.3.75}$$

where

$$B_1 = \sum_{j=1}^{n} \phi_j(\ell)\Delta_j \cos\theta_j \tag{5.3.76}$$

and

$$B_2 = \sum_{j=1}^{n} \phi_j(\ell)\Delta_j \sin\theta_j \tag{5.3.77}$$

From Equation 5.3.75,

$$w(\ell, t) = B \sin(\omega t - \varphi) \tag{5.3.78}$$

where

$$B = \sqrt{B_1^2 + B_2^2} \quad \text{and} \quad \tan\varphi = \frac{B_2}{B_1} \tag{5.3.79a, b}$$

From the MATLAB Program 5.2 with $\omega = 100\,\text{rad/sec}$,

$$B = 0.0155\,\text{m} \quad \text{and} \quad \varphi = 3.1145\,\text{rad}$$

MATLAB Program 5.2

```
%cantilever beam
gamL=[1.875104 4.694091 7.854757 10.995541];%eq.(5.3.49)
E=2e11;%Young's Modulus of Elasticity (N/m^2)
rho=7850;%mass density(kg./m^3)
L=0.8;%length of beam(meter)
b=0.01;%meter
h=0.005;%meter
A=b*h;%cross sectional area
%
P=10;%magnitude of force (Newton)
omega=100;%excitation frequency(rad./sec.)
m=rho*L*A;%mass of beam
Ia=b*h^3/12;
beta=sqrt(E*Ia/(rho*A));
sumcos=0;
sumsin=0;
for i=1:4
omegan(i)=beta*gamL(i)^2/L^2;%Natural Frequencies(rad/sec.),eq.(5.3.50)
  lamd=-(cosh(gamL(i))+cos(gamL(i)))/(sinh(gamL(i))+sin(gamL(i)));%eq.
%(5.3.48)
  ang=gamL(i)*1;
  phiL(i)=(cosh(ang)-cos(ang))+lamd*(sinh(ang)-sin(ang));
  Feq(i)=P*phiL(i)/(m*L);%Equivalent force in mode#i
  den=sqrt((omegan(i)^2-omega^2)^2+(2*0.02 *omegan(i)*omega)^2);
  Amp(i)=Feq(i)/den;%Amplitude of Response in Mode#i
  phase(i)=atan2(2*0.02*omegan(i)*omega,omegan(i)^2-omega^2);%Phase
  sumcos=sumcos+phiL(i)*Amp(i)*cos(phase(i));%B1
  sumsin=sumsin+phiL(i)*Amp(i)*sin(phase(i));%B2
```

```
end
%
Ampw=sqrt(sumcos^2+sumsin^2)%Steady State Amplitude of w(L,t):B
phasew=atan2(sumsin,sumcos)%Steady State Phase of w(L,t)
```

5.4 FINITE ELEMENT ANALYSIS

The number of degrees of freedom of a continuous system is infinite. For simple geometries, such as those considered in Sections 5.2 and 5.3, the governing partial differential equation of motion can be solved analytically to determine the natural frequencies, the mode shapes, and the response of the structure. However, in general, the analytical solution of a partial differential equation is not possible for a real structure, for example, a turbine blade. As a result, the structure is discretized into a finite number of elements and the solution is obtained numerically. This process is known as the finite element method, which has been successfully applied to many real engineering problems. In fact, many commercially available codes, such as ANSYS and NASTRAN, are routinely used in industries. Here, fundamental ideas behind the finite element analysis (Petyt, 1990) will be illustrated via two simple examples: longitudinal vibration of a bar and transverse vibration of a beam.

5.4.1 Longitudinal Vibration of a Bar

Consider the longitudinal bar shown in Figure 5.2.5 again. This bar is divided into n elements of equal lengths (Figure 5.4.1), that is, the length of each element is

$$\ell_e = \frac{\ell}{n} \tag{5.4.1}$$

Each element has two nodes associated with it. For example, the first element has nodes 1 and 2, and the last element has nodes n and

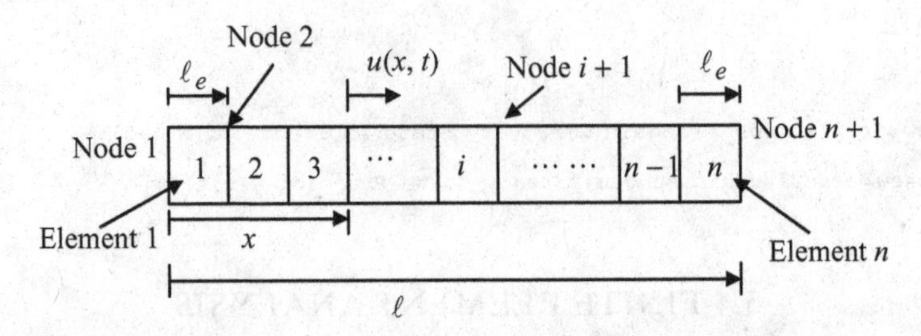

Figure 5.4.1 Discretization of a longitudinal bar by a finite number of elements

$n+1$. In general, the element i will have the node numbers i and $i+1$ (Figure 5.4.2).

Each node has one degree of freedom (axial displacement u). Let $u_i(t)$ be the axial displacement of the node i. Then, a discrete displacement vector $\mathbf{u}(t)$ can be defined as

$$\mathbf{u}(t) = [\, u_1(t) \quad u_2(t) \quad \cdots \quad u_n(t) \quad u_{n+1}(t)\,]^T \tag{5.4.2}$$

Inside each element, the displacement is assumed to be a predefined function of the axial coordinate ξ. Here, this function will be chosen as linear. For example, inside the element i,

$$u(\xi, t) = a_1 + a_2\xi; \quad 0 \le \xi \le \ell_e \tag{5.4.3}$$

Note that

$$\text{At}\, \xi = 0, \quad u(0, t) = u_i(t) \tag{5.4.4}$$

$$\text{At}\, \xi = \ell_e, \quad u(\ell_e, t) = u_{i+1}(t) \tag{5.4.5}$$

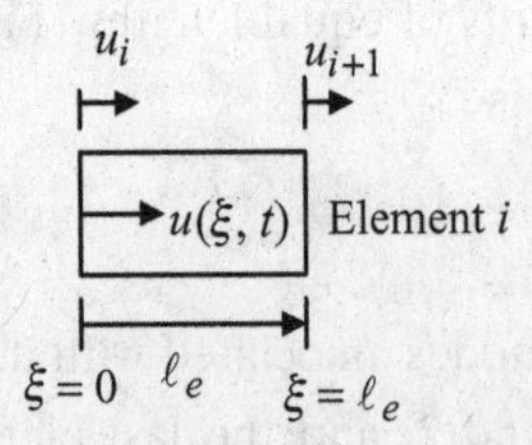

Figure 5.4.2 An element of a longitudinal bar

Substituting Equations 5.4.4 and 5.4.5 into Equation 5.4.3,

$$a_1 = u_i \tag{5.4.6}$$

and

$$a_2 = \frac{u_{i+1} - u_i}{\ell_e} \tag{5.4.7}$$

Substituting Equations 5.4.6 and 5.4.7 into Equation 5.4.3,

$$u(\xi, t) = \left(1 - \frac{\xi}{\ell_e}\right) u_i(t) + \frac{\xi}{\ell_e} u_{i+1}(t) \tag{5.4.8}$$

Differentiating Equation 5.4.8 with respect to ξ,

$$\frac{\partial u}{\partial \xi} = -\frac{1}{\ell_e} u_i(t) + \frac{1}{\ell_e} u_{i+1}(t) \tag{5.4.9}$$

Equation 5.4.9 can be expressed as

$$\frac{\partial u}{\partial \xi} = \boldsymbol{\kappa}^T \mathbf{v}_i(t) \tag{5.4.10}$$

where

$$\mathbf{v}_i(t) = \begin{bmatrix} u_i(t) \\ u_{i+1}(t) \end{bmatrix} \tag{5.4.11}$$

and

$$\boldsymbol{\kappa}^T = \begin{bmatrix} -\frac{1}{\ell_e} & \frac{1}{\ell_e} \end{bmatrix} \tag{5.4.12}$$

Differentiating Equation 5.4.8 with respect to time,

$$\frac{\partial u}{\partial t} = \left(1 - \frac{\xi}{\ell_e}\right) \dot{u}_i(t) + \frac{\xi}{\ell_e} \dot{u}_{i+1}(t) \tag{5.4.13}$$

Equation 5.4.13 can be expressed as

$$\frac{\partial u}{\partial t} = \mathbf{n}^T(\xi) \dot{\mathbf{v}}_i(t) \tag{5.4.14}$$

where

$$\dot{\mathbf{v}}_i(t) = \begin{bmatrix} \dot{u}_i(t) \\ \dot{u}_{i+1}(t) \end{bmatrix} \tag{5.4.15}$$

and

$$\mathbf{n}^T(\xi) = \left[\left(1 - \frac{\xi}{\ell_e}\right) \ \frac{\xi}{\ell_e} \right] \tag{5.4.16}$$

The kinetic energy of the element i is written as

$$T_i = \int_0^{\ell_e} \rho A \left(\frac{\partial u}{\partial t}\right)^2 d\xi = \rho A \dot{\mathbf{v}}_i^T \left[\int_0^{\ell_e} \mathbf{n}(\xi)\mathbf{n}^T(\xi) d\xi \right] \dot{\mathbf{v}}_i \tag{5.4.17}$$

where ρ is the mass density of the material and

$$\int_0^{\ell_e} \mathbf{n}(\xi)\mathbf{n}^T(\xi) dx = \begin{bmatrix} \int_0^{\ell_e} \left(1 - \frac{\xi}{\ell_e}\right)^2 d\xi & \int_0^{\ell_e} \left(1 - \frac{\xi}{\ell_e}\right) \frac{\xi}{\ell_e} d\xi \\ \int_0^{\ell_e} \left(1 - \frac{\xi}{\ell_e}\right) \frac{\xi}{\ell_e} d\xi & \int_0^{\ell_e} \left(\frac{\xi}{\ell_e}\right)^2 d\xi \end{bmatrix} = \frac{\ell_e}{6} \begin{bmatrix} 2 & 1 \\ 1 & 2 \end{bmatrix} \tag{5.4.18}$$

Substituting Equation 5.4.18 into Equation 5.4.17,

$$T_i = \frac{1}{2} \dot{\mathbf{v}}_i^T M_e \dot{\mathbf{v}}_i \tag{5.4.19}$$

where M_e is the mass matrix of the element defined as follows:

$$M_e = \frac{\rho A \ell_e}{3} \begin{bmatrix} 2 & 1 \\ 1 & 2 \end{bmatrix} \tag{5.4.20}$$

Using Equation 5.4.10, the potential energy of the element i is written as

$$P_i = EA \int_0^{\ell_e} \left(\frac{\partial u}{\partial \xi}\right)^2 d\xi = EA \int_0^{\ell_e} \left(\frac{\partial u}{\partial \xi}\right)^T \left(\frac{\partial u}{\partial \xi}\right) d\xi$$

$$= EA \mathbf{v}_i^T \left[\int_0^{\ell_e} \boldsymbol{\kappa}\boldsymbol{\kappa}^T d\xi \right] \mathbf{v}_i \tag{5.4.21}$$

$$\Phi_i = \begin{bmatrix} 0 & \cdots & 0 & 1 & 0 & 0 & \cdots & 0 \\ 0 & \cdots & 0 & 0 & 1 & 0 & \cdots & 0 \end{bmatrix}$$

Column i (arrow pointing down to the 1 in row 1); Column $(i+1)$ (arrow pointing up to the 1 in row 2)

Figure 5.4.3 Matrix connecting $\mathbf{v}_i$ to $\mathbf{u}$

where E and A are the Young's modulus of elasticity and the cross-sectional area of the element, respectively. Furthermore,

$$\int_0^{\ell_e} \boldsymbol{\kappa}\boldsymbol{\kappa}^T d\xi = \begin{bmatrix} \int_0^{\ell_e} \frac{1}{\ell_e^2} d\xi & -\int_0^{\ell_e} \frac{1}{\ell_e^2} d\xi \\ -\int_0^{\ell_e} \frac{1}{\ell_e^2} d\xi & \int_0^{\ell_e} \frac{1}{\ell_e^2} d\xi \end{bmatrix} = \frac{1}{\ell_e} \begin{bmatrix} 1 & -1 \\ -1 & 1 \end{bmatrix} \tag{5.4.22}$$

Substituting Equation 5.4.22 into Equation 5.4.21,

$$P_i = \frac{1}{2} \mathbf{v}_i^T K_e \mathbf{v}_i \tag{5.4.23}$$

where K_e is the stiffness matrix of the element defined as

$$K_e = \frac{2EA}{\ell} \begin{bmatrix} 1 & -1 \\ -1 & 1 \end{bmatrix} \tag{5.4.24}$$

Total Kinetic and Potential Energies of the Bar

From Equation 5.4.11,

$$\mathbf{v}_i(t) = \Phi_i \mathbf{u}(t) \tag{5.4.25}$$

where Φ_i is a $2 \times (n+1)$ matrix defined in Figure 5.4.3. The kinetic and potential energies of the element i are defined as

$$T_i = \frac{1}{2} \dot{\mathbf{u}}^T \Phi_i^T M_e \Phi_i \dot{\mathbf{u}}(t) \tag{5.4.26}$$

$$P_i = \frac{1}{2} \mathbf{u}^T \Phi_i^T K_e \Phi_i \mathbf{u}(t) \tag{5.4.27}$$

The total kinetic energy of the bar is obtained by summing the kinetic energy of each element:

$$T = \sum_{i=1}^{n} T_i = \frac{1}{2}\dot{\mathbf{u}}^T M_t \dot{\mathbf{u}}(t) \tag{5.4.28}$$

where M_t is the mass matrix defined as

$$M_t = \sum_{i=1}^{n} \Phi_i^T M_e \Phi_i \tag{5.4.29}$$

The total potential energy of the bar is obtained by summing the potential energy of each element:

$$P = \sum_{i=1}^{n} P_i = \frac{1}{2}\mathbf{u}^T K_t \mathbf{u}(t) \tag{5.4.30}$$

where K_t is the stiffness matrix defined as

$$K_t = \sum_{i=1}^{n} \Phi_i^T K_e \Phi_i \tag{5.4.31}$$

The differential equations of motion will then be

$$M_t \ddot{\mathbf{u}}(t) + K_t \mathbf{u}(t) = 0 \tag{5.4.32}$$

It should be noted that Equation 5.4.32 refers to a **free–free** bar, as there is no constraint imposed on any nodal displacement.

Example 5.4.1: Natural Frequencies of a Free–Free Bar
Consider a free–free **steel** bar with length = 0.04 m and cross-sectional area = 4×10^{-4} m^2. Determine the first five natural frequencies using 5, 10, and 15 elements, and compare them to the theoretical frequencies obtained in Section 5.2.

Solution

The MATLAB program 5.3 is used. Results are as follows:

First five theoretical frequencies:

$$[0 \quad 0.4075 \quad 0.8150 \quad 1.2226 \quad 1.6301] \times 10^6 \text{ rad/sec}$$

First five natural frequencies from finite element analysis:

Number of elements = 5,

$$[0 \quad 0.4143 \quad 0.8691 \quad 1.3978 \quad 1.9580] \times 10^6 \text{ rad/sec}$$

Number of elements = 10,

$$[0 \quad 0.4092 \quad 0.8285 \quad 1.2682 \quad 1.7382] \times 10^6 \text{ rad/sec}$$

Number of elements = 15,

$$[0 \quad 0.4083 \quad 0.8210 \quad 1.2428 \quad 1.6781] \times 10^6 \text{ rad/sec}$$

As the number of elements increases, the natural frequencies converge to their theoretical values.

MATLAB Program 5.3: Finite Element Analysis of a Longitudinal Bar

```
clear all
close all
%Free-Free Longitudinal Bar Vibration
betal2_ff=[22.4 61.7 121];
nel=10%number of elements
E=210e9;%Young's Modulus of Elasticity (N/m^2)
rho=7.8e3;%mass density (kg./m^3)
lt=4e-2;%Length of Beam (Meter)
Area=4e-4%Cross Sectional Area (m^2)
l=lt/nel;%Element Length (Meter)
ndof=nel+1;%number of degrees of freedom
sum_K=zeros(ndof,ndof);
sum_M=zeros(ndof,ndof);
P=zeros(2,ndof);
```

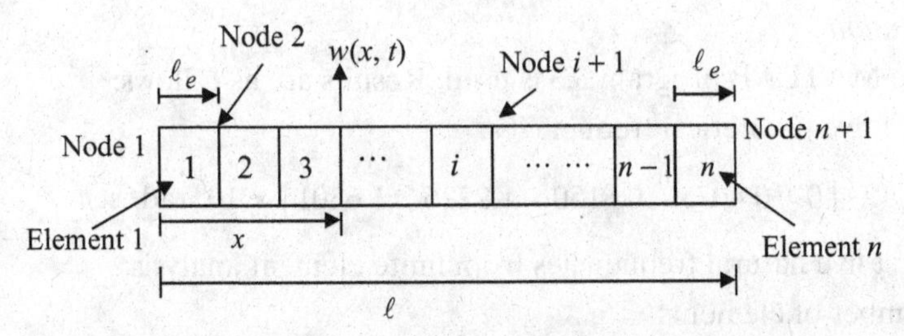

Figure 5.4.4 Discretization of a beam by a finite number of elements

```
P(1:2,1:2)=eye(2);
for i=1:nel
    if (i>1)
    P=zeros(2,ndof);
    P(1:2,i:i+1)=eye(2);
    end
    Ke=(2*E*Area/l)*[1 -1;-1 1];
    Me=(rho*Area*l/3)*[2 1;1 2];
    sum_K=sum_K+P'*Ke*P;% eq.(5.4.31)
    sum_M=sum_M+P'*Me*P;%eq.(5.4.29)
end
K=sum_K;
M=sum_M;
%
%Free-Free
%
ogff_th=sqrt(E/rho)*pi/lt*[1 2 3 4 5]%Theoretical Natural Frequencies
%
[V_ff,D_ff]=eig(K,M);
ogff_fem=sqrt(diag(D_ff))%Natural Frequencies from FEM
%
```

5.4.2 Transverse Vibration of a Beam

Consider the beam shown in Figure 5.3.1 again. This beam is divided into n elements of equal lengths (Figure 5.4.4), that is, the length of

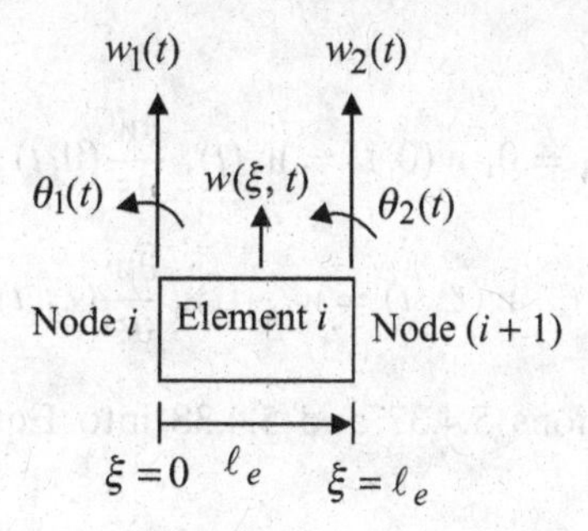

Figure 5.4.5 An element of a beam

each element is

$$\ell_e = \frac{\ell}{n} \tag{5.4.33}$$

Each element has two nodes associated with it. For example, the first element has nodes 1 and 2, and the last element has nodes n and $n + 1$. In general, the element i will have the node numbers i and $i + 1$ (Figure 5.4.5).

Each node has two degrees of freedom (transverse displacement w and the slope $\frac{\partial w}{\partial x}$). Let $w_i(t)$ and $\theta_i(t)$ be the transverse displacement and the slope at the node i. Then, a discrete displacement vector $\mathbf{u}(t)$ can be defined as

$$\mathbf{u}(t) = [\, w_1 \quad w_2 \quad \cdots \quad w_n \quad w_{n+1} \quad \theta_1 \quad \theta_2 \quad \cdots \quad \theta_n \quad \theta_{n+1} \,]^T \tag{5.4.34}$$

Inside each element, the transverse displacement is assumed to be a predefined function of the axial coordinate ξ. Here, this function will be chosen as a cubic polynomial. For example, inside the element i,

$$w(\xi, t) = a_1(t) + a_2(t)\xi + a_3(t)\xi^2 + a_4(t)\xi^3 \tag{5.4.35}$$

Differentiating Equation 5.4.35 with respect to ξ,

$$\frac{\partial w}{\partial \xi}(\xi, t) = a_2 + 2a_3\xi + 3a_4\xi^2 \tag{5.4.36}$$

Note that

$$\text{At } \xi = 0,\ w(0,t) = w_i(t),\ \frac{\partial w}{\partial \xi}(0,t) = \theta_i \tag{5.4.37}$$

$$\text{At } \xi = \ell_e,\ w(\ell_e,t) = w_{i+1}(t),\ \frac{\partial w}{\partial \xi}(\ell_e,t) = \theta_{i+1} \tag{5.4.38}$$

Substituting Equations 5.4.37 and 5.4.38 into Equations 5.4.35 and 5.4.36,

$$w(0,t) = a_1 = w_i(t) \tag{5.4.39a}$$

$$\frac{\partial w}{\partial \xi}(0,t) = a_2 = \theta_i(t) \tag{5.4.39b}$$

$$w(\ell_e,t) = a_1 + a_2\ell_e + a_3\ell_e^2 + a_4\ell_e^3 = w_{i+1}(t) \tag{5.4.39c}$$

$$\frac{\partial w}{\partial \xi}(\ell_e,t) = a_2 + 2a_3\ell_e + 3a_4\ell_e^2 = \theta_{i+1}(t) \tag{5.4.39d}$$

The solutions of Equations 5.4.39a–5.4.39d yield the coefficients a_1, a_2, a_3, and a_4:

$$\mathbf{a}(t) = \Psi \mathbf{q}_i(t) \tag{5.4.40}$$

where

$$\Psi = \begin{bmatrix} 1 & 0 & 0 & 0 \\ 0 & 0 & 1 & 0 \\ -\frac{3}{\ell_e^2} & \frac{3}{\ell_e^2} & -\frac{2}{\ell_e} & -\frac{1}{\ell_e} \\ \frac{2}{\ell_e^3} & -\frac{2}{\ell_e^3} & \frac{1}{\ell_e^2} & \frac{1}{\ell_e^2} \end{bmatrix} \tag{5.4.41}$$

$$\mathbf{a}^T(t) = [\,a_1(t) \quad a_2(t) \quad a_3(t) \quad a_4(t)\,] \tag{5.4.42}$$

$$\mathbf{q}_i^T(t) = [\,w_i(t) \quad w_{i+1}(t) \quad \theta_i(t) \quad \theta_{i+1}(t)\,] \tag{5.4.43}$$

From Equations 5.4.35, 5.4.40, and 5.4.41,

$$w(\xi, t) = [1 \quad \xi \quad \xi^2 \quad \xi^3]\,\mathbf{a}(t) = [1 \quad \xi \quad \xi^2 \quad \xi^3]\Psi\mathbf{q}_i(t) = \mathbf{n}^T(\xi)\mathbf{q}_i(t) \tag{5.4.44}$$

where

$$\mathbf{n}(\xi) = \begin{bmatrix} 1 - \dfrac{3}{\ell_e^2}\xi^2 + \dfrac{2}{\ell_e^3}\xi^3 \\ \dfrac{3}{\ell_e^2}\xi^2 - \dfrac{2}{\ell_e^3}\xi^3 \\ \xi - \dfrac{2}{\ell_e}\xi^2 + \dfrac{1}{\ell_e^2}\xi^3 \\ -\dfrac{\xi^2}{\ell_e} + \dfrac{\xi^3}{\ell_e^2}. \end{bmatrix} \tag{5.4.45}$$

The kinetic energy of the element i is written as

$$T_i = \frac{1}{2}\int_0^{\ell_e} \rho A \left(\frac{\partial w}{\partial t}\right)^2 d\xi \tag{5.4.46}$$

where ρ and A are the mass density of the material and the cross-sectional area of the beam, respectively. Differentiating Equation 5.4.44 with respect to time,

$$\frac{\partial w}{\partial t} = \mathbf{n}^T(\xi)\dot{\mathbf{q}}_i \tag{5.4.47}$$

Therefore,

$$\left(\frac{\partial w}{\partial t}\right)^2 = \left(\frac{\partial w}{\partial t}\right)^T \frac{\partial w}{\partial t} = \dot{\mathbf{q}}_i^T \mathbf{n}(\xi)\mathbf{n}^T(\xi)\dot{\mathbf{q}}_i \tag{5.4.48}$$

Substituting Equation 5.4.48 into Equation 5.4.46,

$$T_i = \frac{1}{2}\dot{\mathbf{q}}_i^T M_e \dot{\mathbf{q}}_i \tag{5.4.49}$$

where M_e is the mass matrix of the element defined as follows:

$$M_e = \rho A \left[\int_0^{\ell_e} \mathbf{n}(\xi)\mathbf{n}^T(\xi) d\xi \right] \tag{5.4.50}$$

Using Equation 5.4.45,

$$\int_0^{\ell_e} \mathbf{n}(\xi)\mathbf{n}^T(\xi) d\xi = \frac{\ell_e}{420} \begin{bmatrix} 156 & 54 & 22\ell_e & -13\ell_e \\ 54 & 156 & 13\ell_e & -22\ell_e \\ 22\ell_e & 13\ell_e & 4\ell_e^2 & -3\ell_e^2 \\ -13\ell_e & -22\ell_e & -3\ell_e^2 & 4\ell_e^2 \end{bmatrix} \tag{5.4.51}$$

Therefore, the mass matrix of the beam element is

$$M_e = \frac{\rho A \ell_e}{420} \begin{bmatrix} 156 & 54 & 22\ell_e & -13\ell_e \\ 54 & 156 & 13\ell_e & -22\ell_e \\ 22\ell_e & 13\ell_e & 4\ell_e^2 & -3\ell_e^2 \\ -13\ell_e & -22\ell_e & -3\ell_e^2 & 4\ell_e^2 \end{bmatrix} \tag{5.4.52}$$

The potential energy of the element i is written as

$$P_i = \frac{1}{2} EI \int_0^{\ell_e} \left(\frac{\partial^2 w}{\partial \xi^2} \right)^2 d\xi \tag{5.4.53}$$

From Equation 5.4.44,

$$\frac{\partial^2 w}{\partial \xi^2} = \left(\frac{\partial^2 \mathbf{n}(\xi)}{\partial \xi^2} \right)^T \mathbf{q}_i(t) \tag{5.4.54}$$

where

$$\frac{\partial^2 \mathbf{n}(\xi)}{\partial \xi^2} = \begin{bmatrix} -\frac{6}{\ell_e^2} + \frac{12}{\ell_e^3}\xi \\ \frac{6}{\ell_e^2} - \frac{12}{\ell_e^3}\xi \\ -\frac{4}{\ell_e} + \frac{6}{\ell_e^2}\xi \\ -\frac{2}{\ell_e} + \frac{6}{\ell_e^2}\xi \end{bmatrix} \tag{5.4.55}$$

From Equation 5.4.54,

$$\left(\frac{\partial^2 w}{\partial \xi^2}\right)^2 = \left(\frac{\partial^2 w}{\partial \xi^2}\right)^T \left(\frac{\partial^2 w}{\partial \xi^2}\right) = \mathbf{q}_i^T(t)\left(\frac{\partial^2 \mathbf{n}(\xi)}{\partial \xi^2}\right)\left(\frac{\partial^2 \mathbf{n}(\xi)}{\partial \xi^2}\right)^T \mathbf{q}_i(t) \tag{5.4.56}$$

Using Equation 5.4.55,

$$\int_0^{\ell_e}\left(\frac{\partial^2 \mathbf{n}(\xi)}{\partial \xi^2}\right)\left(\frac{\partial^2 \mathbf{n}(\xi)}{\partial \xi^2}\right)^T d\xi = \frac{1}{\ell_e^3}\begin{bmatrix} 12 & -12 & 6\ell_e & 6\ell_e \\ -12 & 12 & -6\ell_e & -6\ell_e \\ 6\ell_e & -6\ell_e & 4\ell_e^2 & 2\ell_e^2 \\ 6\ell_e & -6\ell_e & 2\ell_e^2 & 4\ell_e^2 \end{bmatrix} \tag{5.4.57}$$

Substituting Equation 5.4.56 into Equation 5.4.53,

$$P_i = \frac{1}{2}\mathbf{q}_i^T(t) K_e \mathbf{q}_i(t) \tag{5.4.58}$$

where K_e is the element stiffness matrix defined as

$$K_e = EI\int_0^{\ell_e}\left(\frac{\partial^2 \mathbf{n}(\xi)}{\partial \xi^2}\right)\left(\frac{\partial^2 \mathbf{n}(\xi)}{\partial \xi^2}\right)^T d\xi \tag{5.4.59}$$

Substituting Equation 5.4.57 into Equation 5.4.59,

$$K_e = \frac{EI}{\ell_e^3}\begin{bmatrix} 12 & -12 & 6\ell_e & 6\ell_e \\ -12 & 12 & -6\ell_e & -6\ell_e \\ 6\ell_e & -6\ell_e & 4\ell_e^2 & 2\ell_e^2 \\ 6\ell_e & -6\ell_e & 2\ell_e^2 & 4\ell_e^2 \end{bmatrix} \tag{5.4.60}$$

Total Kinetic and Potential Energies of the Beam

$$\mathbf{q}_i(t) = \Gamma_i \mathbf{u}(t) \tag{5.4.61}$$

where Γ_i is a $4 \times (2n+2)$ matrix defined in Figure 5.4.6. The kinetic and potential energies of the element i are defined as

$$T_i = \frac{1}{2}\dot{\mathbf{u}}^T \Gamma_i^T M_e \Gamma_i \dot{\mathbf{u}}(t) \tag{5.4.62}$$

$$P_i = \frac{1}{2}\mathbf{u}^T \Gamma_i^T K_e \Gamma_i \mathbf{u}(t) \tag{5.4.63}$$

$$\Gamma_i = \begin{bmatrix} 0 & \cdots & 0 & 1 & 0 & 0 & \cdots & 0 & 0 & \cdots & 0 & 0 & 0 & 0 & \cdots & 0 \\ 0 & \cdots & 0 & 0 & 1 & 0 & \cdots & 0 & 0 & \cdots & 0 & 0 & 0 & 0 & \cdots & 0 \\ 0 & \cdots & 0 & 0 & 0 & 0 & \cdots & 0 & 0 & \cdots & 0 & 1 & 0 & 0 & \cdots & 0 \\ 0 & \cdots & 0 & 0 & 0 & 0 & \cdots & 0 & 0 & \cdots & 0 & 0 & 1 & 0 & \cdots & 0 \end{bmatrix}$$

Column i; Column $(n + 1 + i)$; Column $(n + 1)$; Column $(2n + 2)$

Figure 5.4.6 Matrix connecting $\mathbf{q}_i$ to $\mathbf{u}$

The total kinetic energy of the beam is obtained by summing the kinetic energy of each element:

$$T = \sum_{i=1}^{n} T_i = \frac{1}{2}\dot{\mathbf{u}}^T M_t \dot{\mathbf{u}}(t) \tag{5.4.64}$$

where M_t is the mass matrix defined as

$$M_t = \sum_{i=1}^{n} \Gamma_i^T M_e \Gamma_i \tag{5.4.65}$$

The total potential energy of the beam is obtained by summing the potential energy of each element:

$$P = \sum_{i=1}^{n} P_i = \frac{1}{2}\mathbf{u}^T K_t \mathbf{u}(t) \tag{5.4.66}$$

where K_t is the stiffness matrix defined as

$$K_t = \sum_{i=1}^{n} \Gamma_i^T K_e \Gamma_i \tag{5.4.67}$$

The differential equations of motion will then be

$$M_t \ddot{\mathbf{u}}(t) + K_t \mathbf{u}(t) = 0 \tag{5.4.68}$$

It should be noted that Equation 5.4.68 refers to a **free–free** beam, as there is no constraint imposed on any nodal displacement or slope.

Example 5.4.2: Natural Frequencies of a Cantilever Beam
Consider a cantilever **steel** beam with length = 0.04 m, cross-sectional area = 4×10^{-4} m^2, and area moment of inertia = $\frac{10^{-8}}{3}$ m^4. Determine the first four natural frequencies using 5, 10, and 15 elements, and compare them to the theoretical frequencies derived in Section 5.3.

Solution
The MATLAB Program 5.4 is used. Results are as follows:
First Four Theoretical Frequencies:

$$[0.0329 \quad 0.2063 \quad 0.5775 \quad 1.1317] \times 10^6 \text{ rad/sec}$$

First Four Natural Frequencies from finite element analysis:
Number of elements = 5,

$$[0.0329 \quad 0.2064 \quad 0.5797 \quad 1.1451] \times 10^6 \text{ rad/sec}$$

Number of elements = 10,

$$[0.0329 \quad 0.2063 \quad 0.5777 \quad 1.1329] \times 10^6 \text{ rad/sec}$$

Number of elements = 15,

$$[0.0329 \quad 0.2063 \quad 0.5776 \quad 1.1321] \times 10^6 \text{ rad/sec}$$

As the number of elements increases, the natural frequencies converge to their theoretical values.

MATLAB Program 5.4: Finite Element Analysis of Beam Vibration

```
clear all

close all

%Free-Free and Cantilever Beam

betal2_ff=[4.730 7.853 10.995 14.137].^2;

betal2_cant=[1.875 4.694 7.854 10.995].^2;
```

```
nel=2%number of elements
E=210e9;%Young's Modulus of Elasticity (N/m^2)
rho=7.8e3;%mass density (kg./m^3)
I=1e-8/3;%Area Moment of inertia (m^4)
lt=4e-2;%Length of Beam (Meter)
Area=4e-4%Cross Sectional Area (m^2)
l=lt/nel;%Element Length (Meter)
ndof=2*(nel-1)+4;
sum_K=zeros(ndof,ndof);
sum_M=zeros(ndof,ndof);
P=zeros(4,ndof);
P(1:4,1:4)=eye(4);
for i=1:nel
    if (i>1)
    P=zeros(4,ndof);
    P(1:4,2*(i-1)+1:2*(i-1)+4)=eye(4);
    end
    Ke=(E*I/(l^3))*[12 6*l -12 6*l; 6*l 4*l*l -6*l 2*l*l;...
    -12 -6*l 12 -6*l;6*l 2*l*l -6*l 4*l*l];
    Me=(rho*Area*l/420)*[156 22*l 54 -13*l;22*l 4*l*l 13*l -3*l*l;...
    54 13*l 156 -22*l;-13*l -3*l*l -22*l 4*l*l];
    sum_K=sum_K+P'*Ke*P;
    sum_M=sum_M+P'*Me*P;
end
K=sum_K;
M=sum_M;
%
%Free-Free
%
ogff_th=sqrt((E*I/(rho*Area*lt^4)))*betal2_ff
%
[V_ff,D_ff]=eig(K,M);
ogff_fem=sqrt(diag(D_ff));
```

```
%
%Cantilever
%
ogcant_th=sqrt((E*I/(rho*Area*lt^4)))*beta12_cant
%
K_cant=K(3:ndof,3:ndof);
M_cant=M(3:ndof,3:ndof);
[V_cant,D_cant]=eig(K_cant,M_cant);
ogcant_fem=sqrt(diag(D_cant));
ogcant_fem(1:4)
```

EXERCISE PROBLEMS

P5.1 Consider the model of a bladed disk (Sinha, 1986) shown in Figure P5.1 where each blade is represented by a single mass. Furthermore, it should be noted that $i+1=1$ when $i=N$ and $i-1=N$ when $i=1$, where N is the number of blades. Model parameters are as follows: $m_t = 0.0114$ kg, $k_t = 430{,}000$ N/m, and $K_c = 45{,}430$ N/m.

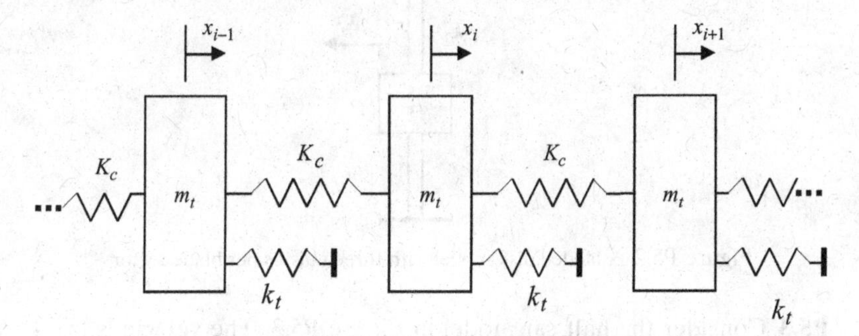

Figure P5.1 A bladed disk model with one mass per blade sector

a. Compute the natural frequencies and the mode shapes when $N=3$. Examine the orthogonality of the mode shapes.
b. Compute the natural frequencies and the mode shapes when $N=10$. Examine the orthogonality of the mode shapes.

P5.2 Consider the model of a turbine blade (Griffin and Hoosac, 1984) shown in Figure P5.2. Model parameters (SI units) are as follows: $m_1 = 0.0114$, $m_2 = 0.0427$, $m_3 = 0.0299$, $k_1 = 430{,}300$, $k_2 = 17{,}350{,}000$, and $k_3 = 7{,}521{,}000$.

a. Compute the natural frequencies and the mode shapes. Examine the orthogonality of the mode shapes.
b. Let $f(t) = \sin \omega t$, where ω is the excitation frequency. Using modal decomposition, find the amplitude and the phase of each mass as a function of the excitation frequencies near the first natural frequency. Assume the modal damping ratio to be 0.01.

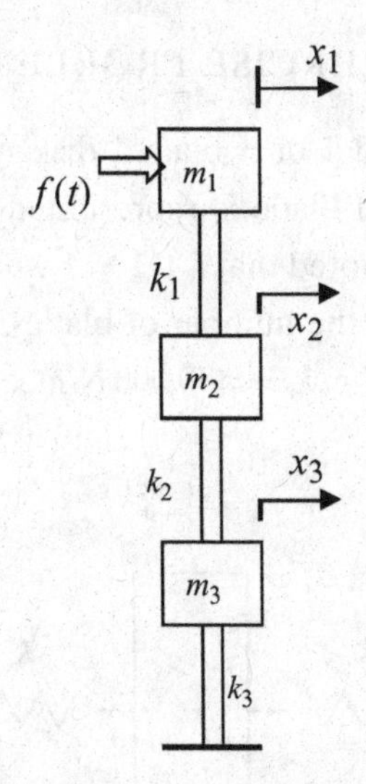

Figure P5.2 A bladed disk model with three masses per blade sector

P5.3 Consider the half car model in Figure P5.3. The vehicle is traveling with a velocity V on a sinusoidal road surface with an amplitude of 0.011 m and a wavelength of 5.3 m.

Parameters of the system are as follows: $\ell_1 = 1.35\,\text{m}$, $\ell_2 = 1.05\,\text{m}$, $I_c = 1{,}556\,\text{kg-m}^2$, $m_1 = 1{,}010\,\text{kg}$, $m_2 = 38\,\text{kg}$, $k_1 = 31{,}110\,\text{N/m}$, $k_2 = 41{,}310\,\text{N/m}$, $k_3 = 321{,}100\,\text{N/m}$, $c_1 = 3{,}980\,\text{N-sec/m}$, and $c_2 = 4{,}980\,\text{N-sec/m}$.

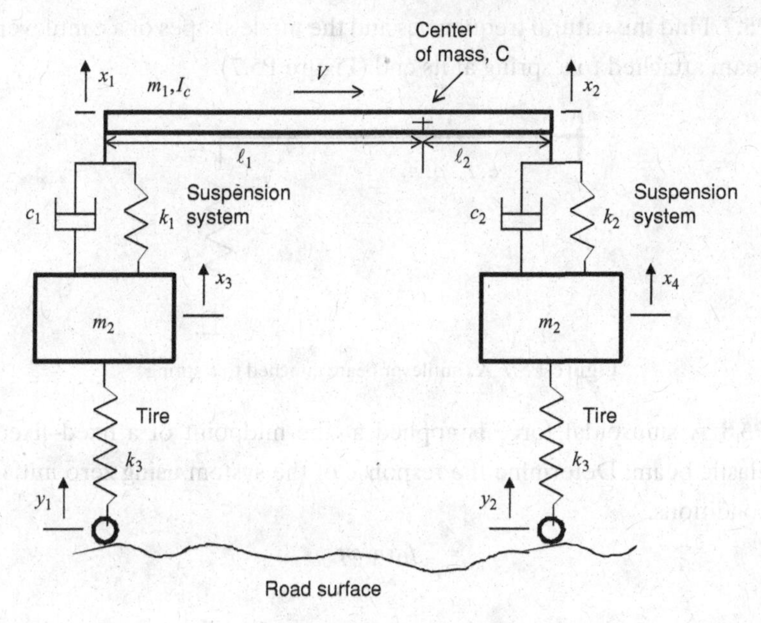

Figure P5.3 A half-car model

a. Compute the natural frequencies and the mode shapes.
b. Compute the modal damping ratios.
c. Find the critical speed of the vehicle. At the lowest critical speed, compute the steady-state response.

P5.4 The steel wire of length 0.9 m and cross-sectional area of 1.3 mm is fixed at both ends in a musical instrument. The tension in the string is 220 N. A musician plucks the string while adjusting the lengths of the strings in the following sequence: 0.5, 0.7, and 0.9 m. Compute the sequence of the fundamental frequencies of the sound generated by the instrument.

P5.5 Find the natural frequencies and the mode shapes of a fixed–fixed longitudinal bar.

P5.6 Find the natural frequencies and the mode shapes of a fixed–fixed torsional shaft.

P5.7 Find the natural frequencies and the mode shapes of a cantilever beam attached to a spring at its end (Figure P5.7).

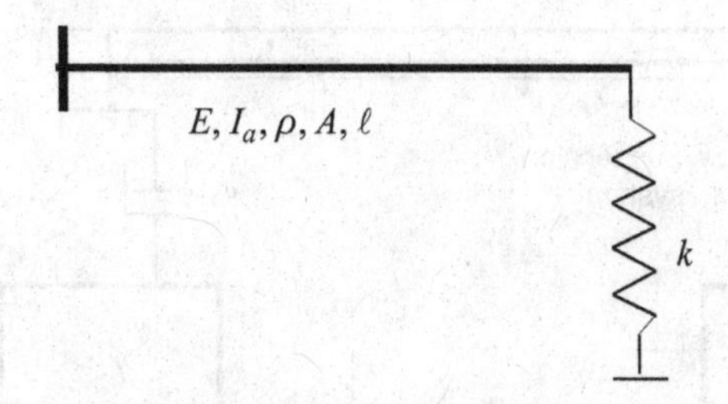

Figure P5.7 A cantilever beam attached to a spring

P5.8 A sinusoidal force is applied at the midpoint of a fixed–fixed elastic beam. Determine the response of the system using zero initial conditions.

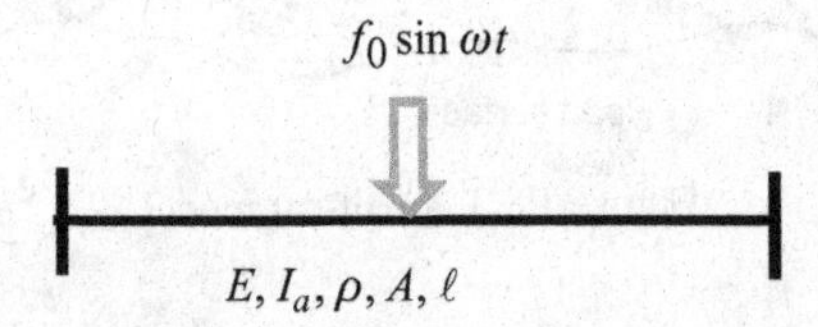

Figure P5.8 A fixed–fixed beam excited by a sinusoidal force

P5.9 Consider a fixed–fixed steel bar with the length = 0.1 m and the cross-sectional area = 4×10^{-4} m^2. Using the finite element method, determine the first three natural frequencies and compare them to the theoretical values.

P5.10 Consider a fixed–fixed steel beam with the length = 0.1 m, the cross-sectional area = 4×10^{-4} m^2 and the area moment of inertia = 0.5×10^{-8} m^4. Using the finite element method, determine the first three natural frequencies and compare them to the theoretical values.

APPENDIX A

EQUIVALENT STIFFNESSES (SPRING CONSTANTS) OF BEAMS, TORSIONAL SHAFT, AND LONGITUDINAL BAR

In this Appendix, equivalent stiffnesses of beams, a torsional shaft, and a longitudinal bar are presented. Derivations of these stiffnesses are based on static deflection of a structure.

A.1 FIXED–FIXED BEAM

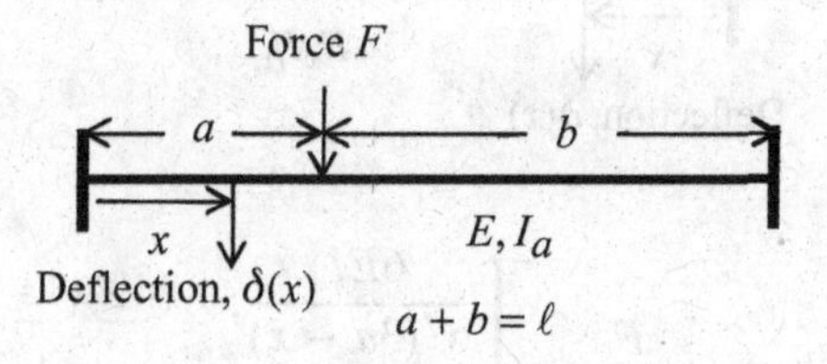

$$k_{eq}(x) = \frac{F}{\delta(x)} = \begin{cases} \dfrac{6EI_a\ell^3}{x^2b^2(3a\ell - x(3a+b))}; & 0 \le x \le a \\ \dfrac{6EI_a\ell^3}{(\ell - x)^2a^2(3b\ell - (\ell - x)(3b+a))}; & a \le x \le \ell \end{cases} \tag{A.1}$$

$k_{eq}(x)$: Equivalent stiffness at a distance x from left end

E: Young's modulus of elasticity

I_a: Area moment of inertia

A.2 SIMPLY SUPPORTED BEAM

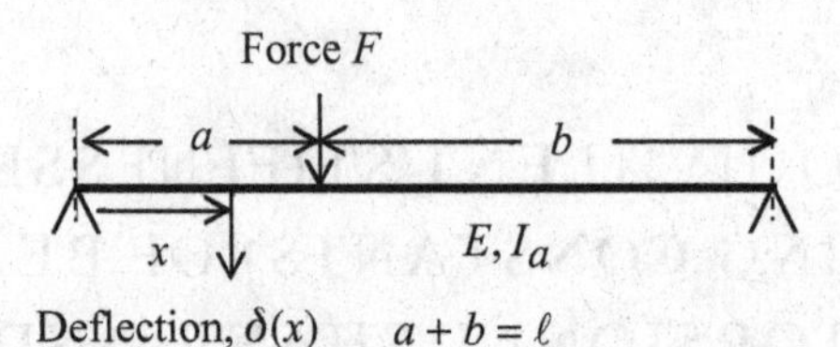

$$k_{eq}(x) = \frac{F}{\delta(x)} = \begin{cases} \dfrac{6EI_a\ell}{xb(\ell^2 - x^2 - b^2)}; & 0 \leq x \leq a \\ \dfrac{6EI_a\ell}{(\ell - x)a(2\ell x - x^2 - a^2)}; & a \leq x \leq \ell \end{cases} \quad \text{(A.2)}$$

A.3 CANTILEVER BEAM

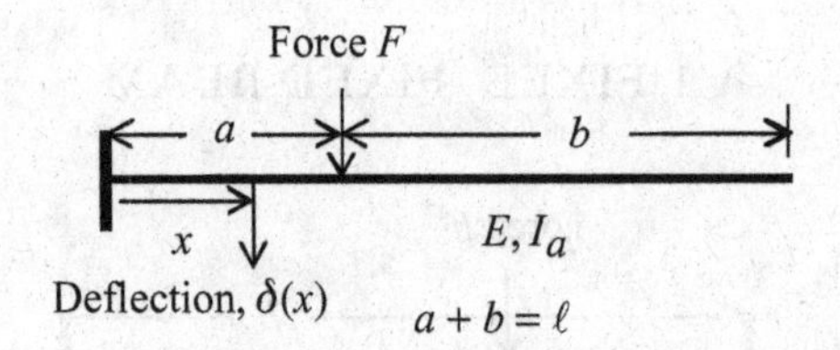

$$k_{eq}(x) = \frac{F}{\delta(x)} = \begin{cases} \dfrac{6EI_a}{x^2(3a - x)}; & 0 \leq x \leq a \\ \dfrac{6EI_a}{a^2(3x - a)}; & a \leq x \leq \ell \end{cases} \quad \text{(A.3)}$$

A.4 SHAFT UNDER TORSION

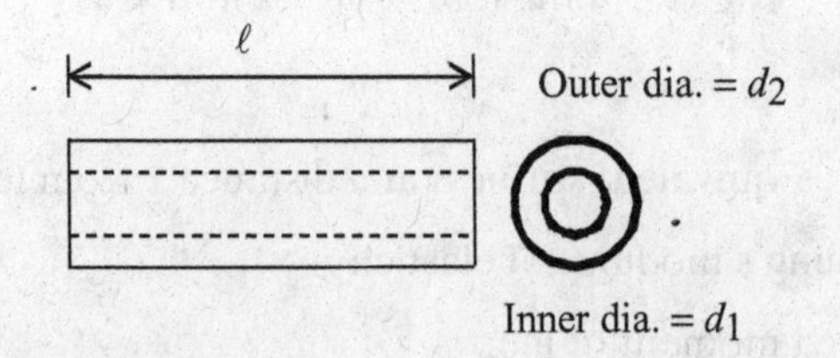

$$k_{eq} = \frac{\pi G(d_2^4 - d_1^4)}{32\ell}; \quad G\text{: Shear modulus of elasticity} \quad \text{(A.4)}$$

A.5 ELASTIC BAR UNDER AXIAL LOAD

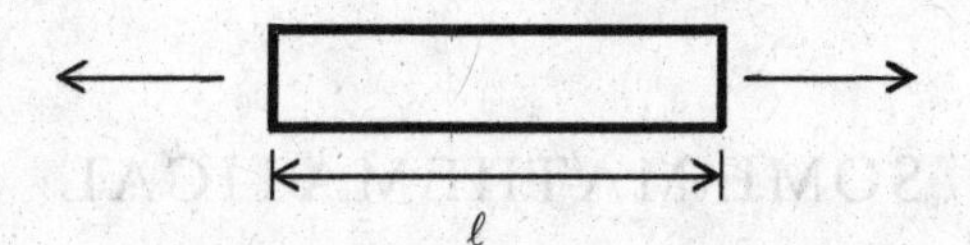

$$k_{eq} = \frac{EA}{\ell}; \quad A\text{: Cross-sectional area} \tag{A.5}$$

APPENDIX B

SOME MATHEMATICAL FORMULAE

B.1 TRIGONOMETRIC IDENTITY

$$\sin(-x) = -\sin x$$

$$\cos(-x) = \cos x$$

$$\sin(x \pm y) = \sin x \cos y \pm \cos x \sin y$$

$$\cos(x \pm y) = \cos x \cos y \mp \sin x \sin y$$

$$\sin x + \sin y = 2 \sin\left(\frac{x+y}{2}\right) \cos\left(\frac{x-y}{2}\right)$$

$$\sin x - \sin y = 2 \cos\left(\frac{x+y}{2}\right) \sin\left(\frac{x-y}{2}\right)$$

$$\cos x + \cos y = 2 \cos\left(\frac{x+y}{2}\right) \cos\left(\frac{x-y}{2}\right)$$

$$\cos x - \cos y = -2 \sin\left(\frac{x+y}{2}\right) \sin\left(\frac{x-y}{2}\right)$$

$$\tan(x \pm y) = \frac{\tan x \pm \tan y}{1 \mp \tan x \tan y}$$

$$e^{jx} = \cos x + j \sin x; \quad j = \sqrt{-1}$$

$$\sinh \gamma x = \frac{e^{\gamma x} - e^{-\gamma x}}{2}$$

$$\cosh \gamma x = \frac{e^{\gamma x} + e^{-\gamma x}}{2}$$

$$\sin \gamma x = \frac{e^{j\gamma x} - e^{-j\gamma x}}{2j}; \quad j = \sqrt{-1}$$

$$\cos \gamma x = \frac{e^{j\gamma x} + e^{-j\gamma x}}{2}; \quad j = \sqrt{-1}$$

$$\cosh^2 x - \sinh^2 x = 1$$

B.2 POWER SERIES EXPANSION

$$e^x = 1 + x + \frac{x^2}{2!} + \frac{x^3}{3!} + \cdots$$

$$\sin x = x - \frac{x^3}{3!} + \frac{x^5}{5!} - \frac{x^7}{7!} + \cdots; \quad x \text{ in radians}$$

$$\cos x = 1 - \frac{x^2}{2!} + \frac{x^4}{4!} - \frac{x^6}{6!} + \cdots; \quad x \text{ in radians}$$

B.3 BINOMIAL EXPANSION

$$(1 - x)^{-p} = 1 + px + \frac{1}{2}p(p + 1)x^2 + \cdots; |x| < 1$$

APPENDIX C

LAPLACE TRANSFORM TABLE

$f(t)$	$f(s)$
$\delta(t)$; unit impulse function	1
$u_s(t)$; unit step function	$\frac{1}{s}$
$\frac{t^{n-1}}{(n-1)!}u_s(t)$; $n = 1, 2, 3, \ldots,$	$\frac{1}{s^n}$
$e^{-at}u_s(t)$	$\frac{1}{s+a}$
$\frac{t^{n-1}e^{-at}}{(n-1)!}u_s(t)$; $n = 1, 2, 3, \ldots,$	$\frac{1}{(s+a)^n}$
$\sin(at)u_s(t)$	$\frac{a}{s^2+a^2}$
$\cos(at)u_s(t)$	$\frac{s}{s^2+a^2}$
$\frac{1}{\omega_d}e^{-\xi\omega_n t}\sin\omega_d t\, u_s(t) \quad \omega_d = \omega_n\sqrt{1-\xi^2}$	$\frac{1}{s^2+2\xi\omega_n s+\omega_n^2}$
$-\frac{\omega_n}{\omega_d}e^{-\xi\omega_n t}\sin(\omega_d t-\phi_1)u_s(t); \quad \omega_d = \omega_n\sqrt{1-\xi^2}$	$\frac{s}{s^2+2\xi\omega_n s+\omega_n^2}$
$\left[1-\frac{\omega_n}{\omega_d}e^{-\xi\omega_n t}\sin(\omega_d t+\phi_1)\right]u_s(t)$	$\frac{\omega_n^2}{s(s^2+2\xi\omega_n s+\omega_n^2)}$
$\omega_d = \omega_n\sqrt{1-\xi^2}; \quad \phi_1 \cos^{-1}\xi; \quad \xi < 1$	

REFERENCES

Billah, K. Y. and Scanlan, R. H., 1991, "Resonance, Tacoma Narrows Bridge Failure, and Undergraduate Physics Textbooks," *American Journal of Physics Teachers*, Vol. 59, No. 2, pp. 118–124.

Binning, G. and Quate, C. F., 1986, "Atomic Force Microscope," *Physical Review Letters*, Vol. 56, No. 9, pp. 930–933.

Boyce, W. E. and DiPrima, R. C., 2005, *Elementary Differential Equations and Boundary Value Problems*, John Wiley and Sons, Hoboken, NJ.

Crandall, S. H., Dahl, N.C. and Lardner, T. J., 1999, *An Introduction to Mechanics of Solids*, McGraw-Hill, New York.

Den Hartog, J. P., 1956, *Mechanical Vibrations*, Dover Publications (1984), New York.

Griffin, J. H. and Hoosac, T., 1984, "Model Development and Statistical Investigation of Turbine Blade Mistuning," *ASME Journal of Vibration, Acoustics, and Reliability in Design*, Vol. 106, No. 2, pp. 204–210.

Griffin, J. H. and Sinha, A., 1985, "The Interaction Between Mistuning and Friction in the Forced Response of Bladed Disk Assemblies," *ASME Journal of Engineering for Gas Turbines and Power*, Vol. 107, No. 1, pp. 205–211.

Hutton, D. V., 1981, *Applied Mechanical Vibrations*, McGraw-Hill, New York.

Newland, D. E., 2004, "Pedestrian Excitation of Bridges," *Journal of Mechanical Engineering Science*, Proc. Instn. Mech. Engineers., Vol. 218, Part C, pp. 477–492.

Oh, J. E., and Yum, S. H., 1986, "A Study on the Dynamic Characteristics of Tennis Racket by Modal Analysis," *Bulletin of JSME*, Vol. 29, No. 253, pp. 2228–2231.

Petyt, M., 1990, *Introduction to Finite Element Vibration Analysis*, Cambridge University Press, Cambridge, UK.

Rao, S. S., 1995, *Mechanical Vibrations*, Prentice Hall, Upper Saddle River, NJ.

Singer, N. C. and Seering, W. P. 1990, "Preshaping Command Inputs to Reduce System Vibration," *ASME Journal of Dynamic Systems, Measurement and Control*," Vol. 112, pp. 76–82.

Sinha, A., 1986, "Calculating the Statistics of Forced Response of a Mistuned Bladed Disk Assembly," *AIAA Journal*, Vol. 24, No. 11, pp. 1797–1801.

Strang, G., 1988, "*Linear Algebra and Its Applications*," Harcourt Brace Jovanovich Publishers, San Diego.

Thompson, J. M. T., 1982, *Instabilities and Catastrophes in Science and Engineering*, John Wiley & Sons, Upper Saddle River, NJ.

Thomson, W. T. and Dahleh, M. D., 1993, *Theory of Vibration with Applications*, 5th edn, Prentice Hall, Upper Saddle River, NJ.

INDEX